STUDIES IN MODERN THERMODYNAMICS 5

STUDIES IN NETWORK THERMODYNAMICS

STUDIES IN MODERN THERMODYNAMICS

1 **Biochemical Thermodynamics** edited by M.N. Jones
2 **Principles of Thermodynamics** by J.A. Beattie and I. Oppenheim
3 **Phase Theory** by H.A.J. Oonk
4 **Thermodynamics** by J.M. Honig

STUDIES IN MODERN THERMODYNAMICS 5

STUDIES IN NETWORK THERMODYNAMICS

L. PEUSNER

181 State Street, Portland, ME 04101, U.S.A.

ELSEVIER
Amsterdam — Oxford — New York — Tokyo   1986

ELSEVIER SCIENCE PUBLISHERS B.V.
Sara Burgerhartstraat 25
P.O. Box 211, 1000 AE Amsterdam, The Netherlands

Distributors for the United States and Canada:

ELSEVIER SCIENCE PUBLISHING COMPANY, INC.
52, Vanderbilt Avenue
New York, N.Y. 10017, U.S.A.

Library of Congress Cataloging-in-Publication Data

```
Peusner, Leonardo, 1943-
    Studies in network thermodynamics.

    (Studies in modern thermodynamics ; 5)
    Bibliography: p.
    Includes index.
    1. Thermodynamics.   2. System analysis.   I. Title.
II. Series.
QC311.P38  1986        536'.7          85-31550
ISBN 0-444-42580-2
```

ISBN 0-444-42580-2 (Vol. 5)
ISBN 0-444-41762-1 (Series)

© Elsevier Science Publishers B.V., 1986

All rights reserved. No part of this publication may be reproduced, stored in a retrieval system or transmitted in any form or by any means, electronic, mechanical, photocopying, recording or otherwise, without the prior written permission of the publisher, Elsevier Science Publishers B.V./Science & Technology Division, P.O. Box 330, 1000 AH Amsterdam, The Netherlands.

Special regulations for readers in the USA — This publication has been registered with the Copyright Clearance Center Inc. (CCC), Salem, Massachusetts. Information can be obtained from the CCC about conditions under which photocopies of parts of this publication may be made in the USA. All other copyright questions, including photocopying outside of the USA, should be referred to the publisher.

Printed in The Netherlands

To my wife Beth

Preface

   This is not a textbook in network thermodynamics, but simply a collection of variations on a single theme. The basic idea is extremely simple: given a physical model, find a potential from which forces can be obtained and, furthermore, make sure that the model is consistent with our fundamental notions of cause-effect. These requirements mean that associated with the set of physical forces there is a set of canonical flows or displacements in the direction of steepest descent--i.e., that of the gradient in the potential surface.

   The most obvious ways to treat these problems is through the use of differential topology and/or tensor algebra. But these techniques are hard and, in the end, they do not add too much to our physical or intuitive understanding. If, on the other hand, we are willing to sacrifice some degree of precision by going to a discrete model, the networks that obey Kirchhoff's laws have a unique place in such physical descriptions. The reason is that Kirchhoff's voltage and current laws lie in orthogonal subspaces, so that i) the cause-effect characteristic of forces and flows being orthogonal are automatically taken care of and ii) they lead to a formulation of the scalar invariance known as Tellegen's theorem. These ideas and other basic concepts in network theory are covered in Chapter 1.

   Chapter 2 deals with the application of these principles to one of the simplest examples provided by physical theory, that of thermostatics. Legendre transformations, Maxwell's reciprocities, the thermodynamic equation of state, jacobian transformations, etc. are derived on the basis of simple network representations and transformations. Moreover, the equivalence between the network properties and those of geometric thermodynamics are pointed out. Chapter 3 turns to a different question: can one find a potential description for chemical kinetics? If one assumes the potentials are concentrations, the well known answer is that this cannot be done. It turns out, however, that such a potential function can be constructed for some special cases. The first order linear reaction coupled to diffusional processes is treated as an example and it is then shown that a convective potential can be developed in the same manner. The more complex problem of reaction-convection-diffusion coupling is then treated for a membrane or slab.

Chapter 4 analyzes the assumptions of Onsager thermodynamics and relates microscopic reversibility to the kinetic potential found in the previous chapter. In particular, it is shown that i) the Onsager construct describes the motion of an incompressible fluid in potential flow and that ii) the assumption of microscopic reversibility is redundant in Onsager´s theory.

Chapter 5 is the steady state counterpart of Chapter 2. It is shown that the network properties lead to Onsager descriptions without the need to go through a dissipation function. Familiar examples --e.g., Kedem-Katchalsky´s equations -- are considered. In addition, previous results related to efficiency of energy conversion are covered.

Chapter 6 makes the transition from steady state to non equilibrium by adding capacitors and inductors to the resistive networks. The properties of these networks are applied to both thermodynamics and kinetics.

Chapter 7 deals with piecewise non-linearities, using resistors, sources and ideal diodes. The Teorell oscillator and some models of muscle contraction are introduced as examples.

Finally, a very short Chapter 8 points out that much remains to be covered in this field.

Although I have attempted to avoid using excessively abstract algebraic ideas, I see, in reading over the manuscript, that the temptation to cover algebra and geometry has taken me off the straight path here and there. I have left the material intact, however, because this "multimedia" characteristic is, in my mind, what completes the Gestalt of network thermodynamics. I hope the reader who is interested enough to open the covers of this book can forgive these imperfections.

Acknowledgments

I would like to thank the people who, directly or indirectly helped to make this monograph possible.

Many people collaborated in the production of the early manuscript versions since 1979: Debbie Crowell, Barbara Lake, Carole Milliken, Pat Maybe, Suzanne Little, Ted Chambers, Elizabeth Stone, Denise Brown Sullivan, Jeanne Ploss and others. Carlota Rutheford typed the final manuscript.

The late John Lennon and Prof. Danielli of Worcester Polytech gave me valuable indirect advice. Both Profs. Alvin Essig and S. Roy Caplan gave me encouragement at different times. Given that most of the problems I deal with in this monograph were identified originally by Prof. Caplan, his associates or students, he deserves credit for having contributed so much of his enthusiasm and knowledge.

Above all I would like to express my deep appreciation for the encouragement D.C. Mikulecky has shown over the years. He has consistently kept an open mind to new ideas and has incorporated my own models to his own theories and models while fighting fiercely with sword, word of mouth and pen to defend my contributions.

Last but not least, I would like to thank my wife Beth who has tolerated a high level of increase in home entropy as little (and relatively large) pieces of paper diffused from room to room...

Leonardo Peusner
Portland, Maine, 1985

TABLE OF CONTENTS

Preface   vii

Chapter 1
GRAPHS AND NETWORKS
1.1   INTRODUCTION   1
1.2   GEOMETRIC GRAPHS   4
1.3   NETWORKS   8
1.4   KIRCHHOFF'S LAWS   10
1.5   VECTOR CALCULUS VS. NETWORK THEORY   13
1.6   TELLEGEN'S THEOREM AND ORIENTABILITY   17
1.7   TOPOLOGICAL PROPERTIES AND GRAPH THEORY   20
1.8   CONTINUOUS ANALOG OF TELLEGEN'S THEOREM   23
1.9   KIRCHHOFF'S LAWS AS VECTOR SPACES   25
1.10 BASIC PROBLEMS OF KIRCHHOFFIAN NETWORKS   27
1.11 NETWORK ANALYSIS AND SYNTHESIS   31
1.12 SIMPLE RESISTIVE ANALYSIS   32
1.13 SIMPLE SYNTHESIS: THE INVERSE PROBLEM   33
1.14 BOUNDARY CONDITIONS: LOAD RESISTANCES   34
1.15 DUAL NETWORK: SYNTHESIS USING A FLOW SOURCE   36
1.16 TWO PORT MACHINES   37
1.17 A WORD OF CAUTION: LINEAR VS. NON-LINEAR   45
1.18 GENERAL RECIPROCITY RELATION   46
1.19 THE T AND PI NETWORKS   48
1.20 LINEAR INDEPENDENCE AND DIMENSION OF THE VECTOR SPACE   52
1.21 THE CUT SET AND TIE SET MATRICES   55
1.22 FUNCTIONAL INVERSION: LEGENDRE TRANSFORMATIONS   57
1.23 DISSIPATION, CONTENT AND COCONTENT   59
Appendix 1 SOME FUNDAMENTAL ALGEBRAIC CONCEPTS   AND THEIR
            NETWORK COUNTERPARTS   60
Appendix 2 NETWORK INCIDENCE MATRICES   66
A.1.1 NETWORK BRANCHES   66
A.1.2 THE NODAL MATRIX   67
A.1.3 MESH EQUATIONS   72
A.1.4 LOOP NETWORK EQUATIONS. THE TIE SET MATRIX   75
A.1.5 NODE PAIR EQUATIONS BASED ON THE CUT SET MATRIX   79
A.1.6 EQUATIONS IN RLC NETWORKS   82

Chapter 2
NETWORK THERMOSTATICS : THERMOSTATICS AND CONNECTED NETWORKS

2.1	INTRODUCTION	86
2.2	WHY NETWORKS?	89
2.3	THERMOSTATIC PORTS	90
2.4	TRANSFORMATIONS AMONG NETWORKS	99
2.5	THE METRIC	104
2.6	THE RECIPROCAL METRICS	105
2.7	CONNECTIVITY: THE T AND PI NETWORKS	106
2.8	THE IDEAL GAS	106
2.9	HYBRID IDEAL GAS DESCRIPTION	108
2.10	AN APPLICATION OF THE TRANSFORMATION MATRICES	109
2.11	EQUILIBRIUM OF HOMOGENEOUS SYSTEMS	110
2.12	DESCRIPTIONS BASED ON ENTROPY NETWORKS	113
2.13	CONNECTIVITY AND RECIPROCITY	115
2.14	THE FUNDAMENTAL METRIC TENSORS	116
2.15	GEOMETRICAL CHARACTERISTICS OF RESISTIVE NETWORKS	117
2.16	GEOMETRICAL PROPERTIES IN GENERAL RESISTIVE NETWORKS	120
2.17	METRIC THERMOSTATICS	122
2.18	RECIPROCITY OF INTERSTATE INNER PRODUCTS AND TELLEGEN´S THEOREM	123
2.19	APPLICATIONS OF THE GENERAL RECIPROCITY CONDITION	125
2.20	RELATIONSHIPS BETWEEN HEAT CAPACITIES AT CONSTANT PRESSURE AND AT CONSTANT VOLUME	128
2.21	ENTROPY LIKE FUNCTIONS FROM GRS	128
2.22	THERMOSTATIC EQUATIONS FROM ENTROPY NETWORKS: THERMODYNAMIC EQUATION OF STATE	129
2.23	ENERGY OF THE IDEAL GAS	130
2.24	UNIQUENESS OF THE EQUILIBRIUM STATE FROM "LOCAL" SECOND LAW ASSUMPTION	131
2.25	THE "MISSING" ENTROPY AND ENERGY RECIPROCITIES CONTAINED IN GRS AND GRE	132
2.26	UNIFICATION OF THERMODYNAMIC CAUSE-EFFECT	133
2.27	CONNECTIVITY AND THE CHEMICAL POTENTIAL	135
Appendix 1	REPRESENTATION IN H(u)	140
Appendix 2	INTEGRATION IN L^2 REDUCES TO A LUMPED PARAMETER NETWORK	142

Chapter 3
ON THE CONSISTENT DEFINITION OF KINETIC POTENTIALS IN DIFFUSION - REACTION - CONVECTION COUPLING AND RELATED TRANSPORT PROBLEMS

3.1	BACKGROUND	144
3.2	THE MEMBRANE INTEGRATION PROBLEM	145
3.3	DIFFUSION AS A DEGENERATE EXAMPLE	149
3.4	FIRST ORDER REACTION DIFFUSION SYSTEM	150
3.5	CONSISTENCY BETWEEN KINETICS AND THERMODYNAMICS	151
3.6	GLOBAL REACTION DIFFUSION EQUATIONS: REACTION IMBEDDING IN DIFFUSION SPACE	152
3.7	FIRST ORDER REACTION DIFFUSION SYSTEM	153
3.8	NETWORK REPRESENTATION	154
3.9	NETWORK EQUATIONS	157
3.10	SOLUTIONS TO THE NETWORK EQUATIONS	159
3.11	DISCUSSION OF THE SOLUTIONS	166
3.12	FACILITATED TRANSPORT	168
3.13	NEED FOR CONNECTIVITY IN PASSIVE PROCESSES	170
3.14	NON-LINEAR CONVECTION- DIFFUSION EQUATIONS	172
3.15	DISCRETE REPRESENTATION OF THE CONVECTION-DIFFUSION PROBLEM	173
3.16	LIMIT OF DOMINANT DIFFUSION RANGE: MANEGOLD AND SOLF EQUATIONS	176
3.17	CONCENTRATION PROFILES	178
3.18	THE CONVECTION REACTION DIFFUSION PROBLEM	178
3.19	DISCUSSION OF SOLUTIONS	183
Appendix	NON-ADDITIVE TERMS	184

Chapter 4
MICROSCOPIC REVERSIBILITY, ONSAGER THERMODYNAMICS AND KINETIC NETWORKS

4.1	INTRODUCTION. RECIPROCITY IN ONSAGER THERMODYNAMICS	191
4.2	MICROSCOPIC FOUNDATION OF ONSAGER THERMODYNAMICS	194
4.3	INTEGRATION OF MICROSCOPIC VARIABLES	197
4.4	MACROSCOPIC EQUATIONS AND TOPOLOGICAL AVERAGING OF THE MICROSCOPIC NETWORK	198
4.5	EUCLIDEAN VECTOR SPACES AND ONSAGER THERMODYNAMICS	200
4.6	MICROSCOPIC PROOF OF ONSAGER RECIPROCITIES	204
4.7	PROBABILITIES OF JOINT NON-EQUILIBRIUM STATES	208
4.8	MICROSCOPIC REVERSIBILITY	211
4.9	KINETIC REVERSIBILITY AND DETAILED BALANCE: THE TRIANGULAR CHEMICAL REACTION	212
4.10	STEADY STATE VS. EQUILIBRIUM	217
4.11	NETWORK EXTENSION FOR THE PROBLEM OF MULTIPLE CHEMICAL REACTIONS WITH LINEAR MASS ACTION	218

4.12 EXAMPLE: TRANSPORT COUPLED TO CHEMICAL REACTION 220
4.13 BALANCED LOOPS, TOPOLOGICAL AND METRICAL REVERSIBILITIES 225
4.14 MACROSCOPIC RECIPROCITY AND CONNECTIVITY 226
4.15 MACROSCOPIC ARGUMENT FOR THE RECIPROCITY OF LINEAR (ONSAGER) SYSTEMS 227
4.16 STATIONARY PROPERTIES OF THE ONSAGER SYSTEM: UNIQUENESS OF FLOW DISTRIBUTION IN THE STATIONARY STATE 228
4.17 FUNCTIONS OF STATE IN THE THERMODYNAMICS OF IRREVERSIBLE PROCESSES 230
Appendix CHANGES IN REFERENCE FRAME WITH RESPECT TO BARYCENTRIC VELOCITY 235

Chapter 5
MACROSCOPIC SYMMETRIC AND HYBRID NETWORKS
5.1 INTRODUCTION 242
5.2 DUALITY BETWEEEN ONSAGER FORMULATIONS AND HYBRID EQUATIONS 243
5.3 THOMSON ANALYSIS OF THE PELTIER HEAT 244
5.4 THE KEDEM-KATCHALSKY EQUATIONS IN ONSAGER THERMODYNAMICS 245
5.5 DERIVATION OF KEDEM-KATCHALSKY EQUATIONS USING NETWORK TRANSFORMATIONS 248
5.6 LIQUID JUNCTION POTENTIALS 252
5.7 ELECTROKINETIC PHENOMENA 255
5.8 GOING BEYOND THE ONSAGER FORMULATION: NON-RECIPROCAL ENERGY CONVERSION 259
5.9 MEASUREMENTS AND PATHS 259
5.10 TWO PORT WITH LINEAR LOAD 261
5.11 EXPERIMENTAL DETERMINATION OF THE OPTIMUM LOAD 264
5.12 EXPERIMENTAL DETERMINATION OF OPTIMUM EFFICIENCY 267
5.13 THE CENTRAL ROLE OF HYBRID PARAMETERS IN ENERGY CONVERSION 268
5.14 NON CYCLIC ENERGY CONVERSION IN THERMOSTATICS 269
5.15 THE COUPLING PARAMETER Q 269
5.16 ENERGY CONVERSION IN MULTIPLE FLOW, LINEAR SYSTEMS 271
5.17 ENERGY CONVERSION FOR ATTACHED IDENTICAL TWO PORTS 277

Chapter 6
TIME BEHAVIOR AND EVOLUTION OF NON-EQUILIBRIUM THERMODYNAMIC AND KINETIC SYSTEMS
6.1 INTRODUCTION: DISSIPATIVE VS. STORING CONSTITUTIVE LAWS 279

6.2	LAGRANGE'S EQUATIONS AND INVARIANCE	281
6.3	MECHANICS AS AN ANALOG COMPUTER FOR NETWORKS	282
6.4	INSTANTANEOUS DISSIPATION AND STORAGE	284
6.5	LAGRANGE'S EQUATIONS ARE IMPLICIT IN TELLEGEN'S THEOREM	288
6.6	EFFECTS OF COMPARTMENTAL STORAGE AND KINETIC TRANSIENTS	291
6.7	TRANSMISSION LINE EQUATIONS DERIVED FROM THE CONNECTED TOPOLOGY	293
6.8	DISSIPATION FUNCTION AND EFFICIENCY IN THE STEADY STATE	298
6.9	GENERALIZED TWO PORT CONVERTER	299
6.10	OPTIMUM LOAD IMPEDANCE	304
6.11	RELATION TO THE STEADY STATE RESULTS	304
Appendix	COMMENTS ON THE STATIONARITY OF THE STEADY STATE AND THE "PRINCIPLE" OF MINIMUM ENTROPY PRODUCTION	306

Chapter 7
PIECEWISE NONLINEAR NETWORKS: STEADY AND OSCILLATORY

7.1	INTRODUCTION	308
7.2	THERMODYNAMIC MODELS OF MUSCLE CONTRACTION	309
7.3	CATASTROPHE THEORY VS. NONLINEAR NETWORK OSCILLATIONS	315
7.4	THE NEGATIVE RESISTANCE	320
7.5	THE TEORELL OSCILLATOR	321
7.6	STATIC NETWORK REPRESENTATION AND DYNAMIC ANALYSIS	327
7.7	INERTIAL DELAYS	332
7.8	PHYSICAL REALIZATION OF THE NEGATIVE RESISTANCE	334
7.9	EXPERIMENTAL TEST OF UNSTABLE STATES	336
7.10	COMPLETE NETWORK MODEL	336
7.11	DISSIPATION FUNCTION	338
7.12	ASTABLE OSCILLATIONS, TRIGGERING AND "ACTION" POTENTIALS	339
7.13	EXPERIMENTAL TESTS	339

Chapter 8
PERSPECTIVES   343

REFERENCES   349

LIST OF SYMBOLS   363

INDEX   367

Chapter 1

GRAPHS AND NETWORKS

1.1 INTRODUCTION

In this monograph we shall only deal with planar networks. These are close relatives of the planar graphs--i.e., the familiar geometric graphs which can be drawn on a sphere by connecting points with non-intersecting lines. Networks are directed graphs in which the connecting edges are assigned arrows representing the flow of matter, information, money, or any arbitrary variable which can be quantitated as the rate of change of an extensive quantity. In addition, the networks considered here also have potentials associated with the nodes or vertices; this property allows the definition of conservative forces.

The use of planar graphs or networks is not as restrictive as it might seem at first. These networks will be used to represent fundamental properties of physical quantities which sometimes precede measurement and are, therefore, based on purely logical constructs. In the absence of other information we cannot decide, a priori, whether forces or displacements should be used as the more primitive, or causative set, leading to the secondary observed effects. As a result, a primitive theory or model must be able to allow **either** set (forces or displacements) to act as the fundamental driving cause. Moreover, both representations should have the same weight and be equally valid. In graph theory, this role can easily be ascribed to the dual transformations between graphs, which allow--from a physical viewpoint -- forces to transform into flows, and vice versa. Given that only planar graphs have duals, the reason for their central role in physical representations should be apparent.

There are two fundamental properties to be considered in a network: its **topology** and its **geometry**. Topological properties depend only on the assignment of connections between different points, or the possible combinations of paths that join a given node with other nodes. These are independent of yardsticks-- i.e., measurements-- and it is irrelevant, in this context whether a graph is drawn on a piece of paper of 1 cm^2 or a globe the size of the earth: if the small graph can be stretched without tearing so that the material points in the imaginary rubber like surface keep the same neighbor associations as the stretching takes place, the two graphs are "equal" from the topological point of view. The mapping or transformation carried out is an **isomorphism** (ref.4).

There is a second equal sign to be introduced, which is related to the more familiar idea of measurement and geometrical congruence: from a geometrical point of view, two graphs are equal if they can be superimposed exactly, so that the corresponding edges have the same length. Two networks have this congruence if both the potentials defined at the vertices or nodes and the flows through their branches are the same for any set of values which can be chosen for either forces or flows. This also requires that there be specific constitutive laws relating the value of a force and a flow at each edge. In some cases, all the geometrical properties of the system can be recovered from a single mathematical expression, the **metric**, which will be defined later on.

Clearly, the topological properties of a physical system or process must be related to the logical, premetrical structure of the theory, while geometrical properties will be related to the phenomenology. Topology (or <u>analysis situ</u>) does not require measurement or calibration; these are geometrical properties. One might expect, then, that physical processes would display very few topological properties as these appear to be unknown until a measurement is performed. In fact, there are so many "physical" results which can be derived purely from abstract topological considerations that one is forced to wonder how much physics is "real" and how much is "logical" and only based on the internal consistency requirements imposed by our own brains and models.

We shall not deal with some of the rather sophisticated points raised by the various branches of topology, both because I am not very proficient in some of these fields and because they tend to obscure, rather than clarify the physical issues. Some primitive ideas should be introduced, however.

A **manifold** is, in very loose terms, a collection of points which are connected in some fashion. We shall deal with n-dimensional, continuous, differentiable manifolds -- the subject of study of differential topology --, in which n numbers, the coordinates, serve to specify a functional property at each point of the manifold. These coordinates can be "smoothly" related from one point to a neighboring point. A manifold defines surfaces of various dimensions.

Using the concept of a continuous manifold one can define curves, vectors, local frames of reference, etc.(ref. 160). There are many simple examples of manifolds. Circles form a one dimensional manifold, as one quantity (the radius) is sufficient to specify the circle. Ellipses form a two dimensional manifold. The equilibrium conditions of an ideal gas in which T, V and the number of moles change, form a three dimensional manifold (ref.264)

The coordinates serve two purposes: the first is to look at changes in direction which take place as a vector moves from point to point in the manifold (a topological property), the second is to define the lengths of vectors in terms of the manifold quantities (a geometrical characteristic). The first problem is dealt with by affine geometry and the second part is considered by differential geometry. The latter looks at small neighborhoods of a point P in which the length of a vector can be defined by means of an Euclidean (flat space) length or metric. The simplest such metric is Pythagoras theorem, which gives the length of a vector $\mathbf{x}$ placed at the origin of the coordinate system in terms of the squares of its n perpendicular components x_i,

$$||\mathbf{x}||^2 = x_1^2 + x_2^2 + x_3^2 + \ldots + x_n^2 . \qquad (1)$$

or, if each axis has a different calibration $\sqrt{a_{ii}}\, x_i$,

$$||\mathbf{x}^2|| = a_{11} x_1^2 + a_{22} x_2^2 + \ldots + a_{nn} x_n^2 . \qquad (2)$$

Moreover, in the case in which the axes are not orthogonal,

$$||\mathbf{x}^2|| = a_{11} x_1^2 + a_{22} x_2^2 + \ldots + a_{nn} x_n^2 \qquad (3)$$
$$+ a_{21} x_1 x_2 + a_{31} x_1 x_3$$
$$\ldots\ldots + a_{ij} x_i x_j ,$$

in which the cross coefficients serve to define the angle θ_{ij} made by axes i and j from the expression

$$\cos \theta_{ij} = a_{ij}/\sqrt{a_{ii} a_{jj}} . \qquad (4)$$

Geometrical theories of physical processes treat invariant physical quantities as lengths which are defined in manifolds.

Networks have equivalent representations for each of the geometrical concepts. The network inputs and outputs, the ports, are the coordinates of the manifold, the internal branches provide a Pythagorean imbedding and the angles between coordinates reflect the "energy" transfer properties between ports.

The invariant characteristics of the Pythagorean length appear as an equality between the power dissipated at the ports and inside the network. But in addition, the network model has several other features: i) it provides a planar representation of an n dimensional manifold, ii) it allows the visualization of indefinite n-dimensional metrics, iii) it separates topological from geometrical considerations, iv) it associates forces and displacements in logically meaningful ways, v) it explains the source of the ubiquitous reciprocities found in the mathematical descriptions of many physical processes --e.g,Maxwell´s, Onsager´s, Betti´s etc.

1.2 GEOMETRIC GRAPHS

The most familiar types of graphs are geometric graphs which consist of points or vertices connected by curves or edges. The planar geometric graphs --the only type of graph we shall deal with in this monograph-- can be represented by two dimensional drawings in which the edges do not cross, such as the example shown in FIGURE 1. By contrast, non-planar graphs--such as the one shown in FIGURE 2--are drawings in which edges cross.

Graph Theory deals with the assignment of associations between pairs of vertices and specific edges, as well as with the more

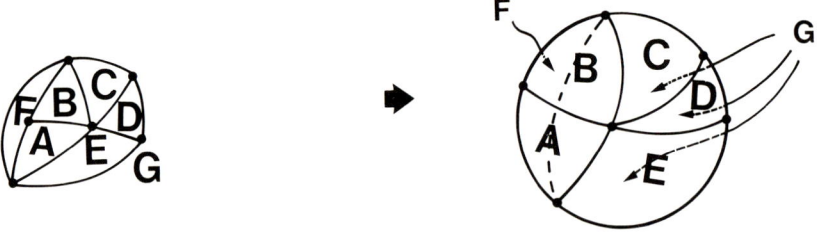

Fig. 1.1 A planar graph is characterized by having no crossing edges. Planar graphs can be drawn on a sphere, as shown. The capital letters denote the faces in the solid. (Note the external face G).

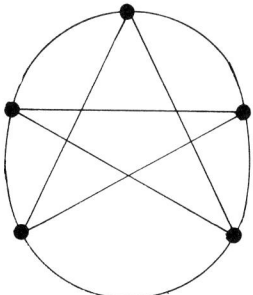

Fig. 1.2 Example of a non-planar graph. These cannot be drawn on a sphere without having crossing edges.

complex combinatorial and metric problems which arise naturally.

An arbitrary graph may be specified by giving two distinct lists which consider all possible pairs of vertices, a list of edges and arrows which associate a given edge with a specific pair of vertices. For example, the geometrical graph of FIGURE 3 can also be represented by the list

EDGES	VERTICES
$E_1 \longrightarrow$	$v_1 v_2$
$E_2 \longrightarrow$	$v_1 v_3$
$E_3 \longrightarrow$	$v_2 v_3$
$E_4 \longrightarrow$	$v_2 v_4$
$E_5 \longrightarrow$	$v_3 v_4$

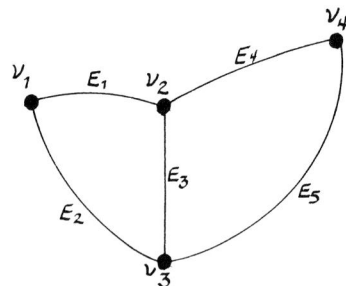

Fig 1.3 Graph corresponding to the mapping given by the list in the text.

while the graph given in FIGURE 4 is represented by the list

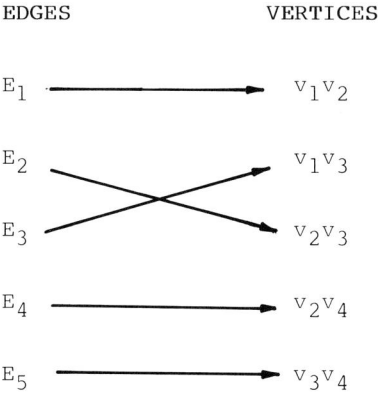

EDGES	VERTICES
E_1 ⟶	$v_1 v_2$
E_2 ⟶	$v_1 v_3$
E_3 ⟶	$v_2 v_3$
E_4 ⟶	$v_2 v_4$
E_5 ⟶	$v_3 v_4$

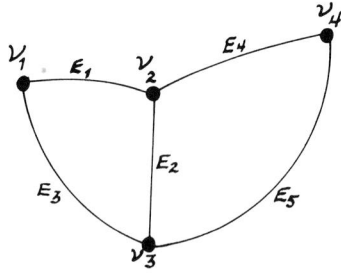

Fig. 1.4 Graph specified by the mapping in the text; this is the same, as the previous one except for the fact that two edges have been interchanged.

which is the same as the previous one, except that different labellings of edges and vertices have been made. By contrast the graph given by the list

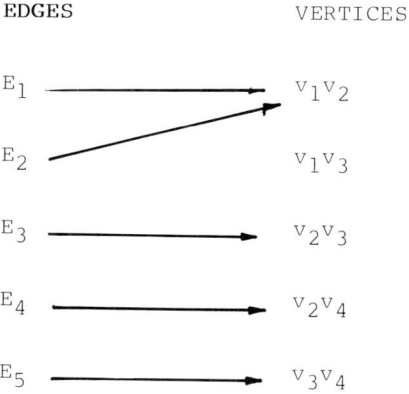

EDGES	VERTICES
E_1	$v_1 v_2$
E_2	$v_1 v_3$
E_3	$v_2 v_3$
E_4	$v_2 v_4$
E_5	$v_3 v_4$

has not only a different assignment of edges, but also a different drawing (FIGURE 5). Although we shall consider here only edges in which the two associated vertices are distinct, it is also possible to have associations of the form

$$E_i \longrightarrow v_j v_j$$

which indicates a loop at node j .

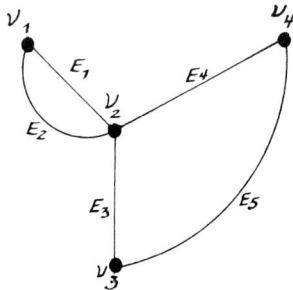

Fig. 1.5 The graph shown has the same number of edges and vertices, but the different assignment or mapping leads to a different drawing.

We can now define a graph, more precisely, in terms of the lists of edges and pairs of vertices and their association or mapping. (ref. 40)

Defintion. A **graph** consists of (disjoint) set of edges E, a (non-empty) set of vertices V and a mapping o of E into V & V, in which V & V is the set of (unordered) distinct pairs of vertices.

An **edge progression**, **edge sequence or branch** sequence (depending on the author) is a finite sequence of consecutive edges. If the edges are all distinct we refer to the sequence as a **chain**, if the initial and final edges are different the path is open, while if the final and initial edges are the same the path defines a circuit or **loop**.

A graph is connected if any pair of vertices is joined by one or more chains, other graphs are disconnected. A connected graph which has no circuits is called a **tree** . Given that there are, in general, many trees for a given graph, the collection of trees belonging to the same graph is sometimes called a **forest**. Clearly, removal of any edge of a tree leads to a disconnected graph.

Given a connected graph G, it can be partitioned into two subgraphs: a tree (which has no circuits or loops, by definition) and edges which are not included in the tree. Edges which belong to the given tree we refer to as **branches** and those which do not belong to the tree are called chords or links. Every time a link is added to a tree, a loop or circuit appears, as shown in FIGURE 6 . The concepts of tree and link are helpful in setting up network equations.

1.3 NETWORKS

The basic graphs considered above are undirected graphs in which there is no directionality assigned to a given edge. In many cases of interest one is concerned with edges which are assigned a positive direction; these are called **directed graphs**.

Definition. A **directed graph** is specified by a non empty set of vertices V, a set of arcs or directed edges E and a mapping of E into V X V--in which V X V is the cartesian product of the set V

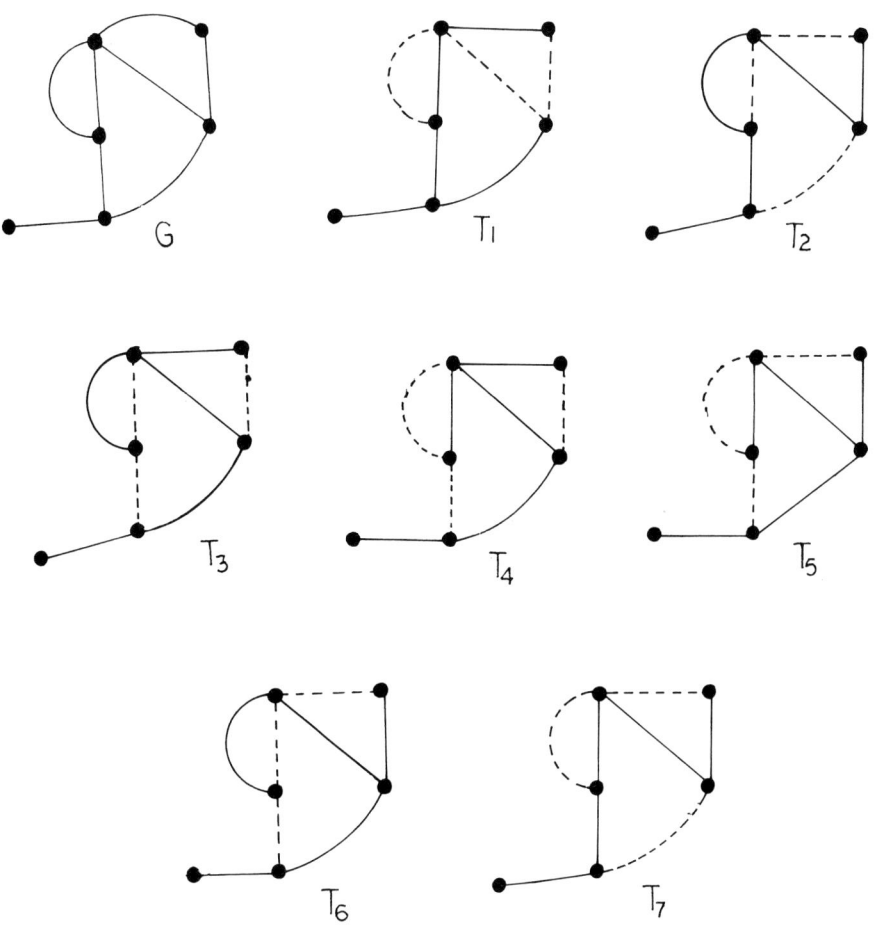

Fig. 1.6 Example of a graph and the trees of its forest. Dotted lines are associated with a given tree. Note that each time a link is added to a tree a loop is restored.

and its own copy and which consists of the ordered pairs belonging to the set V. A directed graph may be drawn by means of edges having associated arrows, as shown in FIGURE 7. The directions assigned to the various edges are clearly arbitrary.

Networks are directed graphs in which the edges are associated

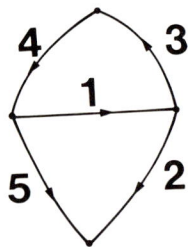

Fig. 1.7 A directed graph has specific, though arbitrary, directions assigned to the edges.

with motion or flow of about anything-- money, matter, information, etc. Two examples are given in FIGURE 8.

1.4 KIRCHHOFF´S LAWS

In the networks found in physical applications, there are two fundamental conservation laws which appear repeatedly.

These are Kirchhoff´s laws. Nodes or vertices are assigned assigned potentials e_i which serve to define a voltage

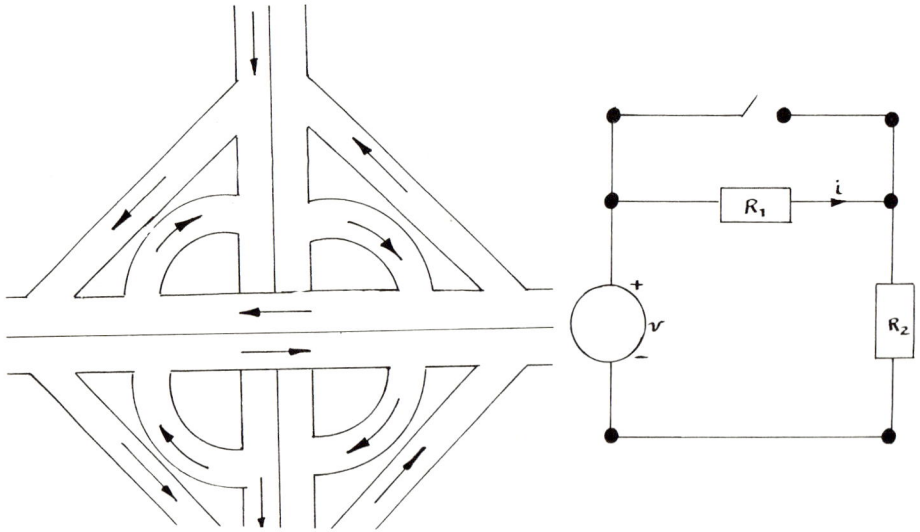

Fig. 1.8 Two examples of networks: a highway and an electrical circuit.

difference or force X_{ij} between two nodes i and j,

$$X_{ij} = e_i - e_j$$

This difference is considered positive if the plus term coincides with the tail of the arrow defining the direction of the edge and the minus term coincides with the head of the arrow. It is simple to show that the sum of all the forces X_{ij} around an arbitrary loop is zero; consider a loop 1,2,...,i,1 and add the forces around the loop to obtain

$$X_{12} + X_{23} + \ldots + X_{-i} + X_{i1} =$$

$$= (e_1 - e_2) + (e_2 - \ldots) + \ldots + (e_i - e_1) = 0 . \quad (5)$$

This is <u>Kirchhoff's " Voltage" Law</u> (K.V.L.). I have introduced quotation marks to stress that the voltage part of the law is purely incidental. The important result is topological : if a potential is given at each point and forces, voltages, or any other physical quantities are defined as differences in potentials K.V.L. will follow.

An example of K.V.L. is given in FIGURE 9. In this case the sum

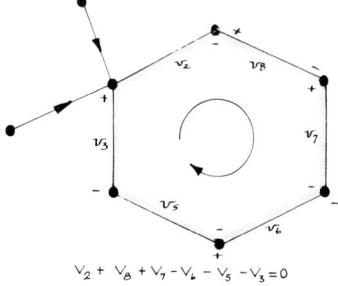

$V_2 + V_8 + V_7 - V_6 - V_5 - V_3 = 0$

Fig. 1.9 Example of Kirchhoff's Voltage Law for an arbitrary node in a network.

of voltages around the loop shown is

$$v_2 + v_8 + v_7 - v_6 - v_5 - v_3 = 0.$$

A second physical assignment made to these networks is that the flowing "material" is conserved--i.e., each edge E_i is assigned a flow J_i in the direction of the given edge and the sum of all flows entering any node is zero. This is Kirchhoff's "Current" Law (K.C.L.).

A network example of K.C.L. is given in FIGURE 10, in which the sum of currents for the node indicated is

$$i_1 + i_4 - i_3 - i_2 = 0.$$

More generally, Kirchhoffs laws are stated as follows:

$$\sum_{\text{loop}} X_k = 0 \quad \text{and} \quad \sum_{\text{node}} J_k = 0$$

Kirchhoff's laws lead to the powerful result known as Tellegen's theorem, which will be considered below.

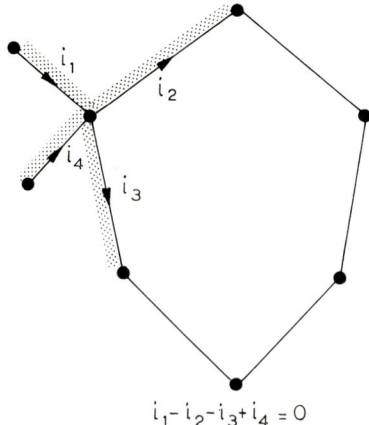

$i_1 - i_2 - i_3 + i_4 = 0$

FIGURE 1.10 An example illustrating Kirchhoff's Current Law.

1.5 VECTOR CALCULUS VS. NETWORK THEORY

There are some interesting and basic relations between the concepts of Vector Calculus and those of Network Theory. These are not coincidental, of course, as they follow from fundamental and rather simple topological concepts.

A field theory is a device to knit cause effect "space" conceptually so that disturbances have an immediate effect on their neighbors rather than act at a distance. Thus, Maxwell's theory of electromagnetism postulates a "knitting" between the electric and magnetic fields to explain how it is possible for electrical disturbances to propagate, while Einstein's general theory of relativity knits space- time to explain how gravitational energy propagates. (In fact, both theories predict the propagation of these disturbances). There is nothing better or worse about a connected, or neighbor theory vs. a disconnected or action at a distance theory, except for the abhorrent nature of the thought that cause effect has long range "jumps"; a view most vehemently expressed by Faraday. Connected topological spaces are therefore a good theatre for our conceptual interpretation of physical phenomena (ref. 160)

Thermodynamics, incidentally provides another example of a field theory in which small displacements or distortions in state space take a system from a set of variables S_1 to a second set of variables S_2 through a continuously connected path. Thermodynamic theory does not consider action at a distance "jumps"--i.e., the irreversible processes. These are represented, instead by some connected path which can go between the same initial and final states. This is, of course, a limitation of the theory because-- unlike the case of gravitation or electromagnetism-- irreversible jumps caused by dissipating processes make perfect physical sense. Thus a desirable goal of a network theory of thermodynamics would be to provide such an irreversible representation.

Networks provide a discrete representation of the same concepts and show the algebraic and topological nature of the field calculus ideas. Kirchhoff's laws knit networks in the same way that fields are knit.In practice one can go back and forth between the two, depending on the nature of the problem. It is mandatory, however, to distinguish network theory as a representation-- what I called "pseudoelectrical" models in my thesis (ref. 182) from electrical models. The latter are only approximations of the field theory equations of electrodynamics in the case in which time variations are slow. By contrast, the networks used here are an exact representation of the algebraic characteristics of the process being described. A third use of networks is in modelling the superficial charactersitics of phenomena. These are useful in some areas of Physiology and Biophysics, for example, but will not be considered in the present context (refs. 182,147-158).

The main difference between field theory and network theory is the specification of the paths of motion: in field theory these are specified by the direction of steepest descent (the gradient), while networks retain this idea by imagining tubes of flow through which motion is trapped while the force is defined as a difference in potential between the ends of the tubes. As we cannot describe the complete potential surface using a network we chose strategic points or nodes where the potential is defined.

The reason that so many problems have network representations is that each field theory operator has a counterpart in network theory, as described below. The further question may arise as to the generality of field theory itself. Is it not possible, for example, to discover a new physical law which cannot be fit into a field theory scheme? If it is required that the observed laws be independent of the frame of reference chosen to describe them -- i.e., if we provide a covariant description, in Einstein's terms-- these laws must be described in terms of entities which have a physical reality independent of the frame of reference. These are the **tensors** and they include scalar invariants (zeroth order tensors) and vectors (first order tensors). They also include certain tensorial operators which are the ones which must be used as building blocks for the physical laws.

The interesting point is that in cartesian frames of reference (euclidean or flat spaces) there are only four invariant operators: the gradient, the curl, the divergence and the Laplacian, all of which have an equivalent expression in terms of Kirchhoff's laws.

These cartesian operators are (ref. 213):

1. The **gradient**, defined as

$$\text{gradient } \phi = \nabla \phi = (\partial \phi / \partial x) \, i_x + (\partial \phi / \partial y) \, i_y + (\partial \phi / \partial z) \, i_z,$$

which points in the direction of steepest descent in a three dimensional surface. This can be easily seen by considering the change in energy (equal to the work done in a conservative system) upon the displacement of the force , $dE = - \text{grad}\phi \cdot dr$, from which it is clear that the maximum decrease in energy takes place when the gradient is a maximum--i.e., in the direction of steepest descent. In network terms, we ignore all the other possible transition paths which are never chosen by nature and trap the motion in a single direction, through a branch. The end points of the branch (+) and (-) correspond to the initial and end points of the path in the potential surface.

(The <u>directional</u> derivative aspect of the gradient should be stressed and this has a close correlation in the network as all voltages (forces) are added in the direction of a flow or flows are added in the direction of the forces.)

2. The **divergence** measures whether there are sources in the space considered. The familiar 3 dimensional formula,

$$\text{divergence } \mathbf{A} = \nabla \cdot \mathbf{A} = (\partial A_x / \partial x) \mathbf{i}_x + (\partial A_y / \partial y) \mathbf{i}_y + (\partial A_z / \partial z) \mathbf{i}_z , \qquad (6)$$

is analogous to specifying, in network terms , how Kirchhhoff´s current law is offset by current sources. Note, however that K.C.L. applies to an arbitrary number of dimensions .

3. The **curl** measures the change in a vector when going around in a closed path and is given, in three dimensional terms by the familiar expression

$$\text{curl} \mathbf{A} = \nabla \times \mathbf{A} = (\partial A_z / \partial y - \partial A_y / \partial z) \mathbf{i}_x + (\partial A_x / \partial z - \partial A_z / \partial x) \mathbf{i}_y + (\partial A_y / \partial x - \partial A_x / \partial y) \mathbf{i}_z . \qquad (7)$$

In network terms, the curl corresponds to a voltage offset from Kirchhoff´s Voltage Law when going around a closed loop. A pictorial example of the curl and gradient is given in FIGURE 11.

When the force is conservative and the curl is zero a potential

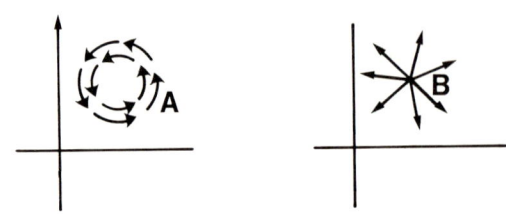

Fig. 1.11 The vector field **A**, described by $\mathbf{A} = -(ay/r)\mathbf{i}_x + (ax/r)\mathbf{i}_y$ has zero divergence, while the field $\mathbf{B} = (ax/r)\mathbf{i}_x + (ay/r)\mathbf{i}_y$, in which $r^2 = x^2 + y^2$, has zero curl (after ref. 213).

can be defined at each point and, moreover, the reciprocities

$$(\partial A_z/\partial y) = (\partial A_y/\partial z),$$
$$(\partial A_x/\partial z) = (\partial A_z/\partial x) \text{ and} \tag{8}$$
$$(\partial A_y/\partial x) = (\partial A_x/\partial y)$$

follow. The network analog is Tellegen's theorem, considered in the next section.

4. The fourth operator, the **Laplacian**, arises by combining the divergence sources with the gradient expression of the force field to give equations of the form

$$\nabla \cdot \nabla \phi = \nabla^2 \phi = \text{sources}, \tag{9}$$

such as Poisson's equation, or Laplace's equation (if the sources are zero).

There are two secondary aspects of the vector calculus view which are not explicitly stated. First, differential calculus as applied to Mechanics or Thermodynamics is usually a linear model in which a given increment is given in terms of first order variations in the coordinates --the reciprocities given above for the curl zero case depend on this fact. The network equivalent of linear calculus corresponds to the introduction of Ohm's laws at the branches, but the Kirchhovian network itself is a coordinate free (geometrical) representation.

In a non-linear coordinate system of arbitrary dimensionality, the operators can be given tensorial generalization (ref. 160), but such situations will only be discussed as required.

1.6 TELLEGEN'S THEOREM AND ORIENTABILITY

Planar graphs can be represented as orientable solids on a sphere (ref. 4). This means that the solid consists of faces-- i.e., the non intersecting loops of the graph-- which can be given the same direction, say counterclockwise, so that each oriented edge sees a loop flowing in the same direction of the edge and a loop flowing against the direction of the edge. If the value of the flow in the given edge is assumed to be the algebraic sum of mesh (i.e., face) flows neighboring the edge (with + or - sign, depending on whether the direction of flow goes with or against the direction of flow of the edge) it is clear that K.C.L. must be obeyed at each node, as whatever flow goes into the node comes out (FIGURE 12).

With this interpretation of K.C.L. a further result is obvious: the product of branch and link voltages and currents for such a closed surface is zero,

$$\sum_{\text{all branches}} x_i y_i = 0 , \qquad (10)$$

because this summation can be decomposed into products of a given mesh current times a sumation of voltages around the given loop or mesh (which is zero, by K.V.L.). This result is one of the many versions of Tellegen's theorem. We now proceed to prove the above result in a simple way.

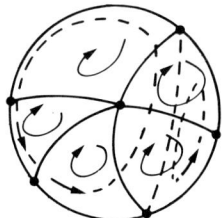

Fig. 1.12 The assumption that there are mesh (or face) flows is equivalent to Kirchhoff's Current Law, as whatever "current" goes into one node comes out of it. Moreover, each branch flow is a linear combination of mesh flows.

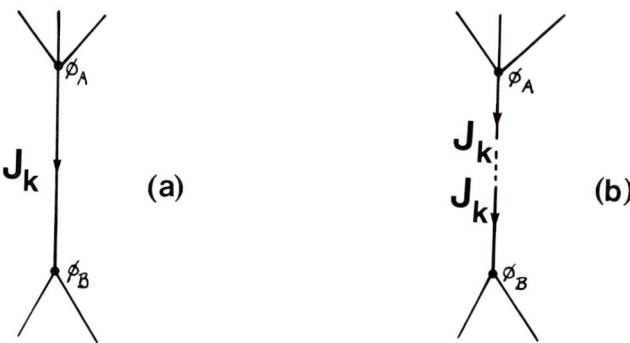

Fig. 1.13 Graphical procedure to split the term $J^k (\phi^A - \phi^B)$, at the left, into two terms $J^k (\phi^A)$ and $-J^k (\phi^B)$, as shown in the right hand drawing.

Consider a branch k in the network, as shown in FIGURE 12. A typical term in the Tellegen summation may be rewritten in the form

$$X^k J^k = J^k \phi^A - J^k \phi^B \qquad (11)$$

in which K.V.L. has been used to express the force as a difference in potentials. We can also use the sign convention for flows by assigning the (+) sign to a flow leaving a node and a (-) sign to the flow entering the node, so that the above equation may be interpreted to be

$$X^k J^k = J^k \phi^A + (- J^k) \phi^B. \qquad (12)$$

When the products $X^k J^k$ are added for all branches, each potential ϕ^i at node i will be multiplied by the sum of flows corresponding to that node, so that the flows leaving the node appear with a (+) sign while those incident appear with a (-) sign. From K.C.L. it then follows that each term multiplying a potential in the summation is zero, so that Tellegen's result follows,

$$\sum X^k J^k = 0, \qquad (13)$$

in which the summation is taken over all branches. By a simple manipulation of terms, Tellegen's theorem may be recast in a more useful way. If the summation is split into terms corresponding to the ports and terms corresponding to the remaining branches inside the network, it follows that

$$\sum_{network} X_k J_k = -\sum_{ports} X_p J_p . \qquad (14)$$

Moreover, it is convenient to redefine the positive direction of flow at the ports to eliminate the (-) sign in the above equality: a port flow is considered positive if it enters a (+) node and leaves a (-) node. The final Tellegen expression therefore becomes

$$\sum_{network} X_k J_k = \sum_{ports} X_p J_p . \qquad (15)$$

Alternatively, we shall also consider the vectors **X** and **J** representing all forces and flows in the network (including the ports). In this notation Tellegen's theorem becomes

$$X^T \cdot J = 0 , \qquad (16)$$

or, alternatively,

$$J^T \cdot X = 0 . \qquad (17)$$

These expressions stress the orthogonal nature of the force and flow vectors which obey Kirchhoff's laws.

From a physical point of view, however, Tellegen's theorem expresses a conservation of a bilinear form:" whatever appears inside came from outside". The most obvious interpretation is, of course the direct application in the conservation of energy, power, or dissipation function. For example, if we can express the internal forces and flows inside a membrane placed between baths in terms of Kirchhoff's laws, the dissipation provided by the baths will equal the dissipation inside the membrane. The typical reaction is: so what is new? What is powerful here is: no physical "laws" need to be postulated beyond K.V.L. and K.C.L. Tellegen's result is a topological concept, the conservation "laws" of the "bilinear" form simply follow from this result.

The point becomes more striking when one notes that no constitutive equations between forces and flows are needed at the branches: the forces and flows could have been measured at different times, or even in different physical systems having the same topology. Moreover, differences in forces or flows will show the same orthogonal characteristics. It therefore follows that the additional equalities

$$\Delta \, J^T \cdot x = 0$$
$$\Delta \, x^T \cdot J = 0 \qquad\qquad (18)$$
and
$$\Delta x^T \cdot \Delta J = 0$$

also hold. These results will be shown to lead directly to Hamiltonian and Lagrangian dynamics.

Converse statements of Tellegen's theorem also follow as corollaries: if a bilinear form is conserved in the sense given above and either K.V.L. or K.C.L. holds the remaining Kirchhoff law will also be obeyed.

1.7 TOPOLOGICAL PROPERTIES AND GRAPH THEORY

Tellegen's theorem can be shown to be more elegantly based on some fundamental concepts of combinatorial topology which have been discussed by Harary (ref. 90), whose presentation we follow here. If the points (vertices, nodes) of a graph are denoted by v_j and the edges (branches) joining these vertices are denoted by e_k, we define a <u>zero chain</u> on a graph G as the summation

$$A = \sum_{j=1}^{p} d_j \, v_j \qquad\qquad (19)$$

and a <u>one chain</u> on the graph G by the summation

$$B = \sum_{k=1}^{q} c_k \, e_k, \qquad\qquad (20)$$

in which q is the number of edges, p is the number of vertices and c_K and d_k are numbers.

The collection of zero chains determine a vector space V^0 having dimension p and the one chains determine a vector space V^1 having dimension q.

It is now possible to map V^1 into V^0 and V^0 into V^1. The mapping functions are called the boundary and coboundary operators, respectively, and are denoted by ∂ and δ:

$$V^1 \underset{\delta}{\overset{\partial}{\rightleftarrows}} V^0 \qquad (21)$$

Given a directed edge with starting point b (or +, say) and ending point a (or -), we define the <u>boundary operator</u> by

$$\partial e = b - a \qquad (22)$$

(this is the reverse of the sign definition given by Harary, so as to keep analogy with branch flow directions). Similarly, we define the <u>coboundary</u> operator by

$$\delta b = \sum e^{\text{towards b}} - \sum e^{\text{away from b}} \qquad (22)$$

in which all edges incident on b are labelled positive and all edges leaving b are labelled negative.

If these definitions are applied to the 1 and zero chains, we obtain

$$\partial B = \partial \sum c^k e^k = \sum c^k \partial e^k \qquad (24)$$

(constant c^k) and

$$\delta A = \delta \sum d^j v^j = \sum d^j \delta v^j \qquad (25)$$

(constant d^j), respectively.

A few additional concepts serve to complete a model for Combinatorial Topology. The <u>inner product</u> of 0 chains and of 1 chains are defined as

$$\langle A, A' \rangle = \sum d^k d'^k \qquad (26)$$

and

$$\langle B, B' \rangle = \sum c^k c'^k, \qquad (27)$$

respectively. Any one chain whose boundary is zero is a topological cycle (or t-cycle), while the collection of all t-cycles is the cycle space of G. A cycle or loop is a closed path which starts and ends at the same point going through oriented edges connected by common points. Every t cycle is a linear combination of cycles. A cocycle is a 1 chain B which is the coboundary of a 0 chain. The following theorems can be proven :

1. Given a 0 chain A and a 1 chain B,

$$< B, \partial A > \; = \; < \delta B, \; A > \qquad (28)$$

2. The inner product of a cycle and a coboundary is zero

3. Any 1 chain orthogonal to every coboundary is a topological cycle.

4. Any 0-chain orthogonal to every cylce is a coboundary

5. Every 1 chain is expressible uniquely as the sum of a cycle and a coboundary.

Kirchhoff's Current law, can be expressed in the language of combinatorial topology by defining the current or flow 1 chain

$$J = \sum_{k=1}^{q} J_k \, e_k \qquad (29)$$

and stating the J is orthogonal to the coboundary of each point--i.e., according to 3, above, J is a cycle on G. The force or voltage 1 chain can be similarly defined by

$$X = \sum X_k \, e_k$$

while the external force chain is given by

$$X^o = \sum X_k^o \, e_k \; .$$

Given Kirchhoff's Voltage law , we can write for an arbitrary cycle Z

$$\langle Z, X \rangle = \langle Z, X^o \rangle \qquad (30)$$

or

$$\langle Z, X - X^o \rangle = 0. \qquad (31)$$

It then follows, from theorem 2, above, that $X - X^o$ is a coboundary on G. Moreover, it also follows from 2 that J is a cycle, the inner product of J and $X - X^o$ is zero,

$$\langle J, X - X^o \rangle. \qquad (32)$$

The above equation is a rigorous expression of the topological basis of **Tellegen's Theorem**.

1.8 CONTINUOUS ANALOG OF TELLEGEN'S THEOREM

A straightforward exercise of vector calculus is the demonstration that the volume integration of the dot product $\mathbf{A} \cdot \mathbf{B}$,

$$\int_V \mathbf{A} \cdot \mathbf{B} \, dV, \qquad (33)$$

in which $\mathbf{A}$ is a solenoidal field ($\text{div} \mathbf{A} = 0$) and $\mathbf{B}$ is a conservative field ($\text{curl} \mathbf{A} = 0$) vanishes inside a closed surface

$$\int_V \mathbf{A} \cdot \mathbf{B} \, dV = 0. \qquad (34)$$

The proof of this statement is simple and requires using the equality

$$\text{div}(\phi \mathbf{B}) = \phi \, \text{div}(\mathbf{B}) + \mathbf{B} \, \text{grad} \, \phi, \qquad (35)$$

which transforms the above integral to

$$\int_V \mathbf{A} \cdot \mathbf{B} \, dV = \int_V [\phi \, \text{div} \, \mathbf{B} - \text{div}(\phi \mathbf{B})] \, dV \qquad (36)$$

The first term vanishes because $\text{div} \, \mathbf{B} = 0$ and the second term transforms to an area integral which also vanishes if there are no perpendicular flows across the boundary area. This proves the original statement.

If there are ports, the area integral is not zero, but leads to

$$\int_V A \cdot B \, dV = \int_V - \text{div}(\phi B) = - \int_A \phi B \, dA. \qquad (37)$$

If the flow like vector B crosses the area only at singular points, the integral on the right hand side reduces to the summation

$$\sum_i \phi^i B^i. \qquad (38)$$

This expression can also be rewritten by expressing n-1 of the n singular potentials in terms of the nth potential,

$$\sum_i (\phi^i - \phi^n) B^i \qquad (39)$$

and, given that the surface must also obey div B = 0, ports can be defined so that a flow component B^i comes out of the surface at point i and returns to the surface at point n. Point n would then return all the flows back into the surface. It may be more convenient to define the flow as positive if it enters the surface and to denote the potential difference

$$\phi^i - \phi^n \qquad (40)$$

as a force X^i. In that case the integral expression reduces to

$$\int_V A \cdot B \, dV = \sum_i X^i B^i. \qquad (41)$$

One of the interesting properties of this result is that no constitutive laws have been used between the X^i and the J^i, so that it is valid for conservative and solenoidal fields which have the same topology, even though they may not be functionally related.

The most useful results are in fact obtained by assuming that the fields correspond to two different sets of cause effect laws or that they correspond to two different times and obey the same laws. For example, consider a set of boundary (port) forces given collectively by the vector $X^{ports}(t)$ and the set of flows measured at a latter time $(t + t^o)$. If we assume that the actual distribution of flows at the latter time goes through some neighborhood of the original flows, both sets of flows have the same topology and we can write

$$\int_V \mathbf{A} \cdot \mathbf{B}' \, dV = \sum_i X_i B'_i, \qquad (42)$$

in which the primed variables reflect the (t+ t_o time) and also

$$\int_V \mathbf{A}' \cdot \mathbf{B} \, dV = \sum_i X'_i B_i. \qquad (43)$$

Moreover, if there is a linear constitutive law of the form

$$\mathbf{B} = a(v) \mathbf{A}$$

at each point inside the volume such that

$$\mathbf{B}' = a(v) \mathbf{A}'$$

is also obeyed, it follows that

$$\sum_i X_i B'_i = \sum_i X'_i B_i.$$

1.9 KIRCHHOFF'S LAWS AS VECTOR SPACES

The set of all flows in the network and, independently, the set of all forces (potential differences) span orthogonal vector spaces before a metric --i.e., cause effect relation --is established. It is easy to verify that all currents and all voltages in the network share the following properties with algebraic vector spaces (the A_i used below are the elements of the vector space considered, either a branch voltage or a branch flow):

1. Commutativity

$$A_i + A_j = A_j + A_i.$$

For example, the order of addition of two currents or two voltages is immaterial.

2. **Associativity**

$$A_k + (A_i + A_j) = (A_j + A_i) + A_k.$$

For example, the order of addition of currents is irrelevant in establishing K.C.L. for a given node or K.V.L. for the voltages in a loop.

3. **Existence of zero element in the set**

$$A_i + 0 = A_i.$$

For example, an additional branch can always be added in a given loop by placing a new node whose potential is the same as one of the neighboring potentials. The voltage across the new branch is zero.

4. **Existence of a negative vector**

$$A_i + (-A_i) = 0.$$

A negative vector to each of the voltages and currents in the network can be found by reversing the direction of a branch.

In addition to the vectors themselves, a vector space also has an associated field of scalars. The vector spaces of K.V.L. and K.C.L. has associated with it the field of numbers +1, -1 and 0, which serve to specify whether there is a positive or negative incidence between, say, a branch and a node or whether there is no relationship between the two graph elements considered. (This will become more explicit in setting up matrices which describe Kirchhoff's laws for specific circuits -- refer to Appendix 2.)

The scalar field has specific properties which are easily verified for Kirchhoff's spaces. Referring to the scalars by c_i, these are:

5. **Distributive property with respect to vector addition**

$$c(A_i + A_j) = c(A_j) + c(A_i)$$

6. Distributive property with respect to scalar multiplication
$(c_i \; c_j) A_k = c_i (c_j \; A_k)$.

7. Existence of the unit scalar
$1 A = A$.

It is interesting to point out that before specific constitutive equations are introduced at the branches, the vector space is affine--i.e., it has built in directions and the concept of parallelism--, but there is no possible way of calibrating a distance (i.e., "power") function . The product of forces (voltages) and flows (currents) at the branches is defined, however.

The introduction of constitutive laws serves to define the specific amounts of power dissipated, which translate, in geometrical terms, into geometrical distances. The purely topological properties of a network are therefore related to Kirchhoff's laws, while the geometrical properties are related to the constitutive relations.

1.10 BASIC PROBLEMS OF KIRCHHOFFIAN NETWORKS

As discussed above, network problems are both topological and geometrical. Topological properties (Kirchhoff's laws) lead to conservation laws and the primitive invariance properties of the network. As Harary has pointed out, Kirchhoff's laws are so fundamental to Algebraic Topology in general, that Kirchhoff can only fairly be considered the founder of this discipline. The basic result that follows from K.V.L. and K.C.L. is Tellegen's theorem and this in turn leads to some powerful results of Classical Mechanics and Thermodynamics.

Geometrical properties of the network--i.e., the metric characteristics-- follow from the constitutive equations which associate specific relations between the flow at a branch and the force at that branch --or forces and flows at other branches. Among the basic constitutive relations is the resistance which represents any of the equations

$F(i,v)=0$
or, in terms of forces and flows, $F(X,J) = 0$. Note that in this definition of a resistance are included the nonlinear X,J type

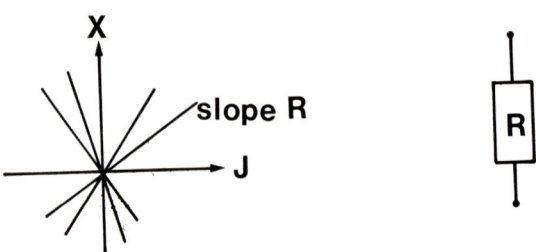

Fig. 1.14 Typical linear resistance behavior in the X-J plane.

equations such as the vacuum tubes, transistors, etc. Ohm's law for a linear resistance is, of course, a very specific case of such an equation and, although most of this monograph will utilize linear resistances, it is worthwhile keeping the general equation in mind. The more familiar Ohmic, or linear resistance has the constitutive law

$$X = R J. \quad\quad\quad (44)$$

In the physical case of positive resistances this law represents a line in the first and third quadrants which goes through the origin, as shown in FIGURE 14.

Negative linear resistances may be realized over a limited range using energy supplying components.

A flow source is a characteristic $F(J) = 0$ and a force source is represented generally by an equation of the form $F(X) = 0$.

A constant flow source J_o is a line parallel to the X axis and a constant force source X_o is a line parallel to the J axis. These are illustrated in FIGURE 15.

By analogy with the resistance we shall also use a second function relating charges and potentials or, more generally, diplacements q and forces X,

$$F(q,X) = 0.$$

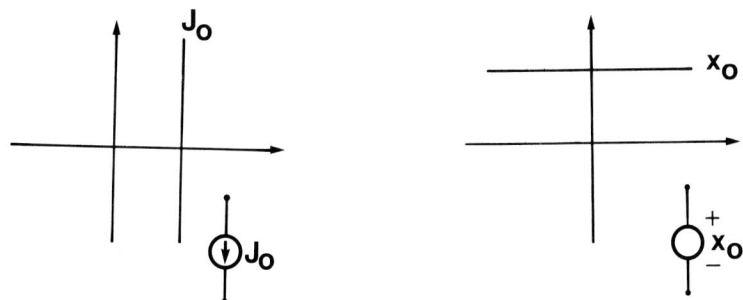

Fig. 1.15 Constitutive relations for a flow source and a force source in the X-J plane.

This is the capacitor, which may either be linear or nonlinear. In particular, we shall consider the linear capacitor for which $C = q/X$.

The inductor is specified as a function which relates magnetic flux and charge or, in the present terminology, momentum and displacement, $F(p, q) = 0$.

There are two basic problems that confront the network worker:

1. Given a network with assigned fluxes and forces at the branches and boundary excitations $e_1, e_2, \ldots, e_n$ --which may be functions of time-- find the global responses at various ports, $r_1, r_2, \ldots, r_n$. This is the problem of Network Analysis.

2. Given a system with known responses to every possible excitation --or an excitation which can represent all others-- find one possible "design" using the simplest possible constitutive relations. This is the problem of network synthesis.

The guiding force behind analysis is simplicity: this translates into using organized techniques of writing equations and chosing very few elements as building blocks. Two secondary problems not considered here are the unicity and existence of the network in analysis, and the realizability of the network in synthesis. These are problems dealt with in network theory books.

It is appropriate, however, to include some of the basic properties some networks may show (ref. 126) :

1. **Linearity**. A system is linear if it obeys the following two principles:

   a) **Superposition**. If an excitation $e_1(t)$ produces a response r_1 and an excitation $e_2(t)$ produces a response r_2, the combined excitation $[e_1(t) + e_2(t)]$ produces a response $[r_1 + r_2]$.

   b) **Proportionality**. If the excitation is $Ae_1(t)$, the response is $Ar_1(t)$.

2. **Passivity**. If forces and flows are imposed on the network terminals energy is always delivered to the network. Moreover, if no excitation has been applied in the past, no response is seen until an excitation is applied. This last requirement is also a causality (or reality) condition.

3. **Time Invariance**. If an excitation $e(t)$ applied at t leads to he response $r(t)$, an excitation $e(t')$ applied at t' leads to the response $r(t')$. This also extends to the derivatives of excitations.

4. **Reciprocity**. If the application of an arbitrary collection of forces $(X_1, X_2, X_3, \ldots, X_n)$ leads to a response $(J_1, J_2, J_3, \ldots, J_n)$ at the corresponding ports, and the set of forces $(X_1', X_2', X_3', \ldots, X_n')$ leads to a response $(J_1', J_2', J_3', \ldots, J_n')$ a network is reciprocal if

$$\sum X_i J_i' = \sum X_i' J_i.$$

In a linear, passive system with time invariant coefficients, this property leads to symmetric matrices--i.e., reciprocal phenomenological equations for the network.

5. **Nonenergenicity**. If the sum of input and output products of forces and flows is zero:

$X_1 J_1 + X_2 J_2 = 0.$

Note that reciprocity and nonenergenicity are not the same. Thus, the network having input

$X_1 = A X_2,$

$$J_2 = - (A) J_1$$

is nonenergenic, but it is not reciprocal.

1.11   NETWORK ANALYSIS AND SYNTHESIS

The two central problems of network theory are the analysis and synthesis of specific circuits. In the first case we are given a network -- which represents here some physical system of interest-- and we are asked to give the equations which describe its behavior, while in the second case we are asked to find a network which obeys a certain set of equations or is isomorphic with a physical system . In the analytical situation we are often interested in finding the behavior of flows and forces across certain pairs of terminals (ports) whose potential difference serves to define a force.

Before a solution is searched for, it is important to have some idea as to whether a solution exists at all. In the analytical case this translates into having some idea as to the behavior of the physical variables before one starts. In the problem of synthesis the question becomes: is the network physically realizable? For example, a response may be required before a signal appears, in which case the network "predicts" the future. This is not "physically reasonable" in most cases; however certain statistical predictors can be built. Moreover, a network need not be physically realizable at all, in general terms, as the physical construction of networks must be distinguished from the mathematical properties (topological and metrical) of networks.

In the context of this monograph we shall restrict our attention to networks which can be physically realized, because of the underlying idea that a system which is isomorphic to a real life process must be representable by real physical processes.

In addition, we shall often search for the simplest network representation of a physical system, provided that the global behavior at the ports agrees with the system being described.

Some of these include: i) resistive networks which obey Kirchhoff's laws, ii) resistive networks with controlled sources and iii) networks with both dissipative (i.e., resistances or other frictional elements) and storing elements ("capacitors" and "inductors").

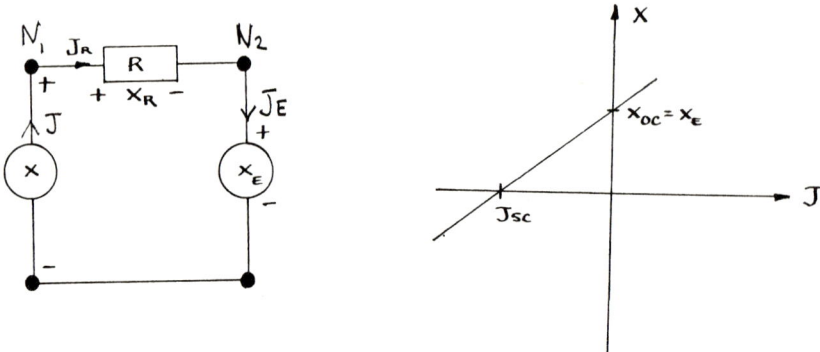

Fig. 1.16 A force source in series with a resistance and the X-J behavior of this circuit.

1.12   SIMPLE RESISTIVE ANALYSIS

As a first example of simple analysis, we place a resistance of value R in series with a force source of value E, as shown in FIGURE 15 , and ask for the set of points which describes the network behavior in the i-v (current-voltage) or X-J (force-flow) planes (note that J has been defined as going into the box). K.V.L. yields

$$X_E - X + X_R = 0 \qquad (45)$$

in which X_R is the force across the resistance and X_E is the force across the force source or "battery". (X is negative, as all the forces are measured in the same direction in the loop). The constitutive relation for the resistance places the following restriction on X_R:

$$X_R = R \, J_R \quad ;$$

i.e., the force across the resistance R is the product of the resistance and its conjugate flow, J_R, by Ohm's law. We then apply K.C.L. to node N_1 in the circuit, to obtain

$$J - J_R = 0 \quad . \qquad (46)$$

From Eqs. (45) and (46) it follows that

$$X = JR + X_E, \quad (47)$$

so that the line describing the behavior of the circuit has an X axis intercept specified by the source, while the slope is specified by the resistance. Clearly, real life, dissipative resistances are positive, so that this type of circuit will cover only the first and third i-v (or X-J) quadrants. However, certain complex elements which have "hidden sources" and feedback may behave, over certain ranges like negative resistances (2nd and 4th quadrants) and actually act as energy supplying, rather than dissipative elements.

1.13 SIMPLE SYNTHESIS: THE INVERSE PROBLEM

Suppose that we are given a physical system which is known to be describable by a single force and a single, conjugate displacement and we are asked to find a network which will be isomorphic--i.e., act as an analog computer-- to this system. Again, we can specify the X intercept of the line and its slope and translate these back into the resistance and source considered above. However, a more physical approach consists in asking for the force (voltage) when there is no flow--i.e., the open circuit force, X_{OC}-- and the flow when there is no force, the short circuit current J_{SC}. Both of these points have the property that their dissipation is zero.

A little thought leads to the two results

$$X_E = X_{OC}$$

(the battery or force source must equal the open circuit force) and

$$R = - X_{OC}/J_{SC}$$

(the resistance is the ratio of the open circuit force to the short circuit current or flow, with appropriate sign). These relations between the open and short circuit characteristics and the networks are shown in FIGURE 17.

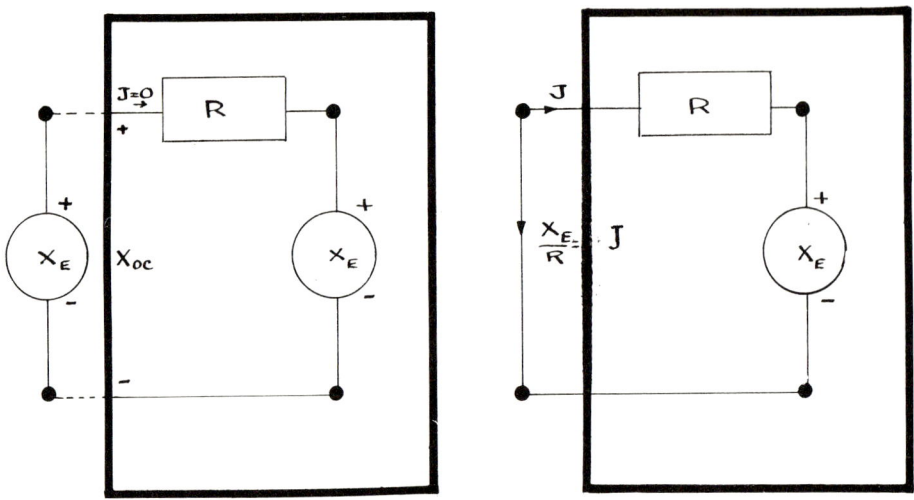

FIGURE 1.17   Network interpretation of the short circuit flow or current and the open circuit voltage or force.

1.14   BOUNDARY CONDITIONS: LOAD RESISTANCES

As in any problem in which different subregions are connected, it is necessary to specify the boundary conditions here. These appear, as long as one deals with only resistive elements, in the form of load resistances which specify constitutive equations at the ports and port sources.

Consider a box with the J-X characteristics—i.e., "current"/"voltage" or "force"/"flow" characteristics shown in FIGURE 17 and let the load resistance R_L be attached to its

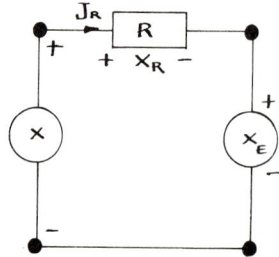

Fig. 1.18   A linear network connected to an external resistor.

terminals. Writing the circuit laws at the port we obtain

$X - X_L = 0$,

$J + J_L = 0$, (48)

$X_L = R_L \, J_L$,

and

$X = - (X_{OC}/J_{SC}) \, J + X_{OC}$. (49)

Equation (49) is the constitutive relation for the black box. From (48) and (49), we obtain the solution point (X_S, J_S) for the circuit:

$J_S = -J_L = X_{OC} / [(X_{OC} / J_{SC}) - R_L]$ (50)

and

$X_S = -R_L \, J_S$. (51)

The point (X_S, J_S) specifies the value of the flow and force which is consistent both with the box equation and with the load conditions. We could have also found the solution graphically, by plotting the X-J relations for the box and the load on the same graph, as shown in FIGURE 19. The intersection of the two lines

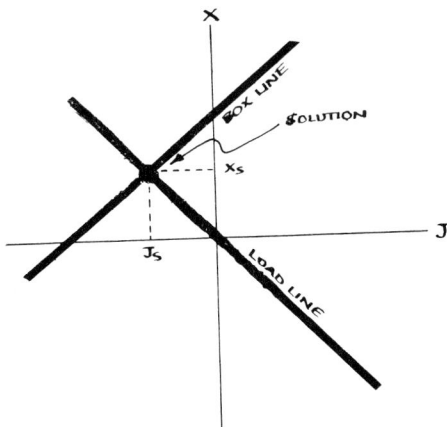

Fig. 1.19 Graphical solution for the boundary value problem shown in Fig. 1.18. The intercept X_S, J_S is the solution.

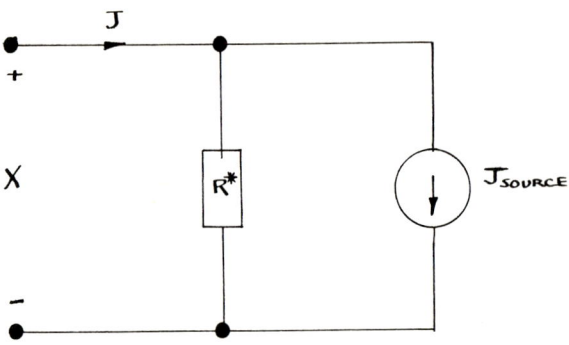

Fig. 1.20 A flow source in parallel with a resistance R´ can be used to model the linear system considered before, provided that $R^* = R$ and $J_{source} = X_E / R$.

gives the solution. Given that the lines either intersect at one point or do not intersect at all, there will be, in general, a unique solution.

1.15 DUAL NETWORK: SYNTHESIS USING A FLOW SOURCE

The previous linear network may also be synthesized using a constant flow source in parallel with a conductance, as shown in FIGURE 20. To show this is the case we first apply K.C.L.:

$$J = J_{R^*} + J_{source} \qquad (52)$$

From Ohm´s law, the flow across the resistance R^* --or conductance $1/R^*$ -- is given by

$$J_{R^*} = X / R^* \qquad (53)$$
$$J_{R^*} = J_{source} \qquad (54)$$

Introducing this value in Equation (52), we obtain

$$J = X/R^* + J_{source}$$

or, rearranging the equation,

$$X = J R^* - J(source) R^* . \qquad (55)$$

Comparing this equation with (47), it follows that R* = R and

$$R J_{source} = X_E. \quad (56)$$

Given that $J_{SC} = -X_E / R$, it also follows that

$$J_{source} = - J_{SC}. \quad (57)$$

It is clear, then that a force source in series with a resistance can be converted to a flow source in parallel with a resistance of the same value, as the linear constitutive equation may be synthesized either way. Note that it is only necessary to find the short circuit flow and the open circuit force:

$$R = - X / J_{SC} \quad (58)$$

while

$$X_E = X_{OC} \quad (59)$$

and

$$J_{source} = -J_{SC}. \quad (60)$$

This is a very useful result usually known as the Thevenin-Norton theorem.

1.16 TWO PORT MACHINES

The same principles can be extended to synthesize multiple port networks.

We can fix ideas by considering some typical phenomenological measurements, in which a force or flow are kept at a fixed value while another force or flow is varied and ratios of the remaining two quantities --in the 2x2 case-- are taken. With two forces and two flows there are 16 covariant measurements, as shown in FIGURE 21 in which at least one of the independent quantities is "thermodynamically" conjugate to one of the dependent quantities. (If we removed this restriction, there would be 8 more measurements in which both of the independent variables are conjugate to each other; this has some practical use, but no physical meaning).

The practical approach to the phenomenological measurements is well known: a force or flow "source is set at one of the inputs while one of the ports is kept at either static head or level head and a meter--e.g., flow meter, height, manometer, etc-- is used to measure the ratio of the driven to the driving variable, under the conditions given.

Eight of the measurements are ratios of forces to flows when the remaining flow is set to zero--or is kept constant if incremental variables are considered-- or ratios of flows to forces when the remaining force is kept to zero. These measurements clearly correspond to the resistive or conductive equations

$$X_1 = R_{11}J_1 + R_{12}J_2$$
$$X_2 = R_{12}J_1 + R_{22}J_1$$

(61)

and

$$J_1 = L_{11}X_1 + L_{12}J_2$$
$$J_2 = L_{21}X_1 + L_{22}J_2 ,$$

(62)

respectively (we do not assume they are necessarily reciprocal).

The remaining eight measurements are hybrid, as they may represent ratios of forces, flows, flows to forces or forces to flows. The ones that have been labelled H measurements lead to the coefficients H_{ik} in the equations

$$X_1 = H_{11}J_1 + H_{12}X_2$$
$$J_2 = H_{21}J_1 + H_{22}X_2 ,$$

(63)

while the set of measurements labelled P yield the coefficients P_{ik} in the set of equations

$$J_1 = P_{11}X_1 + P_{12}J_2$$
$$X_2 = P_{21}J_1 + P_{22}X_2 .$$

(64)

Fig. 1.21 Phenomenological measurements on two port black boxes.

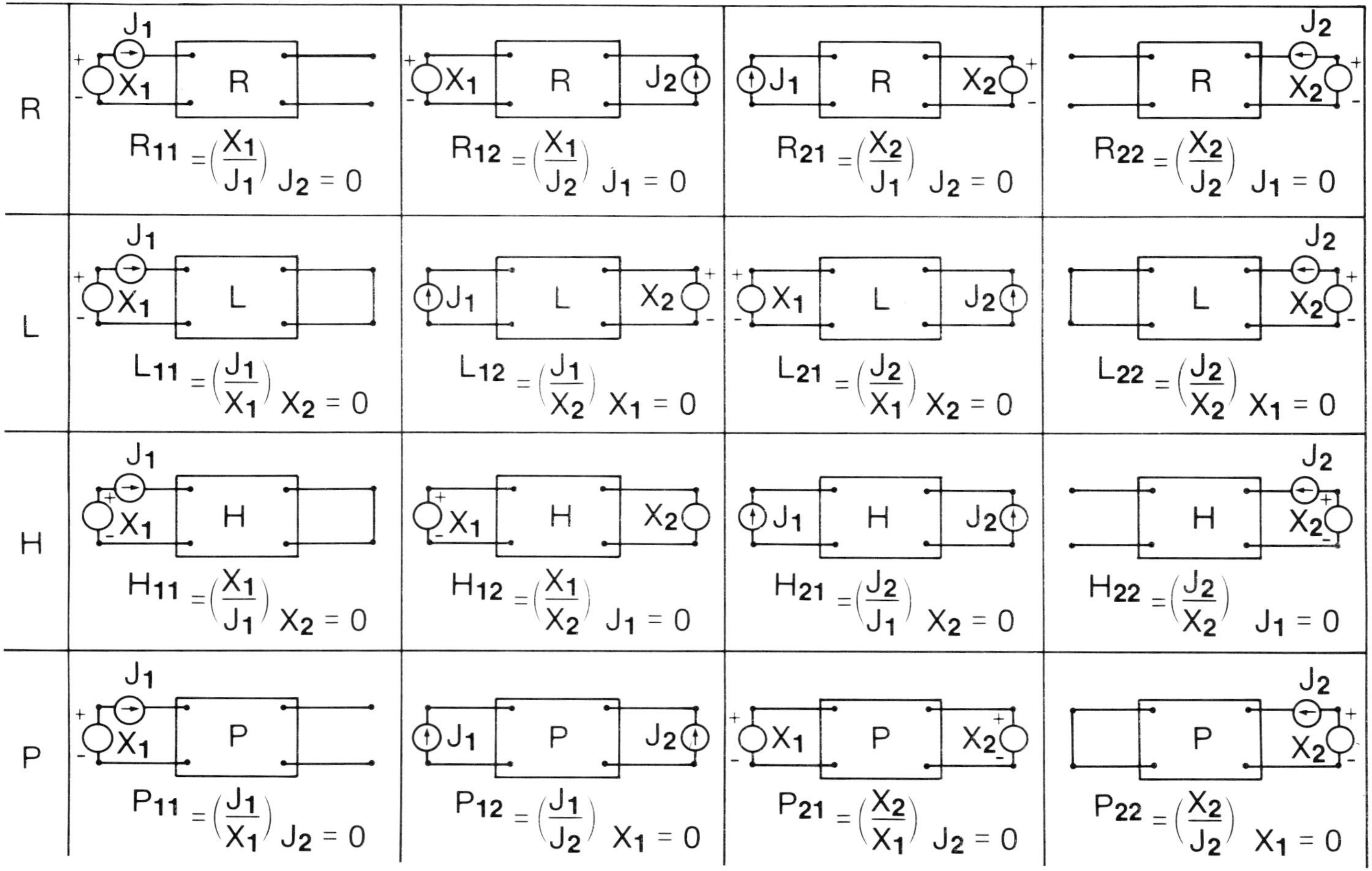

TABLE I. CONVERSIONS AMONG TWO PORT PARAMETERS

To \ From	[R]		[L]		[H]		[P]	
[R]	R_{11}	R_{12}	$\dfrac{L_{22}}{\det L}$	$\dfrac{-L_{12}}{\det L}$	$\dfrac{\det H}{H_{22}}$	$\dfrac{H_{12}}{H_{22}}$	$\dfrac{1}{P_{11}}$	$\dfrac{-P_{12}}{P_{11}}$
	R_{21}	R_{22}	$\dfrac{-L_{21}}{\det L}$	$\dfrac{L_{11}}{\det L}$	$\dfrac{-H_{21}}{H_{22}}$	$\dfrac{1}{H_{22}}$	$\dfrac{P_{21}}{P_{11}}$	$\dfrac{\det P}{P_{11}}$
[L]	$\dfrac{R_{22}}{\det R}$	$\dfrac{-R_{12}}{\det R}$	L_{11}	L_{12}	$\dfrac{1}{H_{11}}$	$\dfrac{-H_{12}}{H_{11}}$	$\dfrac{\det P}{P_{22}}$	$\dfrac{P_{12}}{P_{22}}$
	$\dfrac{-R_{21}}{\det R}$	$\dfrac{R_{11}}{\det R}$	L_{21}	L_{22}	$\dfrac{H_{21}}{H_{11}}$	$\dfrac{\det H}{H_{11}}$	$\dfrac{-P_{21}}{P_{22}}$	$\dfrac{1}{P_{22}}$
[H]	$\dfrac{\det R}{R_{22}}$	$\dfrac{R_{12}}{R_{22}}$	$\dfrac{1}{L_{11}}$	$\dfrac{-L_{12}}{L_{11}}$	H_{11}	H_{12}	$\dfrac{P_{22}}{\det P}$	$\dfrac{-P_{12}}{\det P}$
	$\dfrac{-R_{21}}{R_{22}}$	$\dfrac{1}{R_{22}}$	$\dfrac{L_{21}}{L_{11}}$	$\dfrac{\det L}{L_{11}}$	H_{21}	H_{22}	$\dfrac{-P_{21}}{\det P}$	$\dfrac{P_{11}}{\det P}$
[P]	$\dfrac{1}{R_{11}}$	$\dfrac{-R_{12}}{R_{11}}$	$\dfrac{\det L}{L_{22}}$	$\dfrac{L_{12}}{L_{22}}$	$\dfrac{H_{22}}{\det H}$	$\dfrac{-H_{12}}{\det H}$	P_{11}	P_{12}
	$\dfrac{R_{21}}{R_{11}}$	$\dfrac{\det R}{R_{11}}$	$\dfrac{-L_{21}}{L_{22}}$	$\dfrac{1}{L_{22}}$	$\dfrac{-H_{21}}{\det H}$	$\dfrac{H_{11}}{\det H}$	P_{21}	P_{22}

What is of interest here is to find out if there is any physical significance or practical value in these hybrid forms and how they relate to Onsager's theory.

Given that any set of measurements (R, L, P, or H) must convert to any of the others by usual algebraic procedures, it can be readily shown by direct calculation that the appropriate conversion parameters are those given in TABLE I, in which reciprocity has not been assumed.

Network theory introduces the possibility of synthesizing each of the eight equations given above as follows (refer Fig. 22):

- If the output of the equation is a flow , the two terms added represent two flows in parallel. Moreover, the term proportional to the conjugate force must be a conductance , while the remaining term must be a flow source controlled either by the non conjugate force or flow. Thus,

$$J_1 = L_{11} X_1 + L_{12} X_2 \qquad (65)$$

is a conductance L_{11} in parallel with a controlled flow source of value $L_{12} X_2$ --i.e., proportional to the non conjugate force. The remaining equation for J_2 in the L formulation is the analogous counterpart for the second port. Equation

$$J_1 = P_{11} X_1 + P_{12} J_2 \qquad (66)$$

represents a conductance P_{11} in parallel with a controlled flow source of value $P_{12} J_2$ --i.e., proportional to the non conjugate flow (P_{12} is simply a constant with no units). Similarly, equation

$$J_2 = H_{21} J_1 + H_{22} X_2 \qquad (67)$$

is represented by a conductance H_{22} in paralell with a flow source of value $H_{21} J_1$.

- If the output of the equation is a force, the two terms that appear in the equation are forces which must therefore be in series (because they add). The conjugate term corresponds to a resistance and the non conjugate term corresponds to a

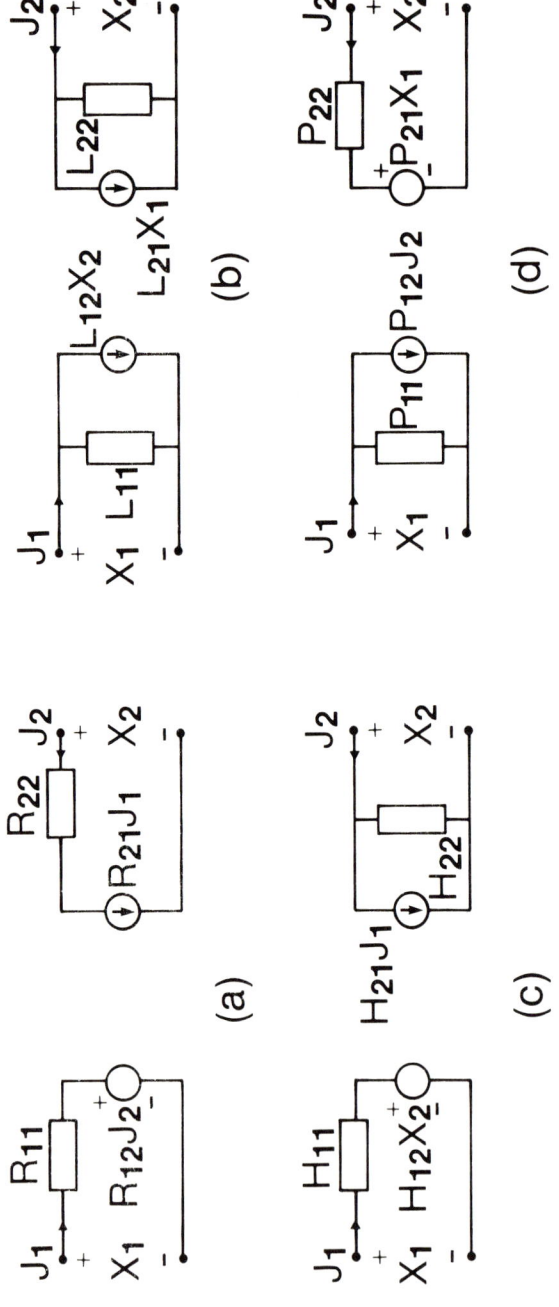

Fig. 1.22 Four basic two port representations. a) R representation of phenomenological two port equations using resistances and series controlled force sources; b) L representation of the phenomenological equations in which force controlled flow sources are placed in parallel with conductances; c) hybrid H representation in which the independent variables are J_1 and X_2; d) P hybrid network has independent variables X_1 and J_2.

force source or battery controlled by either the non conjugate flow or force. Thus,

the equation

$$X_1 = R_{11}J_1 + R_{12}J_2 \quad (68)$$

corresponds to a resistance R_{11} in series with a force source proportional to the non conjugate flow (R_{12} has units of resistance). The companion equation

$$X_2 = R_{21}J_1 + R_{22}J_2 \quad (69)$$

has analogous representation, with the two ports exchanged. Equation

$$X_1 = H_{21}J_1 + H_{12}X_2 \quad (70)$$

is a resistance H_{11} in series with a battery or force source proportional to X_2 (H_{12} has no units), while equation

$$X_2 = P_{21}X_1 + P_{22}J_2 \quad (71)$$

is a resistance P_{22} in series with a source proportional to X_1. (Note that a "resistance" is given as a ratio of a "thermodynamic" force to a thermodynamic flow and a conductance as its inverse)

The equations with corresponding coefficients can be associated back to back in term of the boxes shown in FIGURE 22. The theorem of Norton-Thevenin allows to interconvert between the various port representations, as they all represent the same observable behavior. It is simple to verify that the conversion values given in TABLE I are obtained again. For example, the open circuit value of port 1 can be given either by

$$R_{12}J_2 = X_1^{oc} \quad (72)$$

or by

$$P_{12}J_2 (1/P_{11}) = -X_1^{oc}. \quad (73)$$

The above equations yield

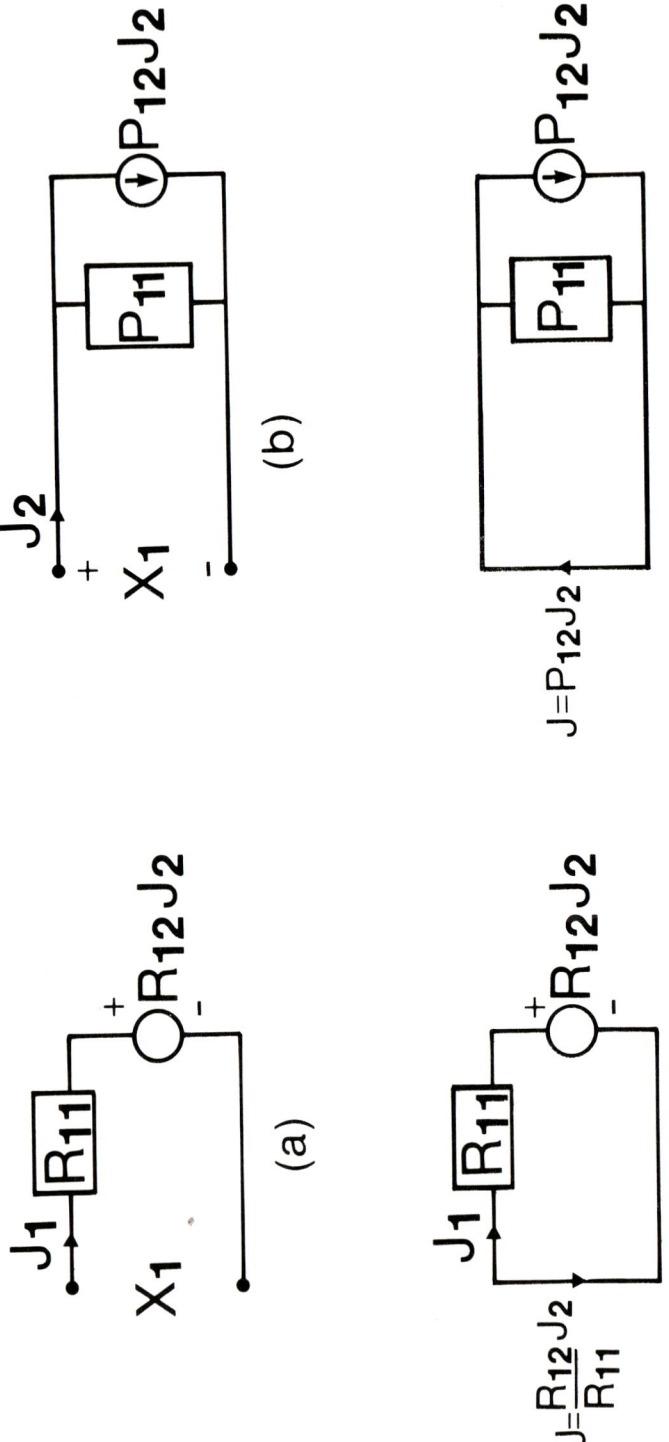

Fig. 1.23 Example showing how transformations between different two port models are found. (a) and (b) represent port 1 in terms of R and P parameters, respectively. By shorting and opening both models the corresponding parameters may be found.

$$(P_{12}/P_{11}) = -R_{12} . \qquad (74)$$

Similar short circuit and open circuit " experiments" yield all the entries of TABLE I .

1.17 A WORD OF CAUTION: LINEAR VS. NON-LINEAR

The terms "linear" and "non-linear" are very losely used throughout the literature. This leads to confusion and it is convenient to make the following distinctions:

A **linear affine vector space** has been defined axiomatically; in network theory terms this is the linearity which appears because all flows are linear combinations of the independent flows (a form of K.C.L.) and all forces or voltages are linear combinations of the independent force (a form of K.V.L.). This linearity remains even when the constitutive relations are nonlinear !

A **linear transformation** is represented by an operator $T(x)$ which obeys the relation

$$T(x_1) + T(x_2) = T(x_1 + x_2)). \qquad (75)$$

According to this definition, the operator

$$T = a x \qquad (76)$$

is linear, but

$$T = a x + b \qquad (77)$$

is not, even though it can be represented by a line ! Thus the addition of constant sources disrupts the linear operator. The repeated multiplication of a linear operator leads to a new linear operator, even in the case where the individual entries of the operator may be polynomials --i.e., "non linear" terms.

Incremental linearity may appear through the linearization of a non linear function, say $f = f(x,y)$ using a Taylor approximation :

$$f(x,y) = f(x_o, y_o) + \partial f / \partial x \ (x_o, y_o) \ dx + \partial f / \partial y \ (x_o, y_o) \ dy$$

in which we use the differential sign to denote either finite or infinitesimal increments in the neighborhood of x_o, y_o.

In the present context, "non linear" will imply that the constitutive relation at one or more of the branches in the network is a non linear function.

1.18 GENERAL RECIPROCITY RELATION

Tellegen's theorem does not depend on any particular constitutive relation between forces and flows at the branches.

However, in the special case in which the branches are Ohmic -- i.e., linear-- resistances, a very useful result is obtained. We can write Tellegen's theorem either as

$$\sum_{\text{ports}} X_p J_p' = \sum_{\text{network}} X_r J_r' \qquad (78)$$

or as

$$\sum_{\text{ports}} X_p' J_p = \sum_{\text{network}} X_r' J_r . \qquad (79)$$

If there network is resistive the first equation becomes

$$\sum_{\text{ports}} X_p J_p' = \sum_{\text{network}} R_r J_r J_r' . \qquad (80)$$

If the resistances are not functions of the flows, the right hand side may be rewritten as

$$\sum_{\text{network}} R_r' J_r J_r' = \sum_{\text{network}} X_r' J_r, \qquad (81)$$

which, according to the second expression given above for Tellegen's theorem can also be given as

$$\sum_{\text{network}} X_r' J_r = \sum_{\text{ports}} X_r' J_r . \qquad (82)$$

From Eqs. (79) and (80) it then follows that

$$\sum_{\text{ports}} X_p' J_p = \sum_{\text{ports}} X_p J_p' . \qquad (83)$$

This is a reciprocity relation which shows the symmetry of the linear phenomenological equations for the resistive network. The converse statement is also true: a time invariant, linear reciprocal system is representable by a network in which all branches are linear resistances. (This is the network equivalent of the continuum statement that an exact differential--i.e., a reciprocal linear system-- leads to a path with a potential function defined at each point).

This can be shown by considering the matrix equation

$$X = R J, \qquad (84)$$

in which R is assumed to be reciprocal. In that case, R may be decomposed in the form

$$R = (a^T r a), \qquad (85)$$

in which r is a diagonal matrix, a result shown in Algebra textbooks. The phenomenological equation (84) may then be decomposed, arbitrarily, as follows

$$X = a^T X_T, \qquad (86)$$

$$X_T = R_T J_T \qquad (87)$$

and

$$J_T = a J. \qquad (88)$$

Therefore, if it is possible to find a matrix **a** such that used as an operator on the port flows it yields a subset of network flows and if, in turn, the transpose of this matrix used as an operator on the corresponding conjugate forces yields the port forces, these forces and flows will be related by a diagonal matrix--i.e., the resistive network will be a synthesis of the reciprocal equations.

A little thought will reveal that matrix **a** must be a statement of Kirchhof´s laws. This is indicated below in more detail.

1.19   THE T AND PI NETWORKS

The mathematical simplicity of the conversion between the symmetric phenomenological matrix and the diagonal matrix --a resistive network-- belies an interesting physical point. We now proceed to show that the assumption of reciprocity leads to the topological connectivity of the network--i.e., if there is reciprocity and K.C.L. is obeyed, K.V.L. will also be obeyed, or viceversa. (This concept will be used in connection with Onsager's reciprocities (ref. 182).

Consider a two port network which might represent the prototypical "Onsager" phenomenology

$$X_1 = R_{11} J_1 + R_{12} J_2$$
$$X_2 = R_{21} J_1 + R_{22} J_2$$
(89)

with the reciprocity $R_{12} = R_{21}$. These equations can be readily represented by two disconnected networks, as follows. The first equation may be thought of as the voltage (force) across two resistances (refer to FIGURE 24): the first resistance has a value $R_{11} - R_{12}$ and a through current (flow) J_1, while the series resistance has a value R_{12} and through current ($J_1 + J_2$) obtained by conceptually injecting an additional current J_2 through the resistance, as shown in the drawing. Clearly, this representation yields the same force as before, as

$$X_1 = (R_{11} - R_{12}) J_1 + R_{12} (J_2 + J_1)$$
$$= R_{11} J_1 + R_{12} J_2 .$$
(90)

The same process may be repeated for the right hand side half of the network by placing a resistance of value ($R_{22} - R_{12}$) having through current or flow J_2 in series with a resistance R_{12} having a through current ($J_1 + J_2$) going through -- again, the J_2 part is "injected" as shown in FIGURE 24. By comparing FIGURES 24a and (b) it becomes clear that, as long as R_{12} and R_{21} are equal the two half networks may be connected as shown in FIGURE 24c. This is the T network.

Is the dissipation the same in both cases? This is easy to show by direct calculation, but the interesting part is that the

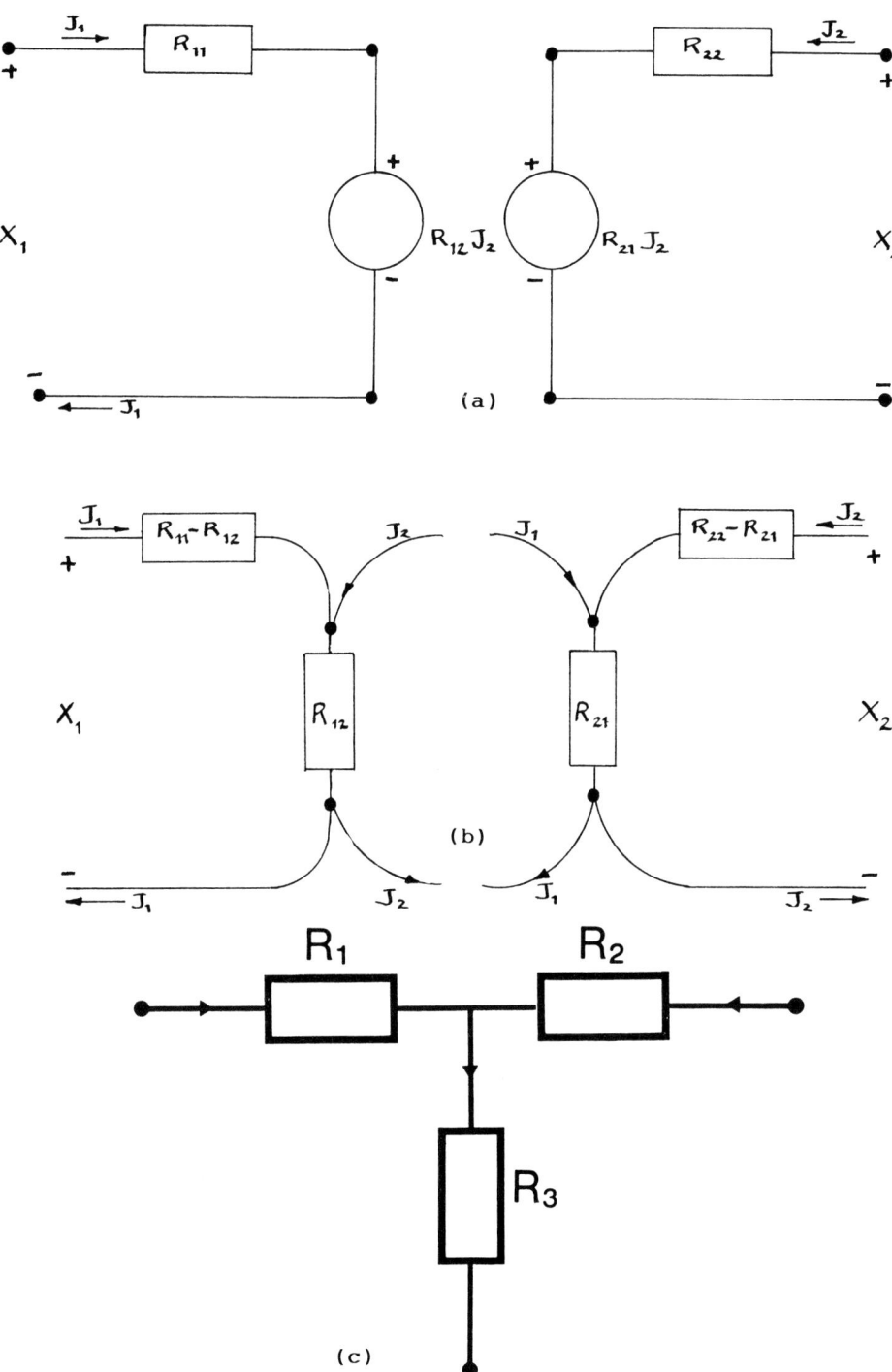

Fig. 1.24 Construction of the T network.

bilinear metric

$$\Phi = R_{11} J_1^2 + 2 R_{12} J_1 J_2 + R_{22} J_2^2 \qquad (91)$$

having a term which involves cross products of flows transforms to the "Pythagorean" form

$$\Phi = (R_{11} - R_{12}) J_1^2 + R_{12} J_3^2 + (R_{22} - R_{12}) J_2^2, \qquad (92)$$

in which $J_3 = J_1 + J_2$ and which has only sums of square terms.

One can view this transformation as taking two oblique axes (the port dissipations) and expressing their resultant length in terms of sums of squares of orthogonal axes in a higher dimensional euclidean space (n= 3 in this case), which corresponds to the "directions" of the resistances in the T network. This property will be discussed in Chapters 2 and 4.

Clearly, the introduction of K.C.L. removes the independence of the third axis. One could then infer that this orthogonality preceeds K.C.L. and that it remains when the K.C.L. restriction is removed --e.g., in irreversible cases.

[There is a further interesting geometrical property which is fundamental to the field of projective geometry and appears here as a network characteristic. A theorem due to Polke (ref.212) says that given 3 arbitrary vectors on a plane, not all equal to zero or dependent, they can be considered projections of three equal orthogonal axes. The vectors considered here along the axes can be further projected on the plane of the manifold. These projections correspond to modulated voltages. I am not developing that construct here because it somewhat removed from the present discussion.]

The same procedure utilized to obtain the T network can be used to consider the connectivity imposed by K.V.L. , rather than K.C.L. Each half of the phenomenological equations

$$J_1 = L_{11} X_1 + L_{12} X_2$$
$$J_2 = L_{21} X_1 + L_{22} X_2 \qquad (93)$$

can now be considered as a conductance ($L_{11} - L_{12}$) or ($L_{22} - L_{12}$) -- for the first and second networks , respectively in

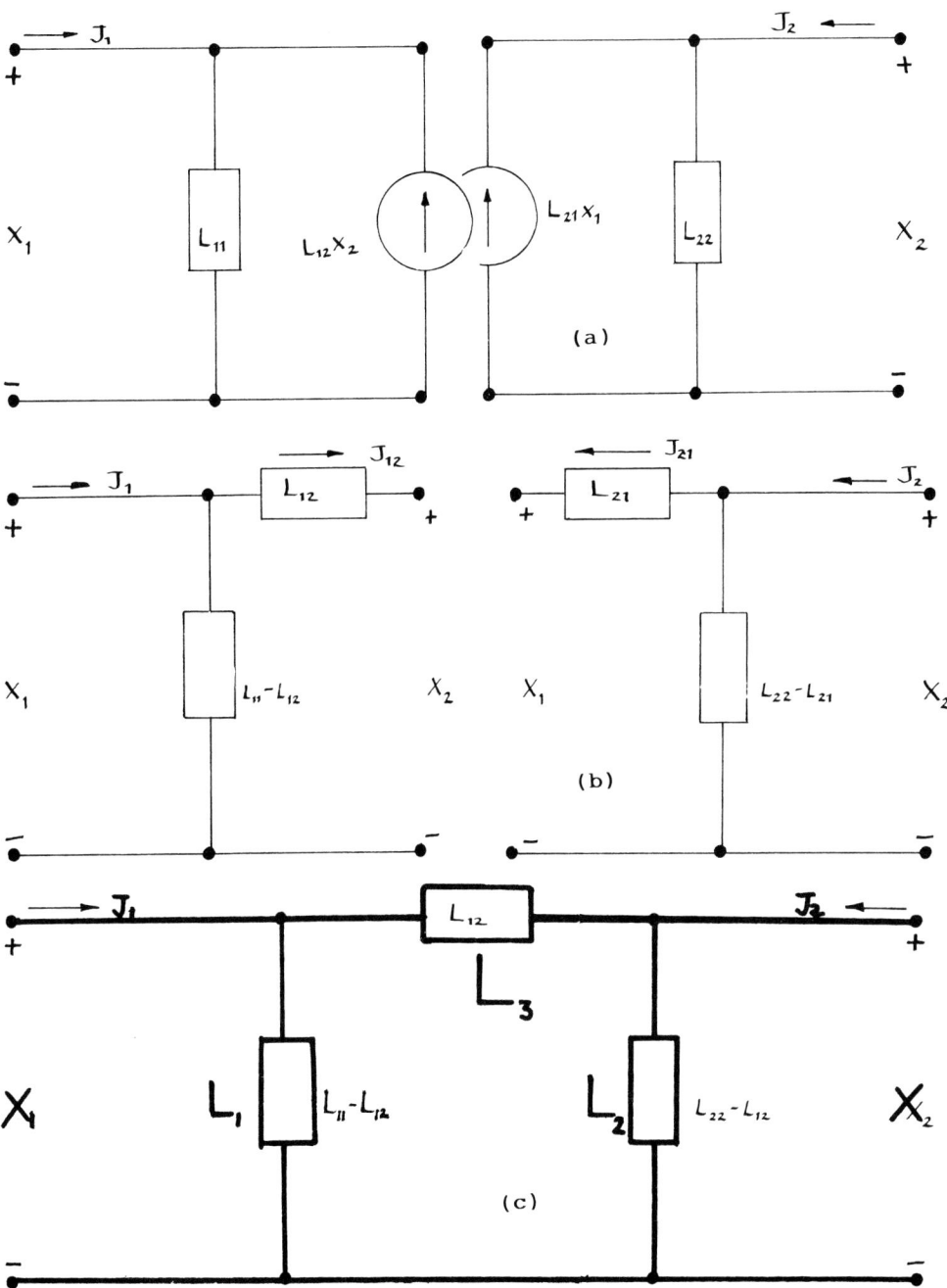

Fig. 1. 25 Stages in the construction of the Pi network.

parallel with a conductance which has the non-conjugate force as an offset "voltage", as shown in FIGURE 25. It is clear that if $L_{12}=L_{21}$ the network can be given in connected form as shown in FIGURE 25 (c). This is the Pi network.

The geometrical representation will be given in Chapter 2. It can readily be shown that the dissipation can either be calculated in the conductances (the "Pythagorean" form) or the ports (the bilinear form with cross coefficients).

It is enlightening to enquire how the process is affected when the equations are not reciprocal. In that case we can represent the situation by a T or Pi resistive network and an additional source. Now the port dissipation will not account for all the dissipation in the resistances, as part of it comes from the source. A reciprocal model, then, has the characteristic that it is completely passive --i.e., it has no sources.

1.20 LINEAR INDEPENDENCE AND DIMENSION OF THE VECTOR SPACE

Given the vectors X_1, X_2,...,X_n in an n-dimensional vector space, these are said to be linearly independent if there is no linear expression

$$k_1 X_1 + k_2 X_2 + ...+ k_n X_n = 0 , \qquad (94)$$

in which the k's are numbers taken from the field associated with the vector space --in the case of networks these are 0, +1 and -1. Conversely if such an expression exists the vectors are linearly dependent. The concept of linear independence appears naturally in establishing trees for a given network. The link flows are the independent vectors, as shown in FIGURE 26, and any branch current can be found by superposition of the link currents.

If the vectors are linearly dependent, there is clearly an operator **k** which maps the vector resultant vector **x** into the zero vector element,

$$\mathbf{kX = 0} . \qquad (95)$$

The maximum number of independent vectors in the vector space is the <u>dimension</u> of the vector space. Since all vectors in the vector space can be expressed as a linear combination of the

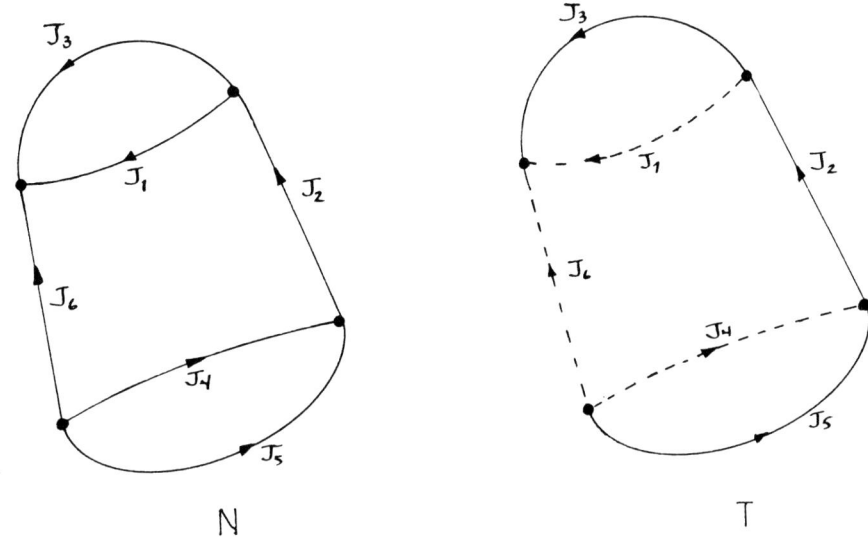

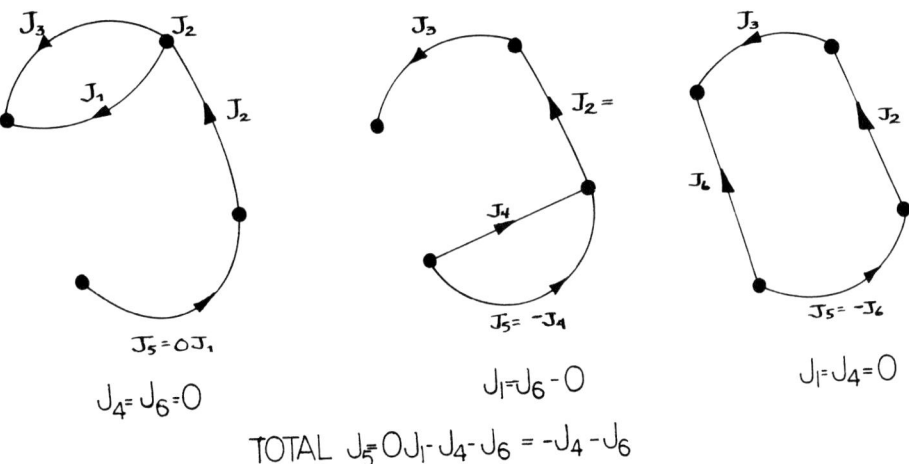

Fig. 1.26 (a) Example showing the independence of the links 1, 4, 6 in a (network) vector space of dimension 3. (b) The total flow through any branch can be calculated by connecting one link at a time and adding the individual contributions (Principle of Superposition).

independent vectors, these are referred to as the <u>basis</u> of the vector space. Clearly, many bases can be chosen for a given vector space.

In the case of connected networks obeying Kirchhoff's laws, the concepts of linear dependence and independence arise naturally. Consider a tree which is, by definition, a connected set of branches through which no flows are present. The addition of any link to the tree completes a loop and allows the link current to flow around the loop. The complete current in any branch is the superposition of the individual link contributions, added one at a time when the other links are zero. The contribution of any link to a given branch is either 0, +1 or -1 (the elements of the field). Thus, the tree flows are linear combinations of the link flows,

$$J_T = a\ J_L, \qquad (96)$$

in which we have explicitly identified the flows as belonging to the tree or the links for that tree. The matrix **a** can be identified as the incidence matrix between loops and the tree branches.

Given that K.V.L. holds, it also follows that any link **force** is a linear combination of the remaining forces in the loop--i.e., the tree forces,

$$X_L = b\ X_T. \qquad (97)$$

Now matrix **b** expresses a loop in terms of the branches intersected by the loop, whereas matrix **a** expresses the branches in terms of the loops intersected. It then follows that one matrix is the transpose of the other, so that

$$b = a^T. \qquad (98)$$

Equation (97) then becomes

$$X_L = a^T\ X_T. \qquad (99)$$

If each branch in the tree has a resistance, the constitutive relation

$$X_T = r\ J_T, \qquad (100)$$

finally leads to the global equations for the links, now identified as ports,

$$X_L = R \; J_L , \tag{101}$$

in which

$$R = a^T \, r \, a. \tag{102}$$

Thus, we have shown, that Kirchhoff's laws provide the desired decomposition and that the reciprocal system can be synthesized by means of connected resistive branches.

Although the matrix **a** used has the all relevant information, it is desirable to express K.V.L. and K.C.L. in the more explicit forms

$$C J_b = 0 \tag{103}$$

and

$$T X_b = 0 , \tag{104}$$

in which J_b and X_b are the complete set of flows and forces in the network, while **C** and **T** are the cut set and tie set matrices, to be defined below.

1.21. THE CUT SET AND TIE SET MATRICES

Equation (96) relating tree flows to the link flows can also be expressed as

$$[\, 0 \mid I \,] \; [\, J_L \mid J_T \,]^T = [\, a \mid 0 \,] \; [\, J_L \mid J_T \,]^T , \tag{105}$$

in which we have defined the extended flow vector

$$J_b = [\, J_L \mid J_T \,]^T . \tag{106}$$

Obviously, the elements of the extended flow vector do not form an independent set of vectors).The above matrix equation can now be rewritten as

$$[\, 0 \mid I \,] \; [\, J_L \mid J_T \,]^T - [\, a \mid 0 \,] \; [\, J_L \mid J_T \,]^T = 0 \tag{107}$$

or, more concisely, in the form

$$[-a \mid I] \, J_b = 0. \tag{108}$$

It therefore follows that the <u>Cut Set matrix</u> defined above is given by $C = [-a \mid I]$.

Each row of the set of equations (108) represents K.C.L. at a node. Moreover, the matrix equations

$$X_L = a^T X_T \tag{109}$$

and

$$X_T = I \, X_T \tag{110}$$

lead to

$$X_b = [X_L \mid X_T]^T = [a_T^T \mid I] X_T = C^T X_T. \tag{111}$$

We can also use the extended force vector to express the dependence of the link forces on the tree forces, as follows:

$$[I \quad 0][X_L \mid X_T]^T = [0 \quad a^T][X_L \quad X_T]^T. \tag{112}$$

After rearranging terms, this expression becomes

$$[I \quad 0][X_L \mid X_T]^T - [0 \quad a^T][X_L \quad X_T]^T = 0 \tag{113}$$

or, equivalently,

$$\{[I \quad 0] + [0 \quad a^T]\} [X_L \quad X_T]^T = 0, \tag{114}$$

which can be expressed more concisely as

$$T \, X_b = 0. \tag{115}$$

Therefore, the Tie Set matrix is given by $T = [I \mid a^T]$.

The transpose of the tie set matrix is $T^T = [I \mid a]^T$

and it leads to the equation

$$[\ I\ |\ a\ ]^T\ J_L = [\ J_L|\ J_T\ ]^T \qquad (116)$$

or, more concisely,

$$T^T\ J_L = J_b\ . \qquad (117)$$

Given that

$$J_b^T\ X_b = (\ J_L^T\ T\ )(\ C^T\ X_T\ ), \qquad (118)$$

and that Tellegen's theorem shows this expression is zero, it follows that

$$C\ T^T = T\ C^T = 0\ . \qquad (119)$$

Tellegen's theorem is therefore equivalent to the orthogonality between the Cut Set and Tie set matrices--i.e., Kirchhoff's Voltage and Current laws. This orthogonality is rooted, in turn, in the topological ideas of boundary and coboundary operators given above.

The question arises as to what is the topological reason these matrices are orthogonal.

When a row a the matrix C is multiplied by a column of the matrix T^T we are essentially taking products of K.C.L. at a node and K.V.L. around a loop. If this product corresponds to a non-neighboring node and loop, all the products that appear in the expression for K.C.L. will be missing in the K.V.L. expression and viceversa. The only non zero products appear when the row in C and the column in T^T correspond to a node that is included in the loop considered. Specific examples using the Tie set and Cut set matrices are given for the interested reader at the end of this Chapter.

1.22 FUNCTIONAL INVERSION: LEGENDRE TRANSFORMATIONS

One problem that often arises consists in finding the force $X(J)$ in terms of the inverse function $J(X)$. If the system is linear, there is little problem and this is easily done for resistors, capacitors and inductors. For non linear systems , on the other hand, the situation is more complex. However, when each variable is uniquely defined in terms of the other variables --

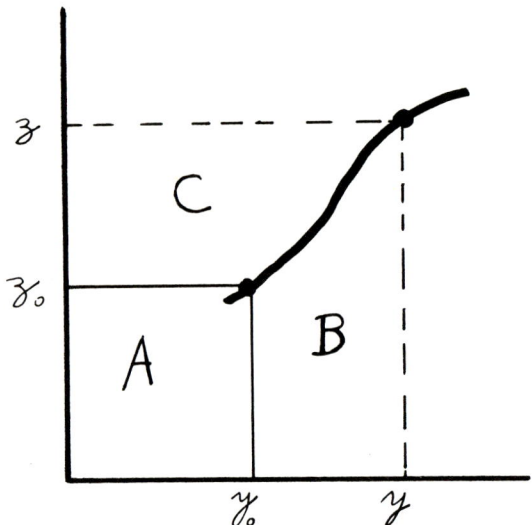

Fig. 1.27 Example of a Legendre transformation.

excluding, then, multiple valued systems--, the technique of Legendre transformations may be used. This technique is basically a form of integration by parts and is familiar to students of classical thermodynamics. The following geometrical approach to the Legendre transformation is useful for introducing the related bilinear forms that appear in all physical theories and, in particular, in network models.

Refer to FIGURE 27, in which two variables are assumed to be related by some function. The areas A, B and C shown obey the relation

$$A + B + C = yz.$$

Moreover, the following integrals can be given:

$$B = \int_{y_0}^{y} z \, dy$$

and

$$C = \int_{z_o}^{z} y \, dz.$$

Clearly, if either $z_o = 0$ or $y_o = 0$, it follows that $A = 0$ and we obtain the (Legendre) transformation

$$\int_{y_o}^{y} z \, dy + \int_{z_o}^{z} y \, dz = yz$$

(in the case of multiple variables the situation is, of course, more complex and the Jacobian formalism is more helpful, but the geometry involved is the same).

1.23 DISSIPATION, CONTENT AND COCONTENT

Various forms of bilinear expressions will be introduced in latter chapters. In an arbitrary resistive element the general form of the dissipation (or power) function is JX. Defining the content

$$G_X = \int_0^J X \, dJ$$

and the co-content

$$G_J = \int_0^X J \, dX$$

it follows that the total dissipation is the sum of the content and cocontent:

$$G_X + G_J = J X.$$

For energy storing elements, similar relations can be given.

Appendix 1

SOME FUNDAMENTAL ALGEBRAIC CONCEPTS AND THEIR NETWORK COUNTERPARTS

Functions of a real variable and general transformations

For our purpose, a function can be thought of as a machine with an input x and an output (ref. 224) y. A function of a real variable is a black box f(x) which gives the real number y as an output when the input is the real number x. The set of inputs is the domain of the function and the admissible outputs are its range. Recall that a function need not be given by a single formula such as $f(x) = x^2$ and that, moreover, it may have different forms for different values of x. Thus, the "machine" may process the inputs differently if $x > 0$ or if $x < 0$, for example.

A transformation F (or function, or operator, or mapping) has an input x which belongs to some arbitrary set X and an output Y which belongs to a second set Y. The domain and the range of the transformation are defined analogously, except that --given that not all the elements of Y may be possible outputs-- we make a distinction depending on whether the range of the function coincides with Y or not. In the first case the transformation is onto, otherwise it is into.

Linear transformations, linear spaces and affine spaces..

Vector spaces (or linear spaces) have been already defined in the text. Two vector spaces V and W over the same field are isomorphic if they have the same dimension.

A transformation T which maps one vector space into another vector space is linear if it obeys the distributive property for addition and if the operator commutes with scalar m so that

$$T(a_1 \mathbf{v}_1 + a_2 \mathbf{v}_2) = a_1 T(\mathbf{v}_1) + a_2 T(\mathbf{v}_2).$$

Matrices

Linear transformations are represented by matrices. An n x m matrix A is a rectangular array of elements $[a_{ij}]$ arranged in n rows and m columns, with i specifying the row number and and j specifying the column number. The transpose matrix A^T is constructed by exchanging rows and columns, so that $[a_{ij}]^T = [a_{ji}]$. The operation of addition between two matrices **A** and **B** having the same number of rows and columns is defined as a matrix **C** whose element in a given row and column is the sum of the corresponding elements in **A** and **B** so that $[c_{ij}] = [a_{ij}] + [b_{ij}]$. A vector is represented as a single column matrix. A square matrix has the same number of rows and columns. The unit matrix, **I**, has all ones in the main diagonal and zeroes elsewhere, while the zero matrix has zeroes for all its entries.

A linear transformation between two vector spaces x and y can be specified by giving the transformation rule between elements x_i and y_j in the form

$$[y_i] = \sum_j [T_{ij}] [x_j],$$

in which $[T_{ij}]$ is the representative matrix element. By extension, if there is a second mapping **M** from the vector space z to x, the two transformations can be compounded to yield

$$[y_i] = \sum_k \sum_j [T_{ij}] [M_{jk}] [x_k],$$

which defines naturally the product $P = T M$ between two matrices as $[P_{ik}] = [T_{ij}] [M_{jk}]$ corresponding to the linear transformation

$$[z_i] = \sum_j [P_{ij}] [z_j].$$

(In terms of the machine view, the product transformation corresponds to hooking the ouptut of the first machine to the input of a second machine).

Metric spaces

An affine space A^n is simply a collection of points defined by n numbers.

A metric space is a collection of points or elements x, y, z... such that to each pair of elements x, y corresponds a number d(x,y), the metric or distance function (refs. 89, 212, 224)

having the following properties:
1) $d(x,y) = d(y, x)$
2) $d(x,y) \geq 0$ [$d(x,y) = 0$ if and only if $x=y$]
3) $d(x,z) \leq d(x,y) + d(y,z)$ ("triangle" inequality).

A special case is the **Euclidean space** R^n which has the metric function

$$d(x,y) = [\sum_n (x_n - y_n)^2]^{1/2}.$$

A **normed linear space** is defined in terms of the absolute value

$$d(x,y) = || x - y ||,$$

which constitutes a "natural" metric. A **Cauchy Sequence** is the converging limit (ref. 224).

$$\lim_{m,p \to \infty} d(x_m, x_p) = 0,$$

in which $\{ x_k \}$ is a sequence of elements in a metric space. A metric vector space is complete if every Cauchy sequence is convergent. A normed linear space which is complete in its natural metric is a **Banach Space**.

The *inner product space* ("dot" product) used in the text is defined in terms of ordered pairs $\langle x,y \rangle$

$$\langle x, y \rangle = x_1 y_1 + x_2 y_2 + \ldots + x_n y_n$$

with the properties:

1) $\langle x,x \rangle \geq 0$ ($=0$ only if $x=0$)   (positive definitness)
2) $\langle x,y \rangle = \langle y,x \rangle$ (commutativity)
3) $\langle (x_1 + x_2), y \rangle = \langle x_1, y \rangle + \langle x_2, y \rangle$ (distributivity)

If the inner product metric generates a complete vector space, the space is a **Hilbert Space**. Any finite dimensional Hilbert space is separable. This means that any vector f in the Hilbert space E_n can be represented as a linear superposition of the spanning set $\{f_k\}$:

$$f = \sum_k z_k f_k.$$

The space of square integrable functions used in the text is also separable.

Inner Product and Power dissipated in a network

The power dissipated in a network has an interesting interpretation in terms of the inner product concept. Consider the simple two dimensional case for which the above expression reduces to

$$\mathbf{a} \cdot \mathbf{b} = a_1 b_1 + a_2 b_2.$$

Referring to FIGURE 1, it is clear that the individual vector components may be expressed as

$$a_1 = ||\mathbf{a}|| \cos \theta_a$$

$$a_2 = ||\mathbf{a}|| \sin \theta_a$$

$$b_1 = ||\mathbf{b}|| \cos \theta_b$$

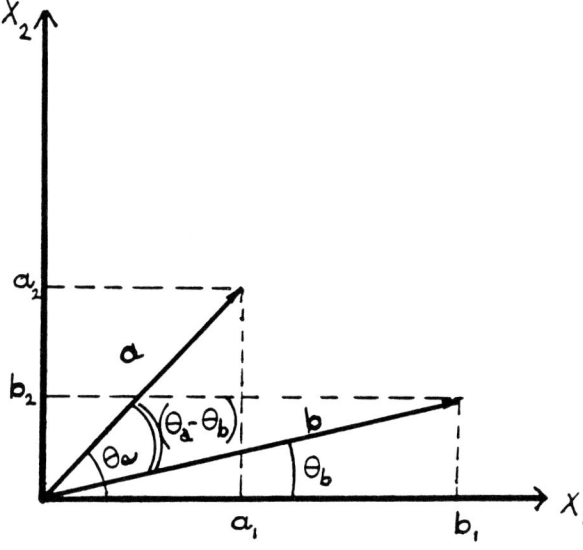

Fig. A.1 Geometrical meaning of the dot product.

and

$$b_2 = ||\mathbf{b}|| \sin \theta_b.$$

It then follows that

$$\mathbf{a} \cdot \mathbf{b} = ||\mathbf{a}||\; ||\mathbf{b}|| \,[\cos \theta_a \cos \theta_b + \sin \theta_a \sin \theta_b]$$

which simply equals the familiar

$$\mathbf{a} \cdot \mathbf{b} = ||\mathbf{a}||\; ||\mathbf{b}|| \cos(\theta_a - \theta_b) = ||\mathbf{a}||\; ||\mathbf{b}|| \cos \theta_{ab}.$$

Clearly, the dissipation function is one such example of a physical dot product. In network terms, there are three ways to visualize this dot products:

1) The dissipation function

$$\Phi = X_1 J_1 + X_2 J_2 + \ldots + X_2 J_2$$

may be considered the products of forces and flows at the ports, as does Onsager thermodynamics.

2) Each of the ports in the disconnected networks R,L,H and P may be thought of as the components of force (or flow) in the direction of the conjugate flow (or force). For example, the input port of the R network gives the component of the force vector $\mathbf{X}$ in the J_1 direction of the flow $\mathbf{J}$. Thus, the complete force ("voltage") vector $\mathbf{X}$ is

$$\mathbf{X} = (R_{11} J_1 + R_{12} J_2)\, \mathbf{j}_1 + (R_{21} J_1 + R_{22} J_2)\, \mathbf{j}_2,$$

in which $\mathbf{j}_1$ and $\mathbf{j}_2$ are unit vectors in "the direction" of J_1 and J_2. By multiplying individual components and adding we obtain the metric

$$\Phi = (R_{11} J_1^2 + 2 R_{12} J_2 J_1 + R_{22} J_2^2,$$

as expected.

3) The metric given above may be considered the self dot product

of a vector so that

$$s^2 = ||s|| \cdot ||s|| = R_{11} J_1^2 + 2 R_{12} J_2 J_1 + R_{22} J_2^2.$$

There is of course an infinite number of ways to obtain such a vector. The connected resistive network--provides a decomposition in which the squares of components of this vector in a higher dimensional space imbedding appear as power dissipated inside the resistances. The decomposition is cartesian, so that the new axes are orthogonal to each other. Clearly, given the restrictions imposed by K.V.L. and K.C.L. , the axes are not linearly independent, even though they are orthogonal.

Appendix 2

NETWORK INCIDENCE MATRICES

A.1.1 NETWORK BRANCHES

In order to write out organized equations for a complex network, it is first necessary to reduce all branches in the network to some standard form. We assume, as before, that the network obeys Kirchhoff's laws and that it only has constant resistances and sources, although the same approach is also valid when storage elements are present (ref. 53, 38).

Each branch will be thought of as either consisting of a force source or battery, E_b, in series with a resistance R_{bb} or a flow source I_b in parallel with a conductance G_{bb}. (Clearly these can be transformed into each other). The equation obeyed by a given branch will therefore either be

$$X_b = R_{bb} J_b + E_b \qquad (1)$$

or

$$J_b = G_{bb} X_b + I_b , \qquad (2)$$

respectively. From our discussion on the substitution theorem, it is clear that the general branch can be expressed in either form; moreover, we can save a lot of space by writing the above equations in matrix form so that the equations for the complete network become

$$\mathbf{X_b} = \mathbf{R_{bb}} \mathbf{J_b} + \mathbf{E_b} \qquad (3)$$

and

$$\mathbf{J_b} = \mathbf{G_{bb}} \mathbf{X_b} + \mathbf{I_b} , \qquad (4)$$

respectively, and now collectively represent the behavior of all the branches in the network with the vectors (written in transpose form, for economy of space)

$$\mathbf{x}^{tr} = (\; x_1 \quad x_2 \quad \ldots\ldots\ldots \quad x_n \;) \qquad (5)$$

and

$$J^{tr} = (\ J_1 \quad J_2 \quad \ldots\ldots\ldots \quad J_n\ ) \qquad (6.)$$

representing the collection of branch forces (voltages) and flows (currents), respectively, the vectors

$$E_b^{tr} = (\ E_1 \quad E_2 \quad \ldots\ldots\ldots \quad E_n\ ) \qquad (7)$$

and

$$I_b^{tr} = (\ I_1 \quad I_2 \quad \ldots\ldots\ldots \quad I_n\ ) \qquad (8)$$

representing the force and flow sources at all the branches and the diagonal matrices

$$R_{bb} = \begin{bmatrix} R_1 & 0 & 0 & \ldots\ldots & 0 \\ & R_2 & & & \\ & & R_3 & & \\ & & & & \\ & & & & R_N \end{bmatrix} \qquad (9)$$

and

$$G_{bb} = \begin{bmatrix} G_1 & 0 & 0 & \ldots\ldots & 0 \\ & G_2 & & & \\ & & G_3 & & \\ & & & & \\ & & & & G_N \end{bmatrix}$$

representing the resistance or conductance of each branch. Clearly $R_{bb} = G_{bb}^{-1}$.

A.1.2 THE NODAL MATRIX

We define the nodal matrix for a network having b branches and n nodes as follows. Assign a matrix column to each branch in the network and a matrix row to each node. The entries of the matrix specify the incidence properties between branches and nodes as follows: if the branch leaves a given node it is assigned the value + 1, if it is incident on the node it is given the value -1 and if it is not incident in the node it is assigned the value 0.

Clearly, the nodal matrix gives one possible practical approach to the mapping between oriented edges and vertices discussed above. The nodal matrix has other properties, which are discussed below.

As an example, consider the network shown in FIGURE 1 which has the associated graph also depicted in the same drawing. The corresponding nodal matrix is

$$N = \begin{array}{c} \text{nodes} \\ \downarrow \end{array} \underset{\text{branches} \longrightarrow}{\begin{bmatrix} 1 & 0 & 0 & -1 & 1 \\ -1 & 1 & 1 & 0 & 0 \\ 0 & 0 & -1 & 1 & 0 \\ 0 & -1 & 0 & 0 & -1 \end{bmatrix}} \qquad (10)$$

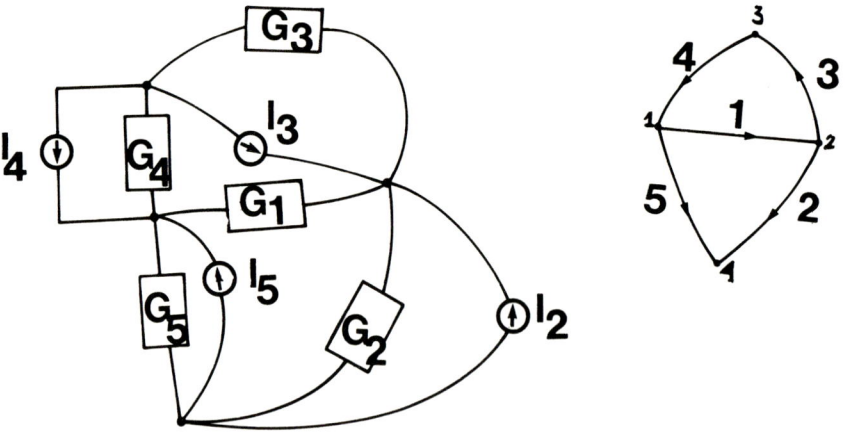

FIGURE A.1   Example used in the text to set up nodal equations.

Note that each column has only two entries (a +1 and a -1) indicating the node at which a branch "enters" (the - node) and the node at which the branch leaves (the +1 node). It is a simple matter to check that the product of the nodal matrix and the branch flow vector is zero,

$$N \, J_b = 0 \qquad (11)$$

--i.e., the nodal matrix takes the flow vector J_b and maps it into the zero vector. The physical counterpart of this topological result is simply Kirchhoff's Current Law (K.C.L.), as can be checked for this specific example by writing Equation (11) explicitly in terms of the individual entries:

$$
\begin{aligned}
J_1 - J_4 + J_5 &= 0 \quad &\text{(NODE 1)} &\quad &(12)\\
-J_1 + J_2 + J_3 &= 0 \quad &\text{(NODE 2)} &\quad &(13)\\
-J_3 + J_4 &= 0 \quad &\text{(NODE 3)} &\quad &(14)\\
-J_2 - J_5 &= 0 . \quad &\text{(NODE 4)} &\quad &(15)
\end{aligned}
$$

It can also be checked by writing the equations explicitly that if a **node potential** vector e_n is defined giving the collection of the potentials at all nodes in the network, the forces ("voltages") across the branches can be obtained from the matrix equation

$$N^T \, e_n = X_b , \qquad (16)$$

thus expressing the fact that the transpose of the matrix N expresses Kirchhoff's Voltage Law. For the example under consideration this fact can readily be checked by direct computation. The transpose matrix is given by

$$N^T = \begin{bmatrix} 1 & -1 & 0 & 0 \\ 0 & 1 & 0 & -1 \\ 0 & 1 & -1 & 0 \\ -1 & 0 & 1 & 0 \\ 1 & 0 & 0 & -1 \end{bmatrix} \qquad (17)$$

and the matrix equation (16) becomes

$$e_1 - e_2 = X_1$$
$$e_2 - e_4 = X_2$$
$$e_2 - e_3 = X_3 \quad (18)$$
$$-e_1 + e_3 = X_4$$
$$e_1 - e_4 = X_5 .$$

We now return to the general branch form given by

$$J_b = G_{bb} X_b + I_b , \quad (19)$$

in which the vector I_b is, for the example considered

$$I_b^{tr} = (I_1 \quad I_2 \quad I_3 \quad I_4 \quad I_5) . \quad (20)$$

(Note that all G´s are given as inverse resistances)

Premultiplying Eq (19) by N, we obtain (ref. 53)

$$N J_b = N G_{bb} X_b + N I_b . \quad (21)$$

The left hand side of this equation is zero because of Eq (11) -- i.e., K.C.L. In addition, Eq. (16) allows to express the forces explicitly in terms of node potentials, so that

$$0 = (N G_{bb} N^T) e_n + N I_b . \quad (22)$$

By defining the conductance matrix

$$G_n = (N G_{bb} N^T) \quad (23)$$

equation (22) becomes

$$G_n e_n = - N I_b , \quad (24)$$

thus yielding an equation that relates the node potentials to the flow sources in the network.

We can calculate G_n and $N I_b$ for the example considered here. First, the product specified by Eq. (23) is set up:

$$(N G_{bb} N^T) =$$

$$
= \begin{bmatrix} 1 & 0 & 0 & -1 & 1 \\ -1 & 1 & 1 & 0 & 0 \\ 0 & 0 & -1 & 1 & 0 \\ 0 & -1 & 0 & 0 & -1 \end{bmatrix} \begin{bmatrix} G_1 & 0 & 0 & 0 & 0 \\ 0 & G_2 & 0 & 0 & 0 \\ 0 & 0 & G_3 & 0 & 0 \\ 0 & 0 & 0 & G_4 & 0 \\ 0 & 0 & 0 & 0 & G_5 \end{bmatrix} \begin{bmatrix} 1 & -1 & 0 & 0 \\ 0 & 1 & 0 & -1 \\ 0 & 1 & -1 & 0 \\ -1 & 0 & 1 & 0 \\ 1 & 0 & 0 & -1 \end{bmatrix}
$$

After the multiplication is carried out, we obtain

$$
G_n = \begin{bmatrix} (G_1+G_4+G_5) & -G_1 & -G_4 & -G_5 \\ -G_1 & (G_1+G_2+G_3) & -G_3 & -G_2 \\ -G_4 & -G_3 & (G_3+G_4) & 0 \\ -G_5 & -G_2 & 0 & (G_2+G_5) \end{bmatrix}
$$

while the product $N\ I_b$ is

$$
N\ I_b = \begin{bmatrix} -I_5 - I_4 \\ -I_2 - I_3 \\ +I_3 + I_4 \\ +I_2 + I_5 \end{bmatrix}
$$

and the resultant equation (24) can be set up.

Moreover, it can be simplified further by arbitrarily assigning the reference potential zero to one of the nodes, say e_4.

The entries of the general G_n matrix can be written by inspection by noting that any off diagonal element is the sum of the G's attached to the nodes indicated by the row and column of the matrix (taken with negative sign), while the diagonal elements ii are the sum of all G's attached to node i.

The matrix $N\ I_b$ is simply the sum of all equivalent flow sources leaving a given node.

A.1.3 MESH EQUATIONS

The planar graphs considered can be drawn on a sphere so that the edges determine faces of a three dimensional solid. As pointed out above, Kirchhoff's current law is consistent with the definition of fictitious mesh currents. From a topological point of view this corresponds to the orientability of the faces of the solid. Denote the mesh currents by j_m, noting that in each planar graph there will be an additional mesh current corresponding to the "hidden face" which will run against the direction of all other faces, when seen from the direction facing the reader.

The directions of flow determined by the mesh currents can be used to write K.V.L. for all faces in a consistent fashion. For example, in the network shown on FIGURE 2 there are three meshes yielding the "force" equations

$$-X_1 - X_3 - X_4 = 0 \quad \text{(MESH 1)}$$

$$X_1 + X_2 - X_5 = 0 \quad \text{(MESH 2)}$$

$$-X_2 + X_3 + X_4 + X_5 = 0 \quad \text{(MESH 3)}.$$

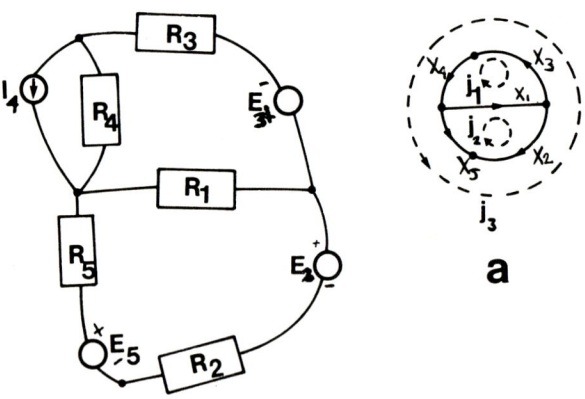

FIGURE A.2   A network example showing mesh flows. The parallel flow/resistance combination must be transformed into a series force source/resistance combination (R_4 I_4).

These equations can also be written in matrix form as

$$\begin{bmatrix} -1 & 0 & -1 & -1 & 0 \\ 1 & 1 & 0 & 0 & -1 \\ 0 & -1 & 1 & 1 & 1 \end{bmatrix} \begin{bmatrix} X_1 \\ X_2 \\ X_3 \\ X_4 \\ X_5 \end{bmatrix} = 0$$

or denoting the multiplying matrix by **M** and the branch force vector by X_b the equations become

$$M \, X_b = 0 , \qquad (25)$$

in which **M** is the mesh matrix.

The branch flows J_k can also be expressed in terms of the mesh flows j_m as follows

$$\begin{aligned} J_1 &= -j_1 + j_2 \\ J_2 &= -j_3 + j_2 \\ J_3 &= -j_1 + j_3 \\ J_4 &= -j_1 + j_3 \\ J_5 &= -j_2 + j_3 \end{aligned} \qquad (26)$$

which can be expressed in matrix form as

$$\begin{bmatrix} J_1 \\ J_2 \\ J_3 \\ J_4 \\ J_5 \end{bmatrix} = \begin{bmatrix} -1 & 1 & 0 \\ 0 & 1 & -1 \\ -1 & 0 & 1 \\ -1 & 0 & 1 \\ 0 & -1 & 1 \end{bmatrix} \begin{bmatrix} j_1 \\ j_2 \\ j_3 \end{bmatrix}$$

and it is clear that these equations correspond to the matrix equation

$$J_b = M^T j_m \quad . \tag{28}$$

Equations (25) and (28) will be valid in general provided that **M** is constructed by assigning mesh flows to all faces and assigning a row to each mesh and a column to each branch. The entries in the matrix have the following entries:

+1 if the direction of the flow in the edge coincides with the direction of mesh flow,

-1 if the flow in the edge has opposite direction to that of the mesh flow, and

0 if the branch does not belong to the mesh considered.

The mesh matrix can now be used to set up network equations. Using the branch form

$$X_b = R_{bb} J_b + E_b$$

and premultiplying by **M** we obtain

$$M X_b = M R_{bb} J_b + M E_b \tag{29}$$

Because of (25) the left hand side of this equation is zero, while J_b may be given in explicit form using Eq (27) to yield

$$0 = (M R_{bb} M^T) j_m + M E_b \quad . \tag{30}$$

Defining the resistance matrix R_m by

$$R_m = (M R_{bb} M^T) \tag{31}$$

and rearranging it follows that

$$R_m j_m = - M E_b \quad . \tag{32}$$

For the specific example considered above, this matrix product is given by

$$\begin{bmatrix} -1 & 0 & -1 & -1 & 0 \\ 1 & 1 & 0 & 0 & -1 \\ 0 & -1 & 1 & 1 & 1 \end{bmatrix} \begin{bmatrix} R_1 & 0 & 0 & 0 & 0 \\ 0 & R_2 & 0 & 0 & 0 \\ 0 & 0 & R_3 & 0 & 0 \\ 0 & 0 & 0 & R_4 & 0 \\ 0 & 0 & 0 & 0 & R_5 \end{bmatrix} \begin{bmatrix} -1 & 1 & 0 \\ 0 & 1 & -1 \\ -1 & 0 & 1 \\ -1 & 0 & 1 \\ 0 & -1 & 1 \end{bmatrix}$$

in which $R_i = 1/G_i$ and the product **M E_{bb}** --which requires transforming each branch to an equivalent force source and resistance in series -- is given by

$$\mathbf{M}\ E_{bb} = \begin{bmatrix} 1 & 0 & -1 & -1 & 0 \\ -1 & 1 & 0 & 0 & -1 \\ 0 & -1 & 1 & 1 & 1 \end{bmatrix} \begin{bmatrix} 0 \\ E_2 \\ E_3 \\ -I_4 R_4 \\ E_5 \end{bmatrix} = \begin{bmatrix} -E_3 + I_4 R_4 \\ E_2 - E_5 \\ -E_2 - E_3 + I_4 + E_5 \end{bmatrix}$$

A.1.4 LOOP NETWORK EQUATIONS. THE TIE SET MATRIX

A third approach to writing network equations consists in relating forces and flows in the tree branches to those in the links. To fix ideas, we refer again to the previous example and chose a tree for this network, as shown in FIGURE 3, in which the tree flows have been dotted.(Although in this particular case the links happen to be J_1 and J_2 as previously labelled, it is always convenient to relabel the links so that they have the lowest subscripts). By definition, when the links are removed no flows take place in the tree; when J_1 and J_2 are replaced, one at a time, each determines a flow pattern in the tree, these will be called <u>loop flows</u> and will be denoted by i_1 and i_2, corresponding to the flows induced in the tree by J_1 and J_2, respectively. Each of the flows in the tree branches can be calculated using linear

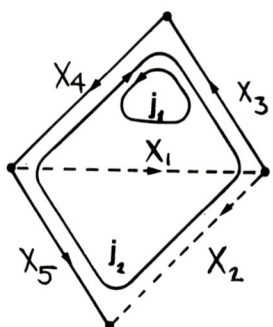

FIGURE A.3   Loop flows for the previous example.

superposition of the two link flows, in this example.(Linear independence can be topologically established by noting that shutting off one link flow does not affect the other).

Given that K.V.L. must be obeyed for each loop we can write

$$X_1 + X_3 + X_4 = 0 \qquad (\text{LOOP 1}) \qquad (33)$$

and

$$X_2 - X_3 - X_4 - X_5 = 0 \qquad (\text{LOOP 2}). \qquad (34)$$

These equations can also be given the following matrix form

$$\begin{bmatrix} 1 & 0 & .1 & 1 & 0 \\ 0 & 1 & .-1 & -1 & -1 \end{bmatrix} \begin{bmatrix} X_1 \\ X_2 \\ \cdots \\ X_3 \\ X_4 \\ X_1 \end{bmatrix} = \mathbf{T}\ \mathbf{X_b} = 0 \qquad (35)$$

in which we have defined the <u>tie set</u> matrix **T** for this example and have subdivided this matrix and the branch force vector into two parts, as follows

$$T = [\ I\ |\ -b\ ]$$

and

$$X_b^{tr} = [\ X_L\quad X_T\ ],$$

in which X_L and X_T are the link and tree force vectors while **b** is the matrix relating the link forces to the tree forces. Note that while K.C.L. established that the tree flows must be a linear combination of the link flows, K.C.L. establishes --through the same tree topology-- that the link forces be linear combinations of the tree forces :

$$X_L = b\ X_T\ .$$

The matrix **I** used above is simply the unit matrix having main diagonal elements equal to 1 and 0 for the remaining entries.
It is also easy to verify that the tree flows are given by

$$T^T i_L = J_b,$$

in which T^T is the transpose of the tie set matrix.
In the example considered these equations are

$$\begin{bmatrix} J_1 \\ J_2 \\ J_3 \\ J_4 \\ J_5 \end{bmatrix} = \begin{bmatrix} 1 & 0 \\ 0 & 1 \\ 1 & -1 \\ 1 & -1 \\ 0 & -1 \end{bmatrix} \begin{bmatrix} i_1 \\ i_2 \end{bmatrix} \qquad (36)$$

Again, Eq (36) follows readily by considering the branch and link flow vector equations

$$J_T = -\ b^T J_L$$

and subdividing the vector of all flows, J_b into two vector

components
$$J_b = [\; J_L \quad J_T \;]. \qquad (37)$$

If each link branch b has an associated resistance R_b, as is the case in the present example, the link and tree flows can be readily related by

$$X_L = b \; X_T$$

$$J_T = -b^T \; J_L$$

and introducing

$$X_T = R_T \; J_T$$

it follows that

$$X_L = -(\; b \; R_T \; b^T) \; J_L \;. \qquad (38)$$

In general, it is more convenient, however, to start with the generalized branch equation

$$X_b = R_{bb} \; J_b + E_b \qquad (39)$$

which yields, after premultiplication by T

$$T \; X_b = T \; R_{bb} \; J_b + T \; E_b \;. \qquad (40)$$

From Eq. (35) it follows that

$$0 = (\; T \; R_{bb} \; T^T) \; i_L + T \; E_b \;.$$

If the resistance

$$R_L = (\; T \; R_{bb} \; T^T)$$

is defined, we finally obtain

$$R_L \; i_L = -T \; E_b \;,$$

an equation which relates the independent loop flows and the force sources in the network.

In the example considered here, the above resistance becomes

$$R_L = \begin{bmatrix} 1 & 0 & 1 & 1 & 0 \\ 0 & 1 & -1 & -1 & 1 \end{bmatrix} \begin{bmatrix} R_1 & 0 & 0 & 0 & 0 \\ 0 & R_2 & 0 & 0 & 0 \\ 0 & 0 & R_3 & 0 & 0 \\ 0 & 0 & 0 & R_4 & 0 \\ 0 & 0 & 0 & 0 & R_5 \end{bmatrix} \begin{bmatrix} 1 & 0 \\ 0 & 1 \\ 1 & -1 \\ 1 & -1 \\ 0 & -1 \end{bmatrix}$$

$$= \begin{bmatrix} (R_1 + R_3 + R_4) & -(R_3 + R_4) \\ -(R_3 + R_4) & (R_2 + R_3 + R_4 - R_5) \end{bmatrix}$$

while the product of matrices $T\ E_b$ yields

$$\begin{bmatrix} 1 & 0 & 1 & 1 & 0 \\ 0 & 1 & -1 & -1 & -1 \end{bmatrix} \begin{bmatrix} 0 \\ E_2 \\ -E_3 \\ I_4 R_4 \\ E_5 \end{bmatrix} \begin{bmatrix} -E_3 + I_4 R_4 \\ E_2 + E_3 - I_4 R_4 - E_5 \end{bmatrix}$$

so that the resultant loop equations are

$$-E_3 + I_4 R_4 = (R_1 + R_3 + R_4) J_1 - (R_3 + R_4) J_2$$

and

$$E_2 + E_3 - I_4 R_4 - E_5 = -(R_3 + R_4) J_1 + (R_1 + R_3 + R_4 - R_5) J_2.$$

A.1.5 NODE PAIR EQUATIONS BASED ON THE CUT SET MATRIX

In addition to the loop equations given above, it is also possible to set equations which relate the tree forces to the flow sources, by defining the **Cut Set Matrix**, as follows. Each branch in the tree is assigned a row and each flow in the network is assigned a column. As above, set up flows in the tree by connecting one link flow at a time and observing the directions of the branch relative to that of the established flow. The entries of the cut set matrix **C** then are:

-1 if the direction of flow through the branch coincides with the direction of the particular link flow,

-1 if the direction of flow is against the direction established by the link flow,

0 if the link flow does not contribute to the flow in that particular branch and

+1 if the branch is the link itself.

For the example given above, the tree flows are related to the link flows by

$$\begin{bmatrix} J_3 \\ J_4 \\ J_5 \end{bmatrix} = \begin{bmatrix} 1 & -1 \\ 1 & -1 \\ 0 & -1 \end{bmatrix} \begin{bmatrix} J_1 \\ J_2 \end{bmatrix}$$

we can write these same equations after rearrangement as

$$\begin{bmatrix} -1 & 1 & . & 1 & 0 & 0 \\ -1 & 1 & . & 0 & 1 & 0 \\ 0 & 1 & . & 0 & 0 & 1 \end{bmatrix} \begin{bmatrix} J_1 \\ J_2 \\ J_3 \\ J_4 \\ J_5 \end{bmatrix} = 0 \quad ,$$

in which we have purposely partitioned the matrix and flow vector. Comparing with the definition of the cut set matrix given above, it is clear that this equation can be expressed as

$$C J_b = 0 \qquad (41)$$

and, furthermore, the previous definition of the matrix b leads to the result

$$C = [\, b^T \, . \, I \,] \qquad (42)$$

This equation is clearly valid for general networks. Moreover, direct multiplication of the cut set transpose by the vector of all network forces leads to

$$C^T X_T = X_b,$$

because

$$b \, X_T = X_L$$

and

$$X_L = X_L .$$

If a general branch in the network is given by

$$J_b = G_{bb} X + I_b \qquad (43)$$

and this expression is premultiplied by C, we obtain

$$C J_b = C G_{bb} X_b + C I_b . \qquad (44)$$

However, the term on the left is zero, because of equation (41), while X_b may be given explicitly as $X_b = C^T X_T$, thus yielding

$$0 = (C G_{bb} C^T) X_T + C I_b .$$

By defining the node pair conductance G_T,

$$G_T = C G_{bb} C^T \qquad (45)$$

and rearranging, we obtain

$$G_T X_T = - C I_b , \qquad (46)$$

which relate tree forces and flow sources. For the present example, the node conductance equation is

$$G_T = C G_{bb} C^T =$$

$$= \begin{bmatrix} -1 & 1 & 1 & 0 & 0 \\ -1 & 1 & 0 & 1 & 0 \\ 0 & 1 & 0 & 0 & 1 \end{bmatrix} \begin{bmatrix} (1/R_1) & 0 & 0 & 0 & 0 \\ 0 & (G_2) & 0 & 0 & 0 \\ 0 & 0 & (1/R_3) & 0 & 0 \\ 0 & 0 & 0 & (1/R_4) & 0 \\ 0 & 0 & 0 & 0 & (1/R_5) \end{bmatrix} \begin{bmatrix} -1 & -1 & 0 \\ 1 & 1 & 1 \\ 1 & 0 & 0 \\ 0 & 1 & 0 \\ 0 & 0 & 1 \end{bmatrix}$$

The product $C I_b$,

$$C I_b = \begin{bmatrix} -1 & 1 & 1 & 0 & 0 \\ -1 & 1 & 0 & 1 & 0 \\ 0 & 1 & 0 & 0 & 1 \end{bmatrix} \begin{bmatrix} 0 \\ E_2 G_2 \\ E_3/R_3 \\ I_4 \\ E_5/R_5 \end{bmatrix}$$

then leads to the final equations

$[(1/R_1) + (1/R_3) + G_2] X_3 + [(1/R_1) + G_2] X_4 + G_2 X_5 = E_2 G_2 - E_3/R_3$

$[(1/R_1) + G_2] X_3 + [(1/R_1) + (1/R_4) + G_2] X_4 + G_2 X_5 = E_2 G_2 - I_4$

$G_2 X_3 + G_2 X_4 + [(1/R_5) + G_2] X_5 = E_2 G_2 - E_5/R_5.$ (47)

A.1.6 EQUATIONS IN RLC NETWORKS

Networks which contain capacitors and inductors--i.e., energy storing or entropy storing elements, depending on the system described-- can be analyzed using the same basic techniques introduced before. I shall only review the bare essentials, as any introductory book on Network theory will cover this subject appropriately and we only have limited use for these equations here.

Analysis of the RLC networks using Kirchhoff's laws leads to differential equations of the form

$$a_n \, dX^n/dt + a_{n-1} \, dX^{n-1}/dt + \ldots + a_0 = b_m \, dJ^m/dt + b_{m-1} \, dJ^{m-1}/dt + \ldots + b_0,$$

for the case of a single port X, J. In this monograph we only deal with equations in which there are only ordinary derivatives of time (no partial derivatives) and which have constant coefficients.

The differential operators

$$L(\ ) = a_n \, d(\ )^n/dt + a_{n-1} \, d(\ )^{n-1}/dt + \ldots + a_0$$

and

$$M(\ ) = b_m \, d(\ )^m/dt + b_{m-1} \, d(\ )^{m-1}/dt + \ldots + b_0,$$

which accept a force X and a flow J in the slots, respectively, are linear because
$L(X_1 + X_2) = L(X_1) + L(X_2)$
and
$L(J_1 + J_2) = L(J_1) + L(J_2),$

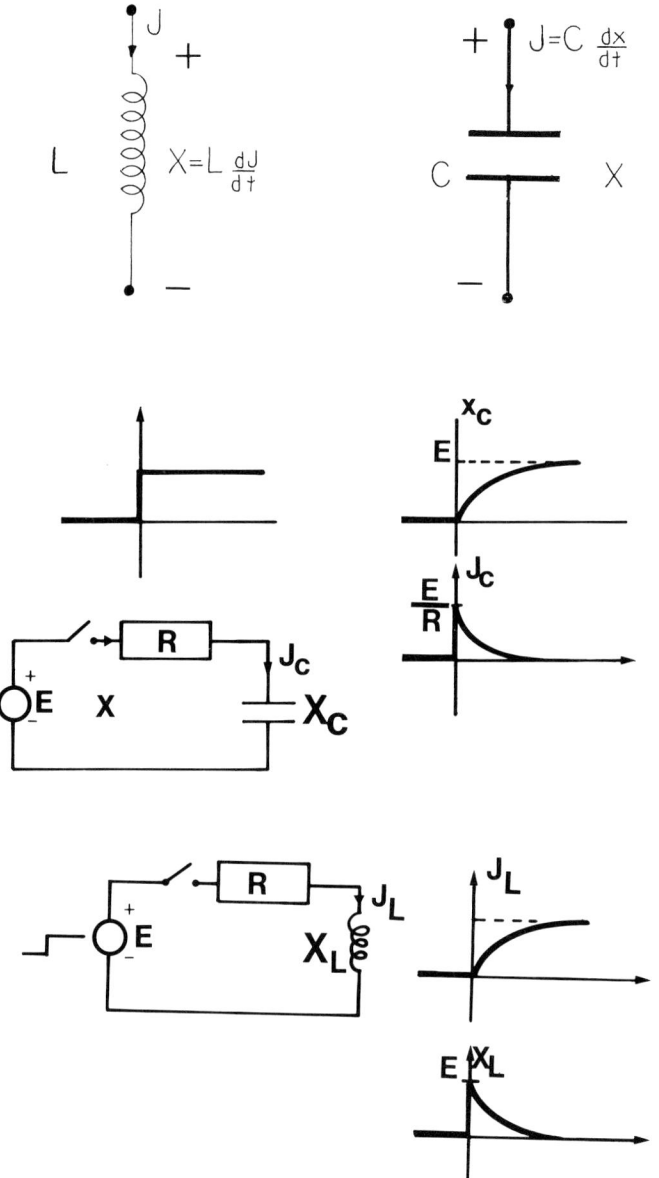

Fig. A.4 The inductor L and capacitor C and responses to a step function.

respectively. In the operator notation, the differential equations may be summarized by L (X) = M (J). A multiport system leads to matrix operators, but has otherwise the same general form.

Solutions to these equations using various excitations such as steps and pulses and various initial conditions are well known and will not be reviewed here. These solutions consist of a homogeneous and a particular solution. For the present purpose we assume that all transients have decayed and that an excitation of the form

X(t) = Real **X(w)** exp (jwt)

forces the system to vibrate "monochromatically" with frequency w. The branch flows will then have the general form

$J(t)^b$ = Real **J(w)**^{b} exp (jwt)

in which **J(w)**^{b} is the complex flow for branch b. All the complex flows and forces at the branches obey an independent form of Kirchhoff's laws, provided that the sinusoidal steady state has been reached (all exponentials drop out of all equations) and the same is true for the complex conjugates $X^*(w)$ and $J^*(w)$. It then follows that Tellegen's theorem can be expressed either as

$$\sum_b X^*_{(w)} J^*_{(w)} = 0$$

or as

$$\sum_b X_{(w)} J_{(w)} = 0 .$$

Each branch can therefore be assigned a complex flow and a complex force and treated as regular forces and flows--i.e., voltages and currents -- provided that the complex "resistances"

Z (w) = $X^b(w)$ / $J^b(w)$

(the impedances of a given branch) are defined. Alternatively, the admittance

Y (w) = $J^b(w)$/ $X^b(w)$

can be specified at each branch. Because the complex forces and flows obey K.V.L. and K.C.L. the impedances and admittances behave like resistances in terms of parallel or series addition. Moreover, all methods for writing network equations can also be restated in terms of the complex quantities provided that the force and flow vectors are replaced by the complex vectors $X_b(w)$, $J_b(w)$ and the resistance and conductance matrices are replaced by the admitance or impedance matrices Y and Z.

With these changes, we formally obtain the following complex equations:

1. Node Equations, $Y_N e_n = -N I_b$
2. Mesh Equations, $Z_m j_m = -M E_b$
3. Loop Equations, $Z_L i_L = -T E_b$
4. Node Pair equations, $Y_T X_T = -C I_b$

with the complex matrices

$Y_N = N Y_{bb} N^T$
$Z_m = M Z_{bb} M^T$
$Z_L = T Z_{bb} T^T$
and
$Y_T = C Y_{bb} C^T$
defined as before.

Chapter 2

NETWORK THERMOSTATICS: THERMOSTATICS AND CONNECTED NETWORKS

2.1 INTRODUCTION

There are two approaches used in the study of thermodynamics: the first analyzes cycles of ideal engines-- such as refrigerators, steam engines,etc-- while the second, invented by Gibbs (ref. 85) , stresses the function of state concept. The former is preferred by engineers, while the latter has the appeal of unification and the possibility of using the powerful ideas of conservative field theory (ref. 132).

The differential change dϕ in any one of the energy potentials ϕ is expressed in concise form as

$$d\phi = \mathbf{X} \cdot \mathbf{dr} \qquad (1)$$

in which $\mathbf{dr}$ is a displacement vector pointing in the direction of the independent thermodynamic variables and $\mathbf{X}$ is a generalized conjugate vector. The statement that dϕ is an exact differential leads to the familiar integral expression $\oint d\phi = 0$, which implies that curl X=0 or, equivalently, that X= grad ϕ . These equations contain Maxwell's reciprocities and express the basic idea that, given a number of independent thermodynamic variables defining a direction of displacement $\mathbf{r}$ in thermodynamic space, there is a function of state ϕ whose change in the direction of <u>steepest descent</u>--i.e.,the direction of the the gradient-- is given by

$$d\phi = \text{grad}_r \cdot \mathbf{dr} \qquad (2)$$

For each independent direction of a displacement in the thermodynamic manifold there is a potential energy surface whose gradient at that point gives the resultant conjugate vector. For example, by looking at the energy surface in entropy volume space, the slope in the direction of maximum change in E at a given point yields the conjugate vector having components in the directions of T and P. This formulation of thermodynamics agrees with the familiar cause effect idea that "motion goes in the direction of the force" and with more subtle geometrical ideas of covariance. Of course, the more familiar definitions of temperature and pressure,

$$T = (\partial E/\partial S) \tag{3}$$

and

$$-P = (\partial E/\partial V)_S \tag{4}$$

are included in the manifold view.

The topological/ vectorial view expresses concisely the first law (existence of the energy potential surfaces) and second law (motion in the direction of steepest descent) of thermodynamics as well as the variational requirement that

$$\delta(\mathbf{x} \cdot \mathbf{dr}) = 0. \tag{5}$$

This potential surface, or function of state approach is pleasing because of its unifying features; but it does not always point the best route that yields a specific set of relationships among thermodynamic variables. Moreover, one loses the visualization of mechanistic relationships which are often suggested by the engine approaches. Available textbooks (refs. 74, 120, 162, 273, etc.) provide a wide range of choices and presentations, and more modern presentations also include "rational" (ref. 250) and logical (ref. 217) expositions of Thermodynamics . The reason for reinterpreting the same subject so many times is two fold.

First, one would like to obtain a better logical understanding of what may be considered one of the most aesthetical scientific structures. Second, one would like to extend its range of applicability to far from equilibrium and to irreversible processes - which has not been fully achieved yet.

But there is also a practical and pedagogical question : what is the most efficient way of summarizing the rich amount of thermostatic information by means of concise, easily understandable expressions? Ideally, one would like to utilize thermodynamic variables as building blocks which could add or multiply like numbers. The early recognition that thermodynamic variables behave more like vectors than numbers, made physical chemists aware of the limitations of such procedures. However, Bridgman succeeded in making a great practical step in such direction by utilizing the Jacobian notation to represent increments in thermodynamic variables taken under specific conditions (ref. 32, 33). Clearly, these approaches, as well as the more fundamental techniques of dealing with Legendre transformations require considerable training in Advanced Calculus and do not provide any further physical insight. In addition, the procedure also requires the availability of adjunctive tables.

The underlying vectorial characteristics of thermodynamic theory were exploited by Hertz in his model of geometric mechanics, which was theoretically powerful, but failed to provide measurable predictions. A more recent revival of such ideas was the metric/geometric model of Thermodynamics proposed by Weinhold, which displayed the full impact of the vectorial properties of the Thermodynamic manifold (refs. 258-260).

Following other branches of Physics, metric/ geometric Thermodynamics assumes that invariant measurements can be equated with the geometric concept of the length of a displacement vector.

These distances are expressed in terms of dot products of vectors and the resultant vector space is an example of an inner product space. If it can be determined that a set of n quantities spans a metric (inner product) vector space, it follows that these quantities can be represented by the physical vectors of n-dimensional euclidean space. Thus, the geometrical view may be extended to processes of any dimensionality.

The vectors themselves are defined, as above, from gradients of functions of state. For example, from the energy E the intensive variable vectors R_i can be defined in the direction of the extensive variable X_i vector,

$$R_i = \left(\partial E / \partial X_i \right)_{X_J} . \qquad (6)$$

The scalar product of these vectors lead to the elements of the Hessian matrix

$$R_i \cdot R_j = \partial^2 E / \partial X_i \partial X_j . \qquad (7)$$

The lengths and relative angles of these vectors can then be identified with specific thermodynamic--i.e., physical --properties. Thus, the analytical problems of thermodynamics are partially reduced to the identical geometrical problem of measuring lengths and angles. In specific situations, the actual geometrical representation is given by drawing the vectors T,P,S,V etc. with appropriate lengths and relative angles shown on the diagram.

The metric/ geometric theory contributes, in principle, a powerful technique, because vectors have an algebraic meaning and physical existence independent of the dimensionality of the process considered or the frame of reference chosen to describe it. However, it has some serious practical limitations, because: i) the representation of n-dimensional processes become harder to visualize for more than two dimensions and ii) the procedure breaks down for irreversible processes, because the representation of physical invariants by means of a distance is only valid for integrable --i.e., reversible -- processes.

2.2 WHY NETWORKS?

This provides a proper background and the rationale for introducing networks, even though from the chronological point of view both models were generated in parallel (refs. 179, 182   ).

The first goal of the network analysis of thermostatics is to provide some freedom from the restrictions inherent to geometrical analyses, while retaining its useful concepts. The most immediate result is heuristic: networks allow a planar representation of n-dimensional geometries and they also yield a proper association between intensive and conjugate extensive variables, as these become "voltages" and "currents" in network branches.

The isomorphism between Network Theory and thermostatics allows the utilization of some powerful theorems which lead to a summary of classical results reciprocities, Legendre transformations,etc.) by means of a single expression, the **General Reciprocity Relation**. Moreover, the analysis shows the close correspondence between the reciprocities of thermostatics and those of steady state Thermodynamics (i.e.,Onsager´s theory), to be considered in CHAPTER 4 and it allows a natural extension to irreversible and non equilibrium processes.

Once the network approach is presented and discussed, we shall consider the network itself in the context of a mapping which takes a Hilbert space into the space of integrable functions-- a subject mathematicians consider in functional analysis.

2.3 THERMOSTATIC PORTS

The most primitive characteristic of a thermodynamic system is that it has ports at which measurements can be performed, heat can be added to the system or work can be done on or by the system.

In order to be consistent with the ideas of cause and effect in the proper covariant framework, each port must have a covariant, or force like quantity and a (conjugate) contravariant or displacement like quantity. In terms of Graph Theory and networks, we have indicated in CHAPTER 1 that these can be associated with input voltages and currents, respectively.

The superficial idea behind the network thermostatic representation is straightforward: the two algebraic structures are isomorphic and corresponding variables can therefore be found in both; in particular, the vector spaces of covariant (force like, intensive) variables and contravariant (extensive, displacement like) variables are represented through the orthogonality of Kirchhoff's current and voltage laws (refs. 182, 189).

Formally, the thermodynamic phase can be represented by black boxes having n-ports corresponding to the n degrees of freedom of the particular problem. The voltages and currents in this Phase Network can then be equated to variations in Entropy, temperature, volume, pressure, etc.

Although any arbitrary assignments of voltages and currents would work from the algebraic point of view, it is most natural to identify the force like quantities (intensive or covariant variables) with the voltages and the displacement like quantities (extensive or contravariant) with the currents in the network. For example, the following obvious identifications may be made in a system with two degrees of freedom :

$$T = v_1 , \qquad (8)$$

$$P = v_2 , \qquad (9)$$

$$S = i_1 \qquad (10)$$

and

$$V = i_2 . \qquad (11)$$

(Note, again, that both S and V are defined as flowing **into** the

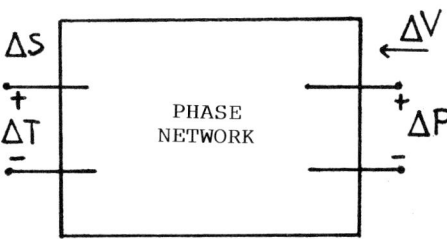

Fig. 2.1 Thermostatic Phase Network for two degrees of freedom having input currents dS, dV and voltages dT, dP.

Phase Network.)

In general, it is more relevant to calculate changes in thermodynamic variables, rather than the variables themselves, so that it is more interesting to represent the incremental network in which the ports represent incremental thermodynamic quantities. --e.g., the port variables dT, dS , dP and dV (for the case of two degrees of freedom), as shown in FIGURE 1.

The end result, from the point of view of the external ports, is that a variation in an intensive quantity is taken across two nodes (labelled + and -), while a variation in a conjugate extensive quantity "flows" into the box at the + terminal and returns, in the same amount, out of the "-" node or terminal, as shown in FIGURE 1. Partial derivatives may then be readily calculated by measuring ratios of the incremental variables dP, dT, etc when other port variables are specified, as indicated in FIGURE 2(a) and (b) . Moreover, in addition to calculating the familiar

$$(\partial P/\partial S)_V \, , \, (\partial T/\partial S)_V, \, (\partial P/\partial S)_V \qquad (12)$$

--which correspond to situations in which the port voltages or currents, now representing differences , are either open or shorted--, it would also be possible to specify more general

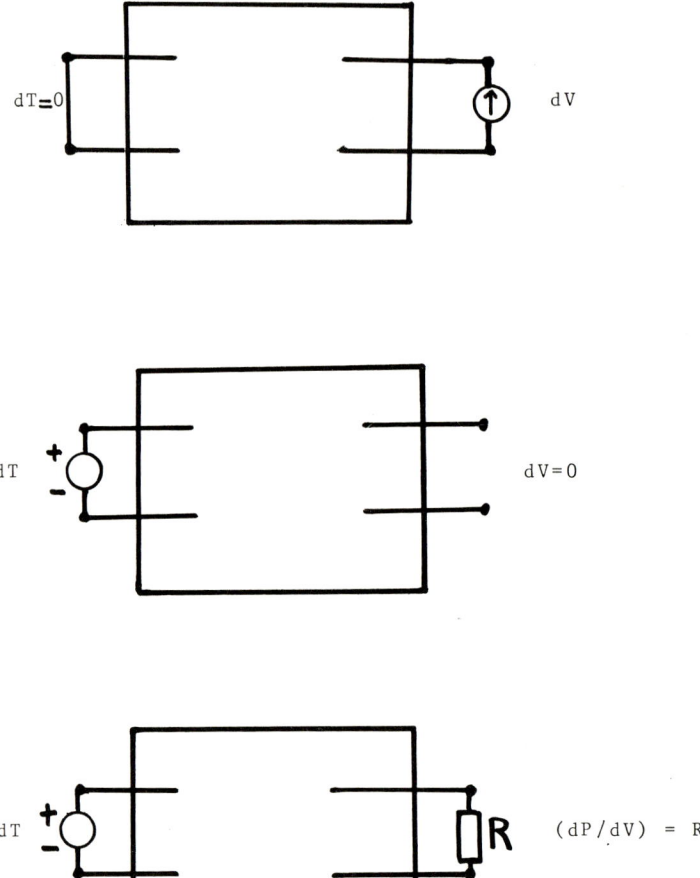

Fig. 2.2 <u>Interpretation of partial derivatives as network measurements.</u> (a) A measurement at constant temperature ($dT = 0$) when V is the independent variable, (b) a measurement at constant V when T is the independent variable and (c) a measurement for which dP/dV = constant and T is the (only) independent variable.

directions of slopes of the form

$$dP/dV = -R, \qquad (13)$$

which, in network terms simply correspond to placing appropriate load resistances R across the mechanical port, as shown in FIGURE 2 (c). Of course, in the special cases in which R= 0 or R = ∞, the familiar partial derivative expressions are recovered.

The network is of course assumed to match the work and heat charactersitics of the ports --i.e., $dW_{re} = -P\,dV$, $dQ_{re} = TdS$ -- and the phenomenological characteristics of the Phase, as given by Eqs.(8) through (11). The actual representation of the internal workings of the network is completely arbitrary; however, all the networks that represent these equations and yield the proper heat and mechanical measurements at the ports are equivalent -- i.e., constitute an equivalence class for this set of measurements. It therefore follows that any network which can be synthesized will do the job.

We now introduce a metric by specifying how variations in thermodynamic variables are calculated. Consider a simple thermo--static system with two degrees of freedom assigned to a thermal port and a mechanical port. If S and V are assumed to be the independent variations, the differential increments in the dependent variables T and P can be expressed as

$$dT = (\partial T/\partial S)_V \, dV + (\partial T/\partial V)_S \, dV \qquad (14)$$

$$dP = (\partial P/\partial S)_V \, dS + (\partial P/\partial V)_S \, dV, \qquad (15)$$

in the usual linear approximation in which higher order differentials of independent variables are ignored. In the following we shall use the differential symbol dx to represent either the infinitesimal or the finite increment Δx; this is done both for economy and to stress the topological isomorphism between the two. In the neighborhood of some equilibrium state, the partial derivatives may be assigned a constant values determined by experiment, so that equations (14) and (15) may be rewritten in the form

$$dT = e_{11}\,dS + e_{12}\,dV \qquad (16)$$

$$dP = e_{21}\, dS + e_{22}\, dV, \tag{17}$$

in which

$$e_{11} = (\partial T/\partial S)_V, \tag{18}$$

$$e_{12} = (\partial T/\partial V)_S, \tag{19}$$

$$e_{21} = (\partial P/\partial S)_V, \tag{20}$$

and

$$e_{22} = (\partial P/\partial V)_S. \tag{21}$$

Note that, in order to keep as much generality as possible, <u>we have not assumed</u> that there exists a function of state which leads to <u>Maxwell's reciprocities</u>.

The network representation of the above equations is straightforward. Each term in Equations (16) and (17) is the sum of two voltage increments: a congugate term, which is proportional to the port current or flow and therefore represents an Ohmic (i.e., linear) resistance e_{11} or e_{22} and a non conjugate term -- e_{12} or e_{21} -- which behaves like a flow (current) controlled force (voltage) source. The resultant two half networks are placed back to back and enclosed by a box to yield the (disconnected) two port shown in FIGURE 3.

The same procedure can now be repeated using the extensive variable measurement and assuming that the increments are given as

$$dS = (\partial S/\partial T)_P\, dT + (\partial S/\partial P)_T\, dP \tag{22}$$

$$dV = (\partial V/\partial T)_P\, dT + (\partial V/\partial P)_T\, dP \tag{23}$$

or in the short hand notation

$$dS = g_{11}\, dT + g_{12}\, dP \tag{24}$$

$$dV = g_{21}\, dT + g_{22}\, dP, \tag{25}$$

with

$$g_{11} = (\partial S/\partial T)_P , \qquad (26)$$

$$g_{12} = (\partial S/\partial P)_T , \qquad (27)$$

$$g_{21} = (\partial V/\partial T)_P , \qquad (28)$$

and

$$g_{22} = (\partial V/\partial P)_T . \qquad (29)$$

Again, no reciprocity or energy function of state is assumed yet.

These equations can now be interpreted as sums of flows, one related to the conjugate force--i.e., a conductance g_{11} or g_{22} -- and one proportional to the non conjugate force. The latter is then a force controlled flow (current) source. When the two half networks are placed back to back the disconnected network in FIGURE 4 is obtained.

Note that these networks contain the following information:

i) The topology indicates the "direction" of the independent variables --i.e., if the independent port variable is a force (voltage) all branches at that port are in parallel and the controlled sources are flows, while if the independent variable at a port is a current or flow the branches are in series.

ii) The networks contain information about the (differential) algebra of the problem--i.e., through the use of K.V.L. and K.C.L. the original equations can be recovered.

[Clearly, these are single representations of an infinite possible number which have "mechanisms" in which each branch has an associated local force and flow. But further decompositions are possible simply by splitting resistances and sources using Kirchhoff's laws. However, the goal of synthesis is to find the simplest network which will do a given job.]

There is no reason to assume that the "batteries" in the first set of networks must be a function of the non conjugate flows (extensive variations) or that the flow sources in the second set

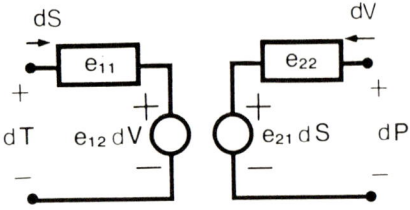

Fig. 2.3   The **e** network.

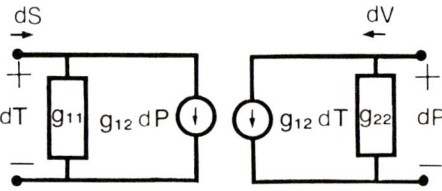

Fig. 2.4   The **g** network.

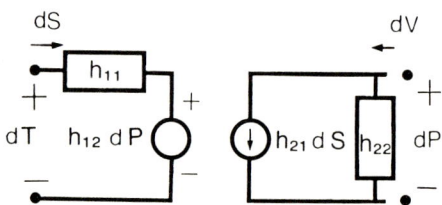

Fig. 2.5   The **h** network.

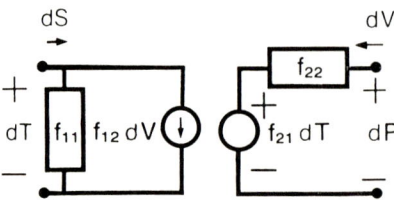

Fig. 2.6 The **f** network.

of networks must be a function of the non conjugate forces. A complete set will allow the first to be alternatively controlled by the non conjugate intensive quantities and the later by non conjugate extensive quantities; this extension leads to four more networks which finally reduce to the two ports shown in FIGURES 5 and 6. These correspond to the additional equations

$$dT = h_{11} dS + h_{12} dP \tag{30}$$

$$dV = h_{21} dS + h_{22} dP, \tag{31}$$

$$h_{11} = (\partial T/\partial S)_P, \tag{32}$$

$$h_{12} = (\partial T/\partial P)_S, \tag{33}$$

$$h_{21} = (\partial V/\partial S)_P, \tag{34}$$

$$h_{22} = (\partial V/\partial P)_S \tag{35}$$

and the set of equations

$$dS = f_{11} dT + f_{12} dV \tag{36}$$

$$dP = f_{21} dT + f_{22} dV, \tag{37}$$

with

$$f_{11} = (\partial S/\partial T)_V, \tag{38}$$

$$f_{12} = (\partial S/\partial V)_T, \tag{39}$$

$$f_{21} = (\partial P/\partial T)_V, \tag{40}$$

$$f_{22} = (\partial P/\partial V)_T. \tag{41}$$

(The question arises as to the possibility of extending the process to replace the resistances and conductances by non conjugate controlled sources; this is possible, but it does not lead to any useful operation with cause and effect variables.)

TABLE I BASIC CHARACTERISTICS OF THE TWO PORT INCREMENTAL ENERGY NETWORKS

NAME	TRANSIMPEDANCE	TRANSCURRENT	TRANSVOLTAGE	TRANSADMITTANCE
DEPENDENT VARIABLES	T P	T V	S P	S V
INDEPENDENT VARIABLES	S V	S P	T V	T P
MATRIX REPRESENTATION	$\begin{bmatrix} e_{11} & e_{12} \\ e_{21} & e_{22} \end{bmatrix}$	$\begin{bmatrix} h_{11} & h_{12} \\ h_{21} & h_{22} \end{bmatrix}$	$\begin{bmatrix} f_{11} & f_{12} \\ f_{21} & f_{22} \end{bmatrix}$	$\begin{bmatrix} g_{11} & g_{12} \\ g_{21} & g_{22} \end{bmatrix}$
EXPLICIT FORM OF MATRIX IN TERMS OF FIRST PARTIAL DERIVATIVES	$\left(\frac{\partial T}{\partial S}\right)_V \quad \left(\frac{\partial T}{\partial V}\right)_S$ $\left(\frac{\partial P}{\partial S}\right)_V \quad \left(\frac{\partial P}{\partial V}\right)_S$	$\left(\frac{\partial T}{\partial S}\right)_P \quad \left(\frac{\partial T}{\partial P}\right)_S$ $\left(\frac{\partial V}{\partial S}\right)_P \quad \left(\frac{\partial V}{\partial P}\right)_S$	$\left(\frac{\partial S}{\partial T}\right)_V \quad \left(\frac{\partial S}{\partial V}\right)_T$ $\left(\frac{\partial P}{\partial T}\right)_V \quad \left(\frac{\partial P}{\partial V}\right)_T$	$\left(\frac{\partial S}{\partial T}\right)_P \quad \left(\frac{\partial S}{\partial P}\right)_T$ $\left(\frac{\partial V}{\partial T}\right)_P \quad \left(\frac{\partial V}{\partial P}\right)_T$
MAXWELL'S RELATIONS WHEN THERMODYNAMIC POTENTIAL EXISTS (NOTICE DEFINITION OF ↓ΔV IN TEXT)	$\left(\frac{\partial T}{\partial V}\right)_S = -\left(\frac{\partial P}{\partial S}\right)_V$	$\left(\frac{\partial T}{\partial P}\right)_S = -\left(\frac{\partial V}{\partial S}\right)_P$	$\left(\frac{\partial S}{\partial V}\right)_T = -\left(\frac{\partial P}{\partial T}\right)_V$	$\left(\frac{\partial S}{\partial P}\right)_T = \left(\frac{\partial V}{\partial T}\right)_P$

The characteristics of the four networks are summarized in TABLE I. Each of the two ports can be represented by a 2 x 2 matrix **e, g, h, or f** because, as it is shown by network techniques below, when specific Maxwell reciprocities hold the matrix entries correspond to the energy, free energy, enthalpy and Helmoltz free energy Hessians. The units of the coefficients are:

i) <u>thermodynamic "resistances"</u>--i.e. ratios of temperature to entropy or pressure to volume--, for e_{11}, e_{12}, e_{21}, e_{22}, h_{11} and f_{22}.

ii) <u>Thermodynamic Conductances</u> --i.e., ratios of entropy to temperature or voolume to pressure--, for g_{11}, g_{11}, g_{11}, g_{11}, f_{11} and h_{22}

iii) <u>Dimensionless quantities</u> for h_{21}, h_{12}, f_{12} and f_{21}.

The heading labels in TABLE 1 referring to the networks as transimpedance, transconductance, transvoltage and transcurrent are used for consistency and possible cross reference with the electrical engineering terminology. This nomenclature relates to the possible measurements which can be performed using the particular network (impedances and admittances are dynamic resistances and conductances , respectively as these networks are more generally used to model time behavior). For example, the **e** description will always yield resistances (impedances in the dynamic case) for any practical measurement. The terms transcurrent and tranvoltage indicate that, in the engineering terminology the left hand side--the heat port in our example-- is always the input, while the right hand side --the mechanical port here-- is always the output.

A question that arises is : given that each of the above networks represents a different mechanism for the same system, is it possible to tell which is the "real" network inside the box. The point is that each of them is <u>the</u> accurate representation for the system when the dependent and the independent variables are those listed in TABLE 1 .

2.4 TRANSFORMATIONS AMONG NETWORKS

Consider now the thermal port of the **h** network , which consists of a resistance in series with a battery or force source. Thevenin-Norton theorems allow this arrangement to be replaced with an equivalent conductance in parallel with a flow source, provided

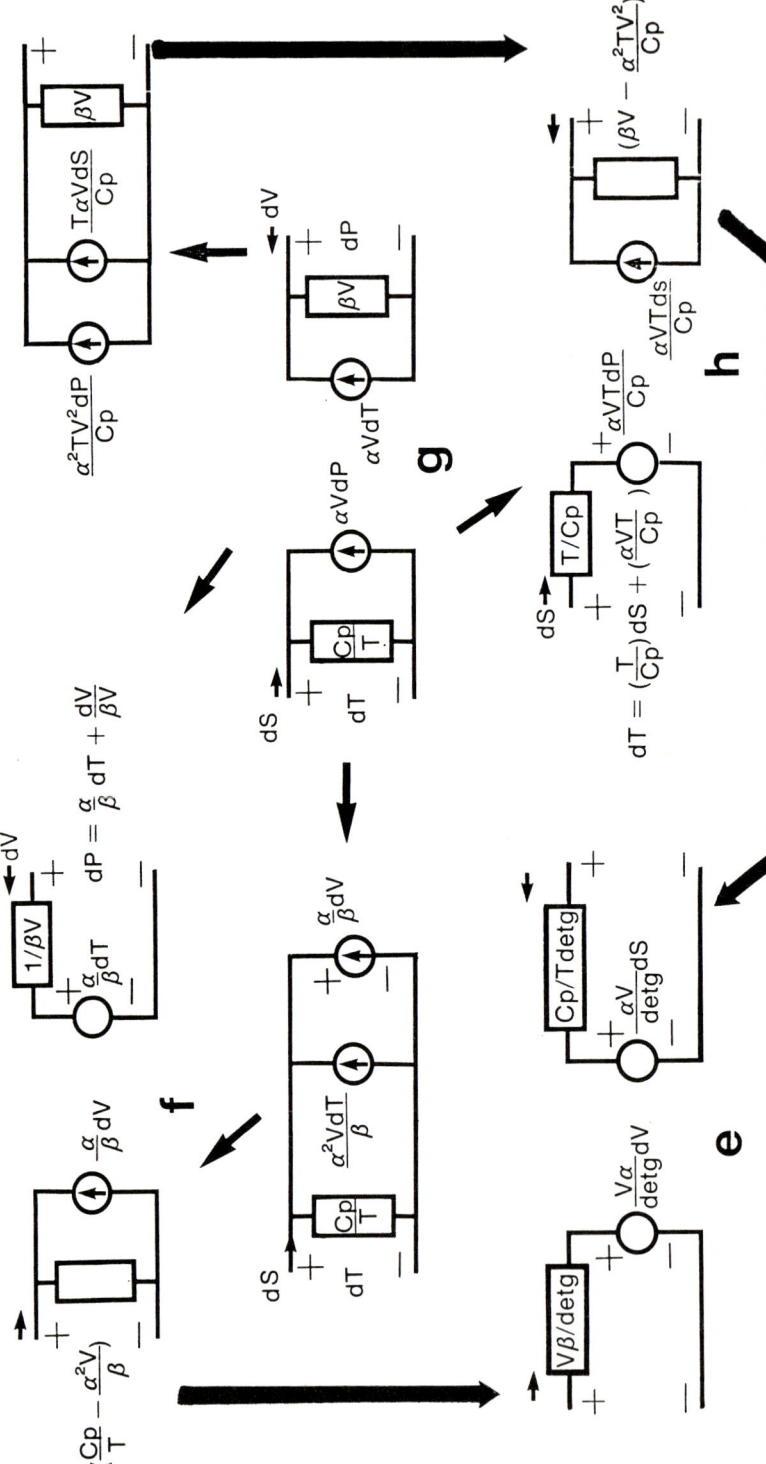

Fig. 2.7 Equivalences among the resistances conductances and sources--i.e., the Hessian entries--in the four energy networks can be readily obtained by using Norton-Thevenin theorem. In this particular example, familiar thermodynamic parameters have been substituted for the more general partial derivative representation and reciprocity is implicit in this parameter representation. The transformations themselves do not require reciprocity.

that they have the values

$1/h_{11}$ (conductance) (42)

and

$(h_{12}/h_{11})dP$, (flow source) (43)

respectively. However, these values must also agree with those of g_{11} and $g_{12}dP$, respectively-- as given by the g network.

Once the identification of the network coefficients is made with the proper partial derivatives, these equalities lead to the familiar results

$$(\partial T/\partial S)_P = 1/(\partial S/\partial T)_P \qquad (44)$$

and

$$(\partial S/\partial P)_T = - (\partial T/\partial P)_S / (\partial T/\partial S)_P . \qquad (45)$$

Similar transformation using Thevenin and Norton's theorem allow interconversion between all networks, as shown in FIGURE 7. These lead to the partial derivative equalities that are shown in TABLE 2. Note again that these do not depend on reciprocal equations or energy functions of state and are therefore more general than the results obtained using Legendre's transformations.

The determinants which appear on this table are the same as the Jacobians (see TABLE 3):

$$J = \begin{bmatrix} (\partial n/\partial x)_y & (\partial n/\partial y)_x \\ (\partial m/\partial x)_y & (\partial n/\partial y)_x \end{bmatrix} \qquad (46)$$

However, no assumption has been made about the existence of an analytical function of the form $\phi = \phi(n, m)$ --i.e., no potential function has been assumed to exist. Thus, these transformations are more general than those based on Legendre transforms, because they admit non reciprocal solutions and are independent of the connectivity of the manifold. Of course, in the special case in which there exists a Pfaffian form

TABLE 2. Interconversions among thermodynamic partial derivatives

FROM / TO	[E]	[G]	[H]	[F]	
[E]	$\left(\frac{\partial T}{\partial S}\right)_V$	$\frac{(\partial V/\partial P)_T}{\det G}$	$\frac{\det H}{(\partial V/\partial P)_S}$	$\frac{1}{(\partial S/\partial T)_V}$	$\frac{-(\partial S/\partial V)_T}{(\partial S/\partial T)_V}$
	$\left(\frac{\partial P}{\partial S}\right)_V$	$\frac{-(\partial V/\partial T)_P}{\det G}$	$\frac{-(\partial V/\partial S)_P}{(\partial V/\partial P)_S}$	$\frac{(\partial P/\partial T)_V}{(\partial S/\partial T)_V}$	$\frac{\det F}{(\partial S/\partial T)_V}$
[G]	$\frac{(\partial P/\partial V)_S}{\det E}$	$\left(\frac{\partial S}{\partial P}\right)_T$	$\frac{1}{(\partial T/\partial S)_P}$	$\frac{\det F}{(\partial P/\partial T)_V}$	$\frac{(\partial S/\partial V)_T}{(\partial P/\partial V)_T}$
	$\frac{-(\partial P/\partial T)_S}{\det E}$	$\left(\frac{\partial V}{\partial T}\right)_P$	$\frac{(\partial V/\partial S)_P}{(\partial T/\partial S)_P}$	$\frac{-(\partial P/\partial T)_V}{(\partial P/\partial V)_T}$	$\frac{-(\partial S/\partial T)_V}{(\partial P/\partial V)_T}$
[H]	$\frac{\det E}{(\partial T/\partial V)_S}$	$\frac{(\partial V/\partial T)_P}{(\partial S/\partial T)_P}$	$\frac{(\partial T}{\partial S)_P$ ... $\left(\frac{\partial T}{\partial P}\right)_S$	$\frac{\det F}{(\partial P/\partial T)_V}$	$\frac{(\partial S/\partial V)_T}{\det F}$
	$\frac{-(\partial P/\partial S)_V}{(\partial P/\partial V)_S}$	$\frac{1}{(\partial T/\partial S)_P}$	$\left(\frac{\partial V}{\partial S}\right)_P$	$\frac{-(\partial P/\partial T)_V}{\det F}$	$\frac{(\partial S/\partial T)_V}{\det F}$
[F]	$\frac{1}{(\partial T/\partial S)_V}$	$\frac{\det G}{(\partial V/\partial P)_T}$	$\frac{-(\partial V/\partial S)_P}{\det H}$	$\frac{(\partial P/\partial V)_T}{\det F}$	$\left(\frac{\partial S}{\partial V}\right)_T$
	$\frac{(\partial P/\partial S)_V}{(\partial T/\partial S)_V}$	$\frac{-(\partial V/\partial T)_P}{(\partial V/\partial P)_T}$	$\left(\frac{\partial T}{\partial S}\right)_P \cdot \left(\frac{\partial V}{\partial P}\right)_S$	$\left(\frac{\partial S}{\partial T}\right)_V$	$\left(\frac{\partial P}{\partial V}\right)_T$
	$\det E = \left(\frac{\partial T}{\partial S}\right)_V \left(\frac{\partial P}{\partial V}\right)_S - \left(\frac{\partial T}{\partial V}\right)_S \left(\frac{\partial P}{\partial S}\right)_V$	$\det G = \left(\frac{\partial S}{\partial T}\right)_P \left(\frac{\partial V}{\partial P}\right)_T - \left(\frac{\partial S}{\partial P}\right)_T \left(\frac{\partial V}{\partial T}\right)_P$	$\det H = \left(\frac{\partial T}{\partial S}\right)_P \left(\frac{\partial V}{\partial P}\right)_S - \left(\frac{\partial T}{\partial P}\right)_S \left(\frac{\partial V}{\partial S}\right)_P$	$\det F = \left(\frac{\partial S}{\partial T}\right)_V \left(\frac{\partial P}{\partial V}\right)_T - \left(\frac{\partial S}{\partial V}\right)_T \left(\frac{\partial P}{\partial T}\right)_V$	

TABLE 3. Interconversions among network determinants (Jacobians)

FROM \ татtoTO	E	G	H	F
det E	$\det E$	$\dfrac{1}{\det G}$	$\dfrac{(\partial T/\partial S)_P}{(\partial V/\partial P)_S}$	$\dfrac{(\partial P/\partial V)_T}{(\partial S/\partial T)_V}$
det G	$\dfrac{1}{\det E}$	$\det G$	$\dfrac{(\partial V/\partial P)_S}{(\partial T/\partial S)_P}$	$\dfrac{(\partial S/\partial T)_V}{(\partial P/\partial V)_T}$
det H	$\dfrac{(\partial T/\partial S)_V}{(\partial P/\partial V)_S}$	$\dfrac{(\partial V/\partial P)_T}{(\partial S/\partial T)_P}$	$\det H$	$\dfrac{1}{\det F}$
det F	$\dfrac{(\partial P/\partial V)_S}{(\partial T/\partial S)_V}$	$\dfrac{(\partial S/\partial T)_P}{(\partial V/\partial P)_T}$	$\dfrac{1}{\det H}$	$\det F$

$$d\phi = m\ dx + n\ dy \qquad (47)$$

in which $d\phi$ is an exact differential the present results reduce to the Legendre transformations.

2.5 THE METRIC

The "power" dissipated inside the network is, according to Tellegen's theorem, equal to the power supplied by the ports. Thus, the sums of products of port intensive and extensive increments or differentials can be expressed as a bilinear form. Moreover, when the networks represent the same system and they therefore obey the transformations given in TABLE 2, this function is the same for all the networks.

$$dT\ dS + dP\ dV = e_{11}\ dS^2 + e_{12}\ dS\ dV + e_{21}\ dV\ dS + e_{22}\ dV^2 \qquad (48)$$

$$= h_{11}\ dS^2 + h_{12}\ dP\ dS + h_{21}\ dS\ dP + h_{22}\ dP^2 \qquad (49)$$

$$= f_{11}\ dT^2 + f_{12}\ dT\ dV + f_{21}\ dV\ dT + f_{22}\ dV^2 \qquad (50)$$

$$= g_{11}\ dT^2 + g_{12}\ dT\ dP + g_{21}\ dT\ dP + g_{22}\ dP^2. \qquad (51)$$

Unlike the usual metrics, these are not necessarily reciprocal, and thus serve to define a geometrical invariant when the analog of distance does not exist. (**e, h, g and f** are of course tensors whether the metric is reciprocal or not). The mathematical definition of a metric with non reciprocal coefficients was considered an aberration by Gauss; however, it allows the physical possibility of defining a scalar invariant analog to the distance in non integrable--i.e., irreversible -- situations.

In the case in which an energy function exists (reciprocal metric) the bilinear form $dTdS + dVdP$ yields the second variations $\Delta_E^2 = \Delta_G^2 = \Delta_F^2 = \Delta_H^2$ within the linear differential approximation of thermodynamics. Thus, for example,

$$d^2E = e_{11}\ dS^2 + e_{12}\ dS\ dV + e_{21}\ dV\ dS + e_{22}\ dV^2 + $$
$$+\ (\partial E/\partial S)\ d^2S + (\partial E/\partial V)\ d^2V.$$

In the linear approximation, the last two terms vanish and $d^2E = dT\ dS + dV\ dP$.

2.6 THE RECIPROCAL METRICS

If an energy function of state is assumed to exist the resultant manifold is connected and a distance can be defined from the invariants given above. In that case the analyticity of the energy functions requires the g and e tensors to be symmetric and the h and f tensors to be antisymmetric. Both of these results induce different forms of orthogonality, in the cartesian sense.

The antisymmetry of f or h has the effect that the contribution to the distance function of the controlled sources --i.e.,the "power dissipated in the transducing part of the network-- is zero. As a result, the invariant scalar given above reduces to either

$$d^2 = h_{11} \, dS^2 + h_{22} \, dP^2 \tag{52}$$

or

$$d^2 = f_{11} \, dT^2 + f_{22} \, dV^2 \tag{53}$$

thus behaving as if the directions of S and P or those of T and V where orthogonal.

In the case in which there is an energy potential function the entries of the tensors are elements of the Hessian matrices

$$\begin{bmatrix} e_{11} & e_{12} \\ e_{21} & e_{22} \end{bmatrix} = \begin{bmatrix} (\partial^2 E/\partial S^2) & (\partial^2 E/\partial S \, \partial V) \\ (\partial^2 E/\partial S \, \partial V) & (\partial^2 E/\partial V^2) \end{bmatrix} \tag{54}$$

$$\begin{bmatrix} h_{11} & h_{12} \\ h_{21} & e_{22} \end{bmatrix} = \begin{bmatrix} (\partial^2 H/\partial S^2) & (\partial^2 H/\partial P \, \partial V) \\ (\partial^2 H/\partial P \, \partial V) & (\partial^2 H/\partial P^2) \end{bmatrix} \tag{55}$$

$$\begin{bmatrix} f_{11} & f_{12} \\ f_{21} & f_{22} \end{bmatrix} = \begin{bmatrix} (\partial^2 F/\partial T^2) & (\partial^2 F/\partial T \, \partial V) \\ (\partial^2 F/\partial T \, \partial V) & (\partial^2 F/\partial V^2) \end{bmatrix} \tag{56}$$

and

$$\begin{bmatrix} g_{11} & g_{12} \\ g_{21} & g_{22} \end{bmatrix} = \begin{bmatrix} (\partial^2 G/\partial T^2) & (\partial^2 G/\partial T \, \partial P) \\ (\partial^2 G/\partial T \, \partial P) & (\partial^2 G/\partial P^2) \end{bmatrix} \tag{57}$$

with the forces and flows defined from the gradients of the energy potentials

$$T = (\partial E/\partial S)_V = (\partial H/\partial S)_P \tag{58a}$$

$$P = (\partial E/\partial V)_S = (\partial F/\partial V)_T \tag{58b}$$

$$S = -(\partial F/\partial T)_V = -(\partial G/\partial T)_P \tag{58c}$$

and

$$V = (\partial H/\partial P)_S = (\partial G/\partial P)_T . \tag{58d}$$

2.7 CONNECTIVITY: THE T AND PI NETWORKS

It was pointed out in Chapter 1 that every connected resistive network which obeys Kirchhoff's laws is reciprocal. Conversely, every linear reciprocal system can be represented by a resistive network. In the 2x2 reciprocal case a resistive network with three resistances is the simplest network which represents the system and the resistances (or conductances) are linear combinations of the matrix coefficients **e, g, h** or **f**.

If the currents (extensive variables) are the independent quantities, the **T** network shown in FIGURE 8 (a) is the simplest to represent the system, while if the intensive quantities are the independent variables, the **Pi** network of FIGURE 8 (b) is the simplest one.

2.8 THE IDEAL GAS

We show an example of the previous results, by applying the network approach to the analysis of an ideal gas. We begin with the Equation of State,

$$P V = n R/ T \tag{59}$$

which can be utilized to find the coefficients f_{21} and f_{22} in the f hybrid network,

$$f_{21} = (\partial P /\partial T)_V = nR/ V \tag{60}$$

and

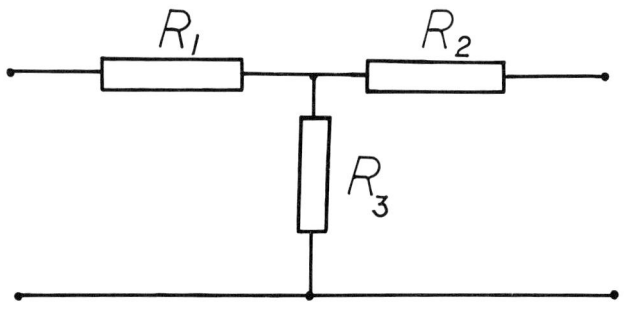

$$R_1 = \left(\frac{\partial T}{\partial S}\right)_V - \left(\frac{\partial T}{\partial V}\right)_S \quad R_2 = \left(\frac{\partial P}{\partial V}\right)_S - \left(\frac{\partial P}{\partial S}\right)_V \quad R_3 = \left(\frac{\partial P}{\partial S}\right)_V - \left(\frac{\partial T}{\partial V}\right)_S$$

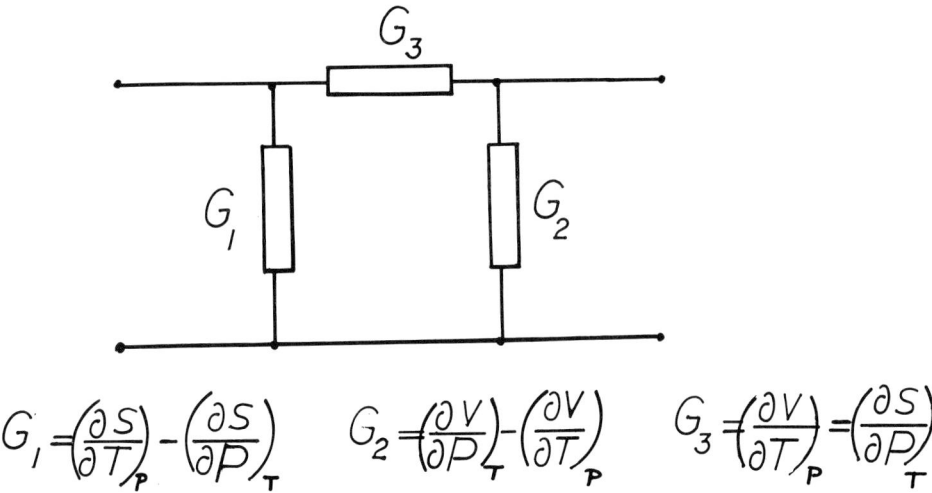

$$G_1 = \left(\frac{\partial S}{\partial T}\right)_P - \left(\frac{\partial S}{\partial P}\right)_T \quad G_2 = \left(\frac{\partial V}{\partial P}\right)_T - \left(\frac{\partial V}{\partial T}\right)_P \quad G_3 = \left(\frac{\partial V}{\partial T}\right)_P - \left(\frac{\partial S}{\partial P}\right)_T$$

Fig. 2.8 Thermostatic T and π networks based on the construction described in Chapter 1.

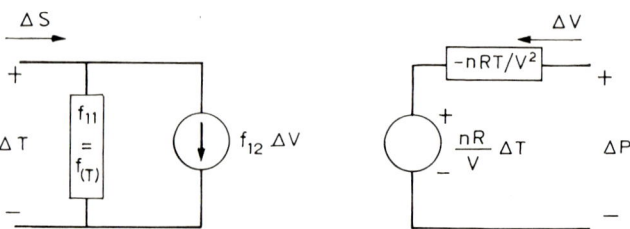

Fig. 2.9 Helmholtz Free energy description of the ideal gas.

$$f_{22} = (\partial P/\partial V)_T = -nRT/V^2. \tag{61}$$

Coefficient f_{12} follows from f_{21}, with a negative sign, while the fact that the f equations express dS as an exact differential in T and V implies that

$$(\partial f_{11}/\partial V)_T = (\partial f_{12}/\partial T)_V \tag{62}$$
$$= -(\partial f_{21}/\partial T)_V$$
$$= (\partial^2 P/\partial T^2)_V \tag{63}$$
$$= 0.$$

2.9 HYBRID IDEAL GAS DESCRIPTION

This last result implies that $f_{11} = f(T)$, as expected. The f equations can now be summarized as (FIGURE 9).

$$\begin{bmatrix} dS \\ dP \end{bmatrix} = \begin{bmatrix} f(T) & -nR/V \\ nR/V & -nRT/V^2 \end{bmatrix} \begin{bmatrix} dT \\ dV \end{bmatrix} \tag{64}$$

Note, in passing, that these equations for the Phase Network summarize the first and second laws, the equation of state of the gas and Maxwell´s energy reciprocities.

2.10 AN APPLICATION OF THE TRANSFORMATION MATRICES

Define the usual heat capacities and expansion coefficents

$$C_P = T\,(\partial S/\partial T)_P$$

$$C_V = T\,(\partial S/\partial T)_V,$$

$$\beta_T = (1/V)(\partial V/\partial P)_T,$$

$$\alpha_P = -(1/V)(\partial V/\partial T)_P,$$

and

$$\alpha_S = (1/V)(\partial V/\partial P)_S$$

and let $\det g = g_{11}g_{22} - g_{12}g_{21}$. From TABLE 1 we readily obtain the matrix equations

$$\begin{bmatrix} g_{11} & g_{12} \\ g_{21} & g_{22} \end{bmatrix} = \begin{bmatrix} (\partial S/\partial T)_P & (\partial S/\partial P)_T \\ (\partial V/\partial T)_P & (\partial V/\partial P)_T \end{bmatrix} = \begin{bmatrix} C_P/T & -V\alpha_P \\ -\alpha_P V & V\beta_T \end{bmatrix} \tag{65a}$$

$$\begin{bmatrix} h_{11} & h_{12} \\ h_{21} & h_{22} \end{bmatrix} = \begin{bmatrix} (\partial T/\partial S)_P & (\partial T/\partial P)_S \\ (\partial V/\partial S)_P & (\partial V/\partial P)_S \end{bmatrix} = \begin{bmatrix} T/C_P & TV\alpha_P/C_P \\ -\alpha_P VT/C_P & T\det g/C_P \end{bmatrix} \tag{65b}$$

$$\begin{bmatrix} f_{11} & f_{12} \\ f_{21} & f_{22} \end{bmatrix} = \begin{bmatrix} (\partial S/\partial T)_V & (\partial S/\partial V)_T \\ (\partial P/\partial T)_V & (\partial P/\partial V)_T \end{bmatrix} = \begin{bmatrix} \det g\,\beta_T V & \alpha_P/\beta_T \\ -\alpha_P/\beta_T & 1/V\beta_T \end{bmatrix} \tag{65c}$$

$$\begin{bmatrix} e_{11} & e_{12} \\ e_{21} & e_{22} \end{bmatrix} = \begin{bmatrix} (\partial T/\partial S)_V & (\partial T/\partial V)_S \\ (\partial P/\partial S)_V & (\partial P/\partial V)_S \end{bmatrix} = \begin{bmatrix} \beta_T V/\det g & \alpha_P V/\det g \\ \alpha_P/\det g & C_P/T\det g \end{bmatrix} \tag{65d}$$

As a straightforward example of the use of these transformations we can refer back to the previously derived equation relating heat capacities at constant temperature and pressure,

$$C_V - C_P = T \, (\partial V/\partial T)_P \, (\partial P/\partial T)_V, \qquad (66)$$

which becomes, after introduction of the appropriate values found above,

$$C_V - C_p = - T \, (\alpha_p V) \, (-\alpha_p/\beta_T)$$
$$= + T V \alpha_p^2/\beta_T \quad,$$

in agreement with common thermodynamics (as before, the sign appears reversed because of the positive definition for dV).

As a second example, we note from the last set of matrices that

$$e_{22} \det g = (C_P/T) \qquad (67)$$

and

$$e_{11} \det g = (\beta_T V) \, ,$$

whence it follows that

$$(e_{22}/e_{11}) = (C_P/\beta_T V T). \qquad (68)$$

With the definitions given above, this equation readily yields

$$(C_V/C_p) = (\beta_s/\beta_T), \qquad (69)$$

which is again in agreement with classical thermodynamics.

2.11 EQUILIBRIUM OF HOMOGENEOUS SYSTEMS

Network interconnections can be used to find the equilibrium properties of thermostatic subsystems. As a simple example, for a gas vessel conceptually subdivided into neighboring regions in equilibrium with each other the temperature and pressure will be the same in all of them. Moreover, any extensive quantity which gets displaced out of one of the regions will enter one of the others, if the system is kept at constant volume. In terms of two

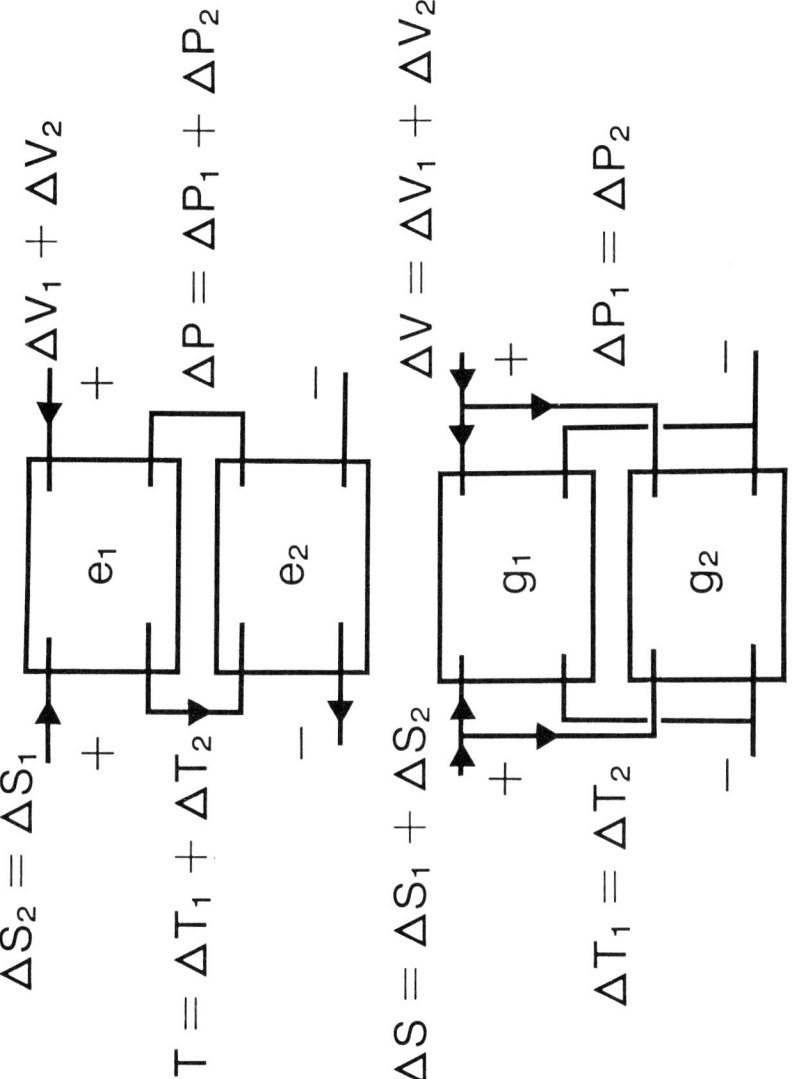

Fig. 2.10 Attachment of thermostatic energy two ports.

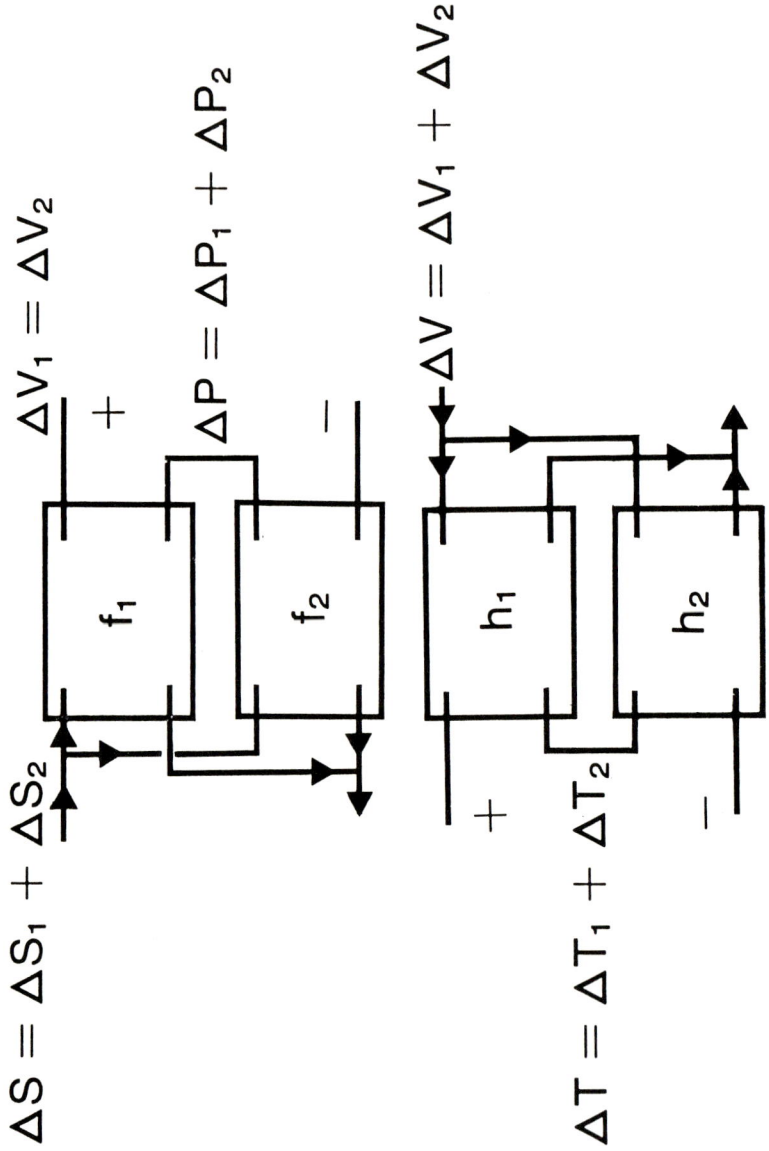

Fig. 2.10 (continued)

ports the situation is that depicted on FIGURE 10, in which variations in pressure and temperature are the same in all ports while variations in extensive quantities add up through K.C.L.

Clearly, the Gibbs' Free energy network provides the correct representation for this situation as the overall g matrix is the sum of the individual g matrices.

The type of boundary matching or network interconnection shown above in which extensive variables add while intensive quantities are matched is called a parallel interconnection. As FIGURE 10 illustrates, other connections are possible, depending on which are the independent variables. Thus, it is also possible to match the extensive quantities and extensiver variables at the same time by using the hybrid networks, for example. By choosing the appropriate matrix form there is always an energy network (or entropy network) which adds up to provide the overall matrix as an output when the correct equilibrium conditions are given.

(A further attachment not considered here is that of both output variables given in terms of both input variables; this is the transmission matrix, which does not have a covariant interpretation although it has some practical uses).

2.12   DESCRIPTIONS BASED ON ENTROPY NETWORKS

The energy networks given above were built on the assumption that entropy and the three other state variables (P,T and V) were the measurable quantities. An alternate approach consists in using an energy function such as E and three other variables which are functions of T,P and V as port variables. The proper distance functions are then based on entropy changes, although it is not required, in the present view, that entropy exists as such in the same way that the "energy" networks requires one to be familiar within the existence of an entropy function, but no energy potential is assumed until reciprocity is introduced. In order to keep the correct units, the port forces and flows would become

$v_1 = d(1/T)$

$i_1 = dE$
$v_2 = d(P/T)$

and

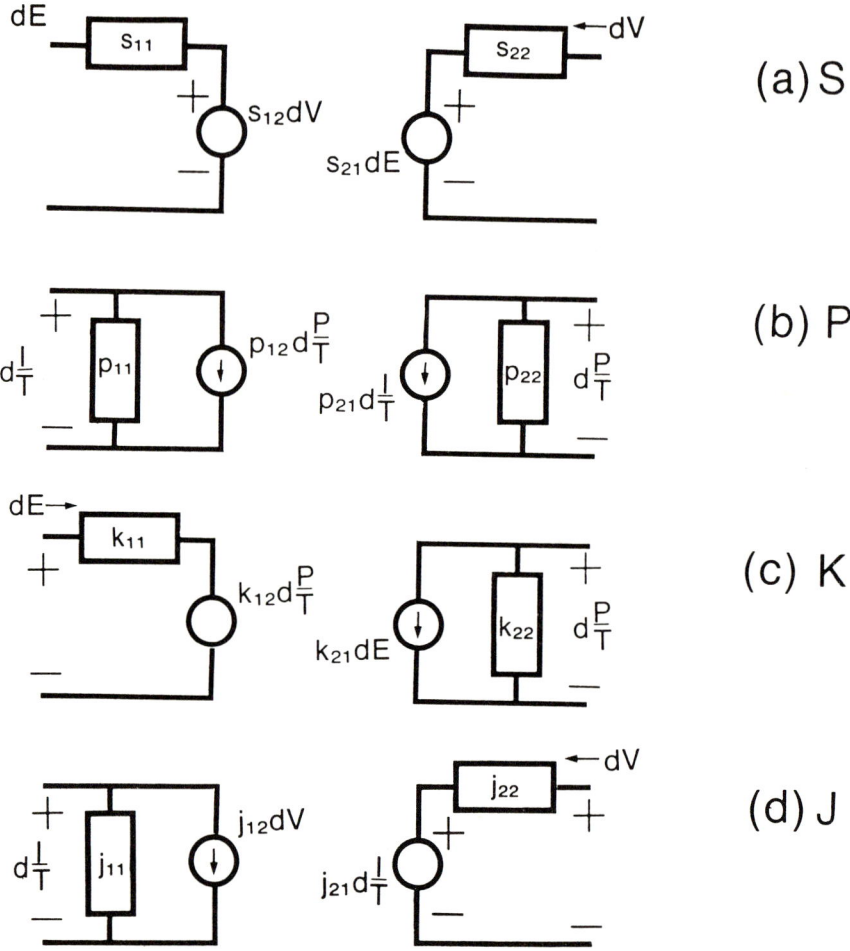

Fig. 2.11 Two ports based on entropy descriptions. When reciprocity is assumed, these lead to the entropy functions S, K, J and P.

$$i_2 = dV. \qquad (70)$$

By analogy with the approach taken before, we can give the four "entropy" networks shown in FIGURE 11, which are labelled S, P, K and J to represent the thermodynamic functions that glue the half networks together--i.e., which provide a continuous manifold--when the networks are reciprocal. The transformation between various partial derivatives follow the same lines given above and are shown in TABLE 2.

These networks also stress the topological approach of Caratheodory's theorem : if the flow dS in the T, S port is equal to zero (adiabatic process) there will be an infinite number of unreachable states which are not spanned by the varying conditions at the remaining independent port.

2.13 CONNECTIVITY AND RECIPROCITY

We now consider some geometrical ideas which are somewhat more abstract in nature but provide some insight on the basis of thermodynamic reciprocities.

The possibility of representing a thermodynamic system by a resistive network is intimately related to the analyticity and, therefore reciprocity of the system. From the network point of view, this was pointed out in Chapter 1 in relation to Tellegen's theorem. It is instructive, however, to pursue this problem further from the thermodynamic point of view.

The exactness of the energy differential,

$$\oint dE = 0$$

gives the analytical equivalent of Kirchhof's voltage law and can be stated in a number of equivalent ways, including:

i) **curl E** $=0$ --i.e., a continuous closed path returns to the original value of the potential; and

ii) $(\partial T/\partial V)_S = (\partial P/\partial S)_V$ --i.e., a Maxwell reciprocity which follows from the connectivity and linearity (first order differential approximation) of thermostatics.

Maxwell's energy reciprocity implies that $e_{12} = e_{21}$, or

$$(e_{12})dV\, dS = (e_{21}\, dS)dV \quad . \tag{71}$$

The terms enclosed in parentheses represent the nonconjugate voltages or forces, while the multiplying term represents the conjugate current or extensive variable at that port. From the two port point of view, it then follows that a reciprocal (i.e., reversible) process is one in which the transducing step of the two port dissipates the same amounts of energy at both input and output. In the case of the hybrid networks representing H and F (A), the opposite signs of the non diagonal terms leads to the result that the bilinear form at the transducing step ("power" dissipated, in network terms) is zero.

2.14 THE FUNDAMENTAL METRIC TENSORS

The bilinear form invariants can also be expressed in terms of classical geometrical (tensorial) concepts. Because d^2 is an invariant, the "matrices" e, g, h and f are second rank tensors which establish a metric for the system. We can distinguish a covariant vector (expressed in transpose form, for convenience)

$$d\mathbf{x}^T = (\, dP \quad dT) \tag{72}$$

from the corresponding contravariant vector

$$d\mathbf{y}^T = (\, dS \quad dV). \tag{73}$$

By using lower and upper indeces for the components of d**x** and d**y**, respectively, the requirement of Tellegen's theorem can be expressed in the form

$$\sum dx_i\, dy^i = \text{invariant} , \tag{74}$$

which is consistent with the usual tensor notation. The fundamental metric tensors **e** and **g** are related by

$$\mathbf{e}\, \mathbf{g} = \mathbf{I} \quad , \tag{75}$$

which can be checked by direct multiplication of the matrices, while the hybrid tensors are related by

$$h\ f = I. \tag{76}$$

These multiplications correspond to index contracting the second order tensors. Moreover, note that

$$a^i{}_j = g^{ik}\ e_{kj}$$

(in which a represent components of h or f), in agreement with the operations of raising tensor indeces. A similar expression in which the order of multiplication of e and g is reversed leads to lowering of indeces. Note that the components of g and h always have one upper index and a lower index.

2.15 GEOMETRICAL CHARACTERISTICS OF RESISTIVE NETWORKS

We pointed out above that by equating the "power" provided by external sources (and dissipated by loads) at the ports the interesting geometrical result emerges that the resistive currents and voltages--i.e., the internal extensive and extensive variables of the thermodynamic system-- behave as orthogonal directions in a space of higher dimension than the original problem. These additional dimensions collapse, of course, once the requirements of K.V.L. and K.C.L. are introduced. We proceed to show the one to one correspondence between the geometrical specifications of the thermodynamic problem and those of the network.

In order to consider a specific example first, we return to the T and Pi networks obtained before for the representation of the homogeneous fluid with two degrees of freedom. The orthogonal directions which act as eigenvectors in the higher dimensional problem (3 dimensions for the T or Pi networks) are, in the case of the T network

$$d_1{}^2 = \left(\sqrt{e_{11} - e_{12}}\ (dS)\right)^2 \tag{77}$$

$$d_2{}^2 = \left(\sqrt{e_{22} - e_{12}}\ (dV)\right)^2 \tag{78}$$

and

$$d_3{}^2 = \left(\sqrt{e_{12}}\ (dS + dV)\right)^2, \tag{79}$$

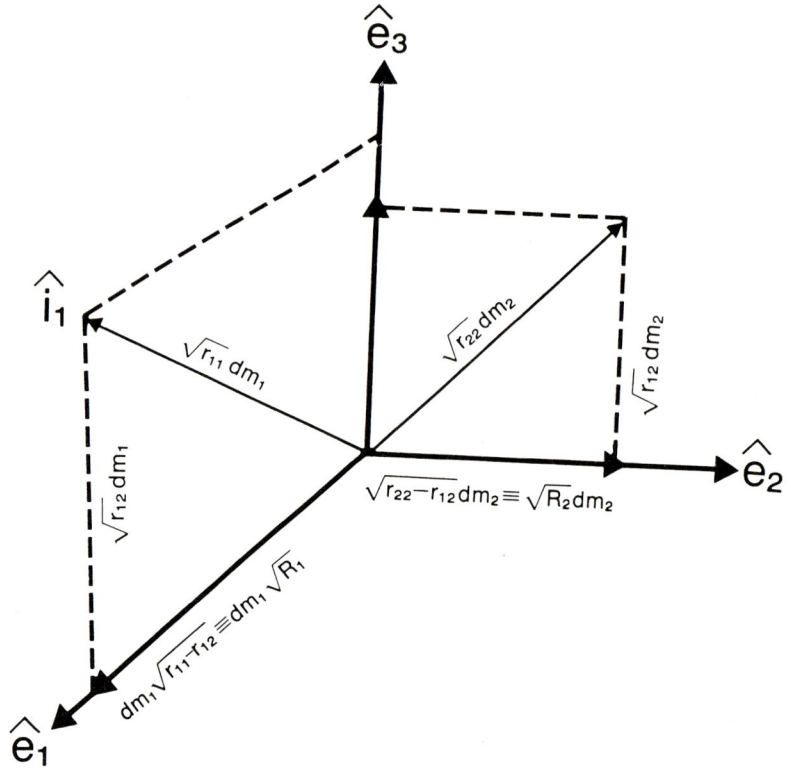

Fig. 2.12 The distance resultant can be expressed in terms of three orthogonal components. The coefficients r_{ij} represent the matrix entries in the **e** or **g** matrices and m_i represent the independent variables.

in which d_1, d_2 and d_3 are the distances or dissipations in the directions of resistances R_1, R_2 and R_3, respectively. These distances can be used to form a cartesian, orthogonal coordinate system. If the vector in the direction of R_3 is split into two (parallel) contributions, e_{12} dS and e_{21} dV --with $e_{12}=e_{21}$--, it follows that the 1-3 plane has the resultant (refer to FIGURE 12) vector

$$\mathbf{v}_1 = \sqrt{e_{11}} \, dS \qquad (80)$$

and plane 2-3 has the resultant vector

$$\mathbf{v}_2 = \sqrt{e_{22}} \, dV , \qquad (81)$$

which are the physical component vectors describing the two dimensional system. The distance invariant can now be expressed in one of three ways:

i) <u>in terms of the physical components</u>:

$$d^2 = e_1 \cdot e_1 + 2 e_1 \cdot e_2 + e_2 \cdot e_2 \qquad (82)$$

$$= v_1^2 + 2 v_1 v_2 q_{12} + v_2^2, \qquad (83)$$

in which q is the cosine of the angle made by the physical components and is given by

$$q = e_{12}/\sqrt{e_{12} \, e_{12}} \qquad (84)$$

and represents the degree to which energy can be transferred from one port to another, as will be discussed in more detail in Chapter 5.

ii) <u>In terms of the orthogonal directions</u>

The distance vector is clearly also determined by the components along the orthogonal directions, in the form

$$\mathbf{d} = \sqrt{e_{11} - e_{12}} \, dS \, \mathbf{i}_1 +$$

$$+ \sqrt{e_{22} - e_{12}} \, dV \, \mathbf{i}_2 + \sqrt{e_{12}} \, (d\dot{S} + dV) \, \mathbf{i}_3 \qquad (85)$$

The self dot product $\mathbf{d} \cdot \mathbf{d}$ yields the square of the distance or thermodynamic invariant. Clearly this corresponds in network terms to adding power (square distance) contributions in the individual resistances.

iii) <u>Dot Product of covariant and contravariant port vectors</u>

It can easily be checked by direct calculation that the dot product between the port forces (intensive variables) and the vector of port displacements (currents, extensive variables) leads to the distance directly,

dx · dy $= d^2$ \qquad (86)

This is the way to calculate the power supplied or dissipated at the port of a network and it is clearly the most direct way of

2.16 GEOMETRICAL PROPERTIES IN GENERAL RESISTIVE NETWORKS

Given a system of higher dimensionality, similar geometrical properties hold for the representative resistive network.

The resistances are again linear combinations of the Hessian

$$e_{ik} = \partial^2 E / \partial x_i \partial x_j \qquad (87)$$

while in a conductance network the conductances are linear combinations of the Free Energy Hessian

$$g_{ik} = \partial^2 G / \partial y_i \partial y_j . \qquad (88)$$

The squared distance can be found, as above, by any of the three techniques (dot product of the covariant and contravariant vectors at the ports, by inner product of the physical components or by adding the distance contributions in the resistive, orthogonal directions).

The angles between the various physical vectors are generally given by their cosines

$$\cos \theta_{ij} = g_{ij} / \sqrt{g_{ii} \ g_{jj}} \qquad (89)$$

$$= e_{ij} / \sqrt{e_{ii} \ g_{jj}} \qquad (90)$$

$$= q_{ij}$$

which have energy coupling properties among the various ports. The individual q_{ij}'s defined this way can be used to measure the hyperangle formed by all physical components as

$$\sin \theta = \begin{bmatrix} 1 & q_{12} & q_{13} & \cdots\cdots\cdots q_{1n} \\ q_{21} & 1 & q_{23} & \cdots\cdots\cdots q_{2n} \\ q_{31} & q_{32} & 1 & \cdots\cdots\cdots q_{3n} \\ \cdots & \cdots & \cdots & \cdots \\ q_{n1} & q_{n2} & q_{n3} & \cdots\cdots\cdots 1 \end{bmatrix} \qquad (91)$$

The area, volume, or higher dimensional content of the polytope defined is given by (ref. 54)

Content = $\sin \theta \prod^{n} g_{ii} \, dy_i$, (92)

in which θ is the hyperangle defined from the matrix given above.

The angles given above can be readily calculated for the two dimensional example of the homogeneous fluid with two degrees of freedom. The cosine made between the S and V axes is given by

$$\cos \theta_{S,V} = e_{12}/\sqrt{e_{11} e_{22}}$$ (93)

while the cosine of the angle subtended by the T and P axes is

$$\cos \theta_{T,P} = g_{12}/\sqrt{g_{11} g_{22}} \; .$$ (94)

Making use of the conversion table, one concludes that these two angles are equal. Note that this result also agrees with our previous conclusion that the V and T axes are orthogonal and that the S and P axes are orthogonal (see FIGURE 13). Geometric thermodynamics also obtain the same results.

The departure between this model and geometric thermodynamics becomes clear when the cosines are written explicitly in the form

$$\cos^2 \theta_{S,V} = (\partial T/\partial V)_S (\partial P/\partial S)_V / (\partial T/\partial S)_V (\partial P/\partial V)_S$$ (95)

and

$$\cos^2 \theta_{T,P} = (\partial S/\partial V)_T (\partial P/\partial T)_V / (\partial S/\partial T)_V (\partial P/\partial V)_T .$$ (96)

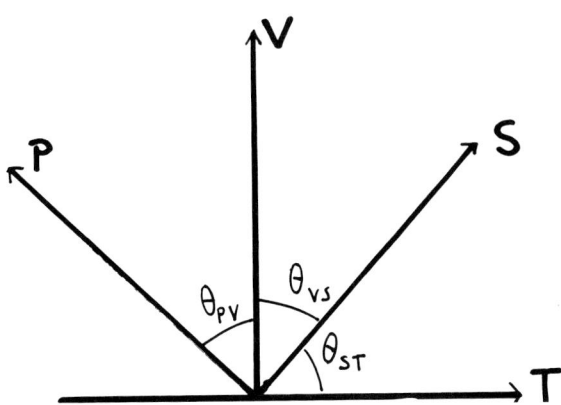

Fig. 2.13   Relative directions of thermostatic axes.

In geometrical terms these angles have sense only if there Maxwell's reciprocities hold--i.e., if there is a unique energy potential function.

In network terms these ratios have a meaning even when the angles cannot be defined--i.e., when there is no unique potential function or when there is more than one potential, such as is the case with an irreversible process which shares the energy potential with a different reversible process. The exact nature of these ratios will be considered in Chapter 3. It will be seen that these ratios are coupling constants which have a more general meaning than the cosine of an angle.

2.17 METRIC THERMOSTATICS

In the appendix to Chapter 1 it was indicated that there are several approaches to setting up global thermodynamic equations based on network theory. These follow one of the standard organized analyses which relate "currents" and "voltages" at nodes, branches or meshes in the network. The four fundamental techniques use the nodal, mesh, tie set and cut set matrices described in Chapter I. From these one can see the close relationship between metric thermostatics and network thermodynamics.

In a simplified analysis in which the independent quantities are the extensive variables (energy function of state) we can visualize some internal tree whose displacements dm_b are linearly related to the external (port) displacements dy_j; this can be expressed, in matrix notation as

$$d\mathbf{m} = \mathbf{T}\, d\mathbf{y} \,. \tag{97}$$

By contrast, the port forces x_j (voltages) are linear functions of the tree forces f_b or, in matrix notation,

$$d\mathbf{x} = \mathbf{T}^T \tag{98}$$

in which T is the tie set matrix specifying K.V.L. and K.C.L. in the network.

If one assumes that at each branch in the tree the equations are linear and of the form ("Ohm's Law")

$$dx_b = R_{bb}\, dm_b \tag{99}$$

the port equations become

$$d\mathbf{x} = (\mathbf{T}^T \mathbf{R} \mathbf{T}) \, d\mathbf{m} \tag{100}$$

in which **R** is a diagonal matrix and the equations are therefore symmetric. Clearly, a free energy analysis in which the independent variables are the forces (intensive variables), rather than the displacements will lead to phenomenological equations of the form $d\mathbf{m} = (\mathbf{C}^T \mathbf{G} \mathbf{C}) d\mathbf{x}$ in which **G** is a diagonal conductance matrix and **C** is the Cut set matrix. As we pointed out before, **C** and **T** are orthogonal, so that $\mathbf{C}\mathbf{T}^T = 0$. Because the matrices **R** and **G** are diagonal, the resultant equations are reciprocal and therefore obey Maxwell's reciprocities.

The simplest resistive networks for dimensions n=1, 2, and 3 which have the port displacements as links of internal trees are given in FIGURE 14. These networks can represent many dissimilar processes and constitute equivalence classes for processes of a given dimensionality. Through the use of Tellegen's theorem (or simply using the analog idea that the power supplied at the ports of a network equals the power dissipated in the internal resistances) one readily obtains the result

$$d^2 = \left(\sqrt{R_b} \, dm_b\right)^2 = a^2, \tag{101}$$

2.18 RECIPROCITY OF INTERSTATE INNER PRODUCTS AND TELLEGEN'S THEOREM

As we discussed in Chapter I, the orthogonality of the cut and tie set matrices which follows from Kirchhoff's laws leads directly to the reciprocity of the port equations when the system is linear-- i.e., a resistive network having time invariant resistances.

In the thermodynamic case there is no euclideanism in the large, in general. However, it can be argued that, as long as the region of the manifold considered is small, an euclidean neghborhood can always be found. (This is similar to the way general relativity reduces to special relativity in the same situation). As a result, if two states are sufficiently close to each other --i.e., within a differential neighborhood which

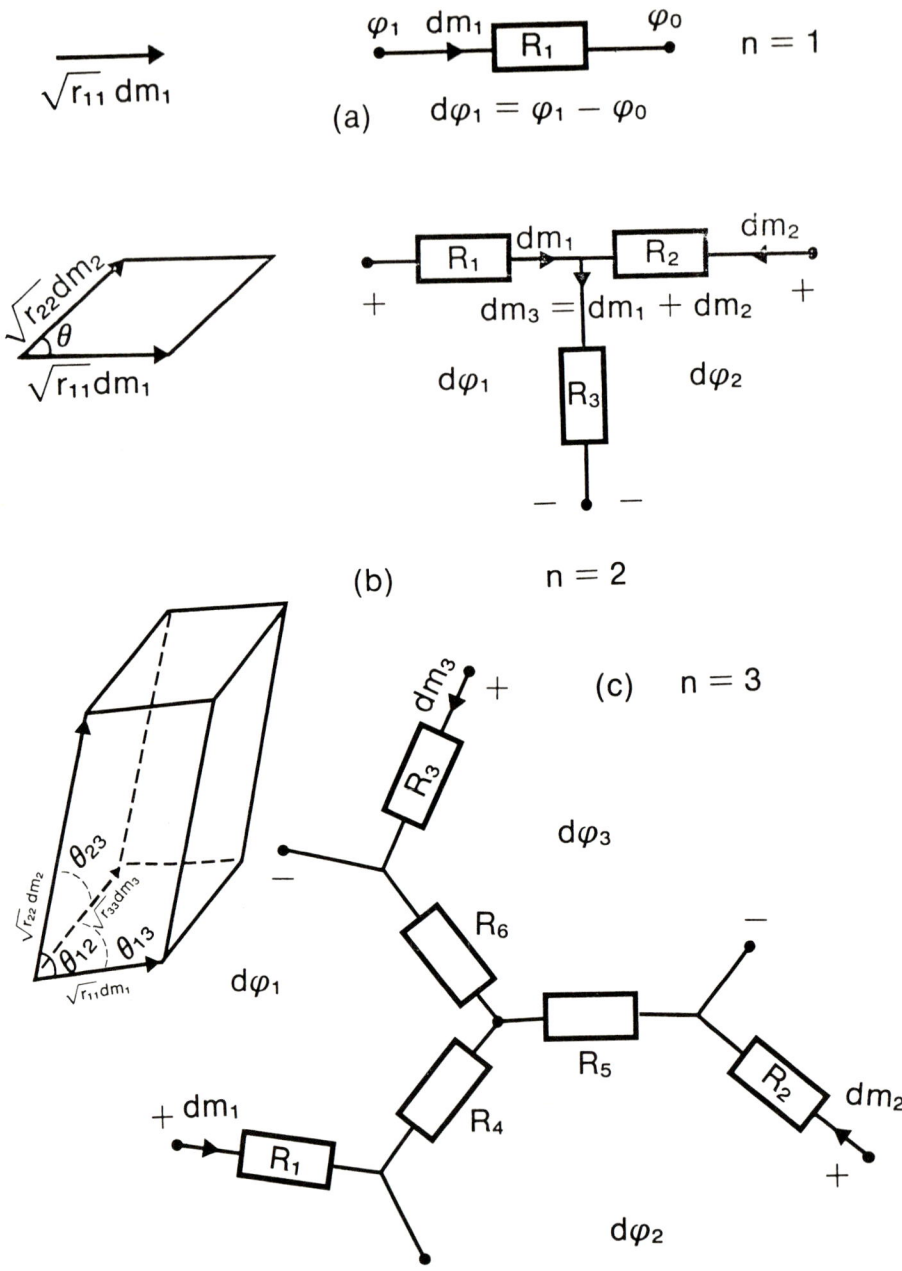

Fig. 2.14 Geometrical equivalences for networks n=1, n=2 and n=3.

appears to be euclidean, or flat in the mainfold-- will behave as if they could both be represented by the same resistive network. The actual dependent and independent variables will of course be different in both states. Tellegen's theorem then yield immediately reciprocities of the form

$$dx'\cdot dy = dx\cdot dy' , \qquad (102)$$

in which the primed and unprimed variables represent measurements made in the neighborhoods of each of the states considered.

This is a **General Reciprocity Relation** which expresses the "conservation" of a bilinear form analogous to a bilinear energy differential, with measurements drawn from two different states. The General Reciprocity Relation serves to unify many thermodynamic results, as will be shown below. From a geometrical point of view it is a direct result of the connectivity of the manifold (Kirchhoff's laws) and the property of uniform calibration of the euclidean geometry in the neighborhood considered.

Two basic families of reciprocities can be derived: if the basic metric chosen uses the energy measure we shall refer to the reciprocity as the GRE (General reciprocity for Energy quantities) while, if it involves the entropy network port variables, we shall refer to it as the general Reciprocity for Entropy quantities (GRS).

In the special case of a homogeneous fluid with two degrees of freedom, these become

$$\boxed{dT'dS + dP'dV = dT\, dS' + dP\, dV'} \qquad \text{(GRE)} \qquad (103)$$

and

$$\boxed{dE'd(1/T) - dV'd(P/T) = dE\, d(1/T)' - dV\, d(P/T)'} \qquad \text{(GRS)}. \qquad (104)$$

2.19   APPLICATIONS OF THE GENERAL RECIPROCITY CONDITION

Given the General Reciprocity condition for Energy Functions derived above (GRE), it is possible to obtain readily some familiar results of Thermodynamics. Thus, we can perform the following thought incremental experiments on the system:

1. Keep S constant in the first experiment (unprimed variables) and V constant in the second experiment (primed variables). With $dS=0$ and $dV'=0$, Eq. (103) reduces to

$$dP'\,dV = dT\,dS' \qquad (105)$$

or

$$(dP'/dS')_{V'} = (dT/dV)_{S} \,, \qquad (106)$$

in which we interpret the differentials which appear in the GRE expression as either differentials or finite differences which can be multiplied and divided like regular numbers. When the infinitesimal limit is approached, Eq. (106) becomes the partial differential equality

$$(\partial P'/\partial S')_{V'} = (\partial T/\partial V)_{S} \,. \qquad (107)$$

Physically, this corresponds to two processes which takes place in a differential neighborhood of each other--i.e., "almost" the same process, but not quite the same. It must then be concluded that if the exact limit could be reached Eq. (107) would reduce to the familiar Maxwell reciprocity

$$(\partial P/\partial S)_{V} = (\partial T/\partial V)_{S} \qquad (108)$$

which can be seen to be valid only as a thought experiment involving neighboring euclidean processes--i.e., linear. The local euclideanism is required, because without linear constitutive relations GRE would not hold.

In the classical derivation of thermodynamic reciprocities this local flatness is implied in the fact that thermodynamics is a first order linear approximation in which second and higher order variations of the independent variables are ignored, so that any dependent variable can be expanded in terms of first variations of the independent variables.

2. Keep P constant in the first experiment and S constant in the

second experiment.

The General Reciprocity condition now reduces to

$$dT' \, dS + dP' \, dV = 0 \, , \tag{109}$$

which, by the same general argument given above reduces, in the limit, to

$$(\partial T/\partial P)_S = -(\partial V/\partial S)_P \tag{110}$$

thus yielding a second Maxwell reciprocity.

3. Keep T constant in the first experiment and V constant in the second experiment. Equation (109) then becomes

$$dT' \, dS + dP' \, dV = 0, \tag{111}$$

which leads to

$$(\partial S/\partial V)_T = -(\partial P/\partial T)_V \tag{112}$$

and, finally,

4. Keep T constant in the first experiment and P constant in the second experiment.

The GRE reduces to $dT' \, dS = dP \, dV'$, thus leading to the fourth reciprocity

$$(\partial S/\partial P)_T = (\partial V/\partial T)_P . \tag{113}$$

This utilization of GRE to obtain formulas relating various experiments is analogous to the classical approach used by Lord Raleigh to obtain reciprocities in the equations of sound (ref. 200).

(Note: In all of the above partial differential equalities the signs appear reversed with respect to the familiar textbook expressions, because of the convention chosen here for the positive sign for dV).

2.20 RELATIONSHIPS BETWEEN HEAT CAPACITIES AT CONSTANT PRESSURE AND AT CONSTANT VOLUME

Using the GRE and dividing each term by $dT\, dT'$, we obtain

$$(dV/dT)(dP'/dT') + (dS/dT)(dT'/dT') = \qquad (114)$$
$$= (dV'/dT')(dP/dT)+(dS'/dT')(dT/dT')$$

which, at constant P and V′ --i.e., when we consider two separate experiments for which $dP=0$ and $dV'=0$ -- becomes

$$(\partial V/\partial T)_P (\partial P'/\partial T')_{V'} + (\partial S/\partial T)_P = (\partial S'/\partial T')_{V'} . \qquad (115)$$

Taking the limit in which the primed and unprimed states are the same, multiplying by T and making the familiar identifications

$$c_p = T(\partial S/\partial T)_P \qquad (116)$$

and

$$c_V = T(\partial S/\partial T)_{V'} \qquad (117)$$

we obtain the relationship between the heat capacities at constant pressure and at constant temperature:

$$c_V - c_p = T\, (\partial V/\partial T)_P\, (\partial P/\partial T)_V . \qquad (118)$$

2.21 ENTROPY LIKE FUNCTIONS FROM GRS

The reciprocity condition for entropy functions (GRS) defines all entropy potentials, even though only S was used to construct this equation. Given GRS, we can perform the following though experiments:

1. Keep E constant in the first experiment and V constant in the second. After the appropriate limits are taken, in the same

manner as in the above exercise, one obtains

$$(\partial (1/T)/\partial V)_E = -(\partial (P/T)/\partial E)_V. \qquad (119)$$

2. Keep T´ and V constant. Following the same procedure, we obtain

$$(\partial E/\partial V)_{1/T} = (\partial (P/T)/\partial (1/T))_V. \qquad (120)$$

3. Keep T´ and (P/T) constant. This yields

$$-(\partial E/\partial (P/T))_{1/T} = (\partial V/\partial (1/T))_{P/T}. \qquad (121)$$

4. Finally, run two experiments at constant E´ and constant P/T, respectively to obtain from GRS

$$(\partial V/\partial E)_{P/T} = (\partial (1/T)/\partial (P/T))_E. \qquad (122)$$

Equations (119), (120), (121) and (122) define the exact differentials

$$dS = -(P/T)\, dV + (1/T)\, dE \qquad \text{(Entropy)}, \qquad (123)$$

$$dJ = -(P/T)\, dV - E\, d(1/T) \qquad \text{(Massieu)}, \qquad (124)$$

$$dY = V\, d(P/T) - E\, d(1/T) \qquad \text{(Planck)}, \qquad (125)$$

and

$$dK = V\, d(P/T) + (1/T)\, dE, \qquad (126)$$

in agreement with the results which would be obtained by using Legendre transformations. (Note, again, the difference in the signs whenever dV appears, because of the positive direction chosen for dV).

2.22 THERMOSTATIC EQUATIONS FROM ENTROPY NETWORKS: THERMODYNAMIC EQUATION OF STATE

We can now derive the well known thermodynamic equation of state,

$$-(\partial E/\partial V)_T = T(\partial P/\partial T)_V - P \qquad (127)$$

using the GRS expression. The GRS is given by

$$dE'\, d(1/T) - dV'\, d(P/T) = dE\, d(1/T') - dV\, d(P'/T'). \qquad (128)$$

Let two experiments be performed: in the first one, the temperature, T, remains constant while in the second the volume, V´, is kept constant. This changes GRS into

$$0 = dE\, d(1/T') - dV\, d(P'/T'). \qquad (129)$$

Moreover, the pressure to temperature ratio can be expressed in terms of independent increments of pressure and temperature as

$$d(P'/T') = P'\, d(1/T) + (1/T')\, dP' \qquad (130)$$

and

$$d(1/T) = -dT/T^2. \qquad (131)$$

Introducing (130) and (131) into (129) yields

$$0 = dE\,(dT'/T'^2)_V \qquad (132)$$
$$+ dV(dP'/T')_V - (P'dT'/T'^2)_V\, dV.$$

Multiplying through by T'^2, dividing by $dT'dV$, and taking the limit of coincident primed and unprimed variables we obtain

$$(-\partial E/\partial V)_T = T(\partial P/\partial T)_V - P, \qquad (133)$$

as expected. Although it can be argued that the number of steps required to derive this equation does not represent considerable savings over the standard techniques, the important point to stress is that <u>all the information required was contained in the GRS itself</u> --i.e., a single equation.

2.23 ENERGY OF THE IDEAL GAS

The GRS leads directly to the independence of the energy of

the ideal gas and the volume it occupies. Assume that in a first experiment T is constant and that in the second experiment V´ is constant. GRS yields

$$0 = dE\, d(1/T) + dV\, d(P´/T´) \qquad (134)$$

which may be rewritten, after introducing the limiting conditions T=T´, P=P´ and V=V´ as

$$(\partial E/\partial T)_T = -(\partial(P/T)/\partial(1/T))_V . \qquad (135)$$

Given the ideal equation PV= nRT, the right hand side is zero, so that

$$(\partial E/\partial V)_T = 0 , \qquad (136)$$

thus showing again that the reciprocity conditions are consistent with other results of thermodynamics and that, furthermore, they summarize a great deal of scattered information.

Of course, application of GRE with dT=0 and dV´=0 yields in the limit considered here

$$(\partial S/\partial V)_T = -(\partial P/\partial T)_V = nR/V \qquad (137)$$

or, upon integration,

$$S = nR \ln V + C , \qquad (138)$$

as expected.

2.24 UNIQUENESS OF THE EQUILIBRIUM STATE FROM "LOCAL" SECOND LAW ASSUMPTION

The connected network splits a bilinear form into the into the individual contributions of (fictitious) dissipations in internal resistances and, as discussed above, corresponds to an equivalent Pythagorean decomposition into sums of squares. Given that the bilinear law is a measurable observable, it makes sense to view the individual directions in internal space as obeying the same macroscopic laws as the complete system.(This is an assumption which only has statistical basis, but cannot be justified from a purely thermodynamic point of view). If this is correct, the second law applied to the individual component resistances requires that the direction of force and displacement always coincide, so that each resistance in the network must be positive.

An immediate consequence, if this view is correct, is that the equilibrium state must be stable as follows readily by appplying Tellegen's theorem, so that

$$\mathbf{dx} \cdot \mathbf{dy}_{ports} = \mathbf{dx} \cdot \mathbf{dy}_{internal\ resistances} \qquad (139)$$

--i.e., the power dissipated in the resistances equals the power supplied at the ports, in network terms. Introducing the constitutive law for each resistance we obtain

$$\mathbf{dx} \cdot \mathbf{dy} = R_b (dy_b)^2 \qquad (140)$$

Consider a situation in which all the extensive quantitites are kept constant and assume the intensive quantities could vary. It then follows that

$$0 = R_b (dy_b)^2 \qquad (141)$$

and in order for the second term to be zero given positive resistances, all variations must independently vanish--i.e., the equilibrium value is unique.

2.25 THE "MISSING" ENTROPY AND ENERGY RECIPROCITIES CONTAINED IN GRS AND GRE

The results derived above can be derived also by the more familiar Legendre transformations and other approaches of differential calculus-- although there is no classical equivalent to GRS and GRE in terms of the amount of information they contain.

GRE and GRS are unique, however, in that they yield two reciprocities which are not directly derivable from the classical formalism, unless sufficient algebraic gymnastics are invoked.

From GRE one obtains, after setting $dT' = 0$ and $dS = 0$ in two separate experiments and then reaching the limit of coincident states:

$$(\partial T/\partial S)_V (\partial S/\partial T)_P + (\partial v/\partial T)_P/(\partial S/\partial P)_V = 1 \qquad (142)$$

and

$$(\partial v/\partial P)_T (\partial P/\partial v)_S + (\partial S/\partial P)_T/(\partial v/\partial T)_S = 1. \qquad (143)$$

These two equations can be consolidated in the form

$$(\partial T/\partial S)_V (\partial S/\partial T)_P + (\partial v/\partial T)_P / (\partial S/\partial P)_V = \qquad (144)$$
$$= (\partial v/\partial P)_T (\partial P/\partial v)_S + (\partial S/\partial P)_T / (\partial v/\partial T)_S$$

Similarly, the use of GRS in two experiments in which dE= 0 and dV´= d(P/T) = 0 leads to:

$$(\partial v/\partial (P/T))_T (\partial (P/T)/\partial v))_E +$$
$$+ (\partial (1/T)/\partial v))_E / (\partial (P/T)/\partial E)_T = \qquad (145)$$
$$= (\partial E/\partial (1/T))_{P/T} (\partial (1/T)/\partial E))_V +$$
$$+ (\partial v/\partial (1/T))_{P/T} / (\partial E/\partial (P/T))_V.$$

2.26   UNIFICATION OF THERMODYNAMIC CAUSE-EFFECT

The network view presents four practical ways to display cause and effect among thermodynamic variables in the system with two degrees of freedom. Thus the E network shows explicitly that the forces (intensive variables) are defined in the direction of the displacements (extensive variables), for example. Similarly, the G network indicates that displacements can be defined in the direction of the intensive quantities. The network paths are, of course, the same as the directions of the gradients when the specific energy function is considered. This suggests a simple nomenclature to unify the cause effect characteristics of the networks in the spirit of Gibbsian thermodynamics.

Consider the change in Energy in a system with two degrees of freedom having a thermal port and a mechanical port only. The differential change

$$dE = T\,dS + P\,dV \qquad (146)$$

(with the sign convention adopted here) can also be rewritten in the form

$$dE = (\mathbf{grad}_{S,V}\, E) \cdot d\mathbf{z}_E \qquad (147)$$

in which the gradient can be explicitly written out in the form

$$\mathbf{grad}_{S,V} = (\partial/\partial S)\mathbf{i}_S + (\partial/\partial V)\mathbf{i}_V \qquad (148)$$

to indicate that the resultant vector

$$\begin{aligned}\mathbf{grad}_{S,V}\, E &= (\partial E/\partial S)\mathbf{i}_S + (\partial E/\partial V)\mathbf{i}_V \\ &= T\,\mathbf{i}_S + P\,\mathbf{i}_V\end{aligned} \qquad (149)$$

has components in the "directions" of the vectors $\mathbf{i}_S$ and $\mathbf{i}_V$ with component magnitudes given by T and P, respectively. while the displacement vector **dz** is given by

$$d\mathbf{z}_E = dS\,\mathbf{i}_S + dV\,\mathbf{i}_V . \qquad (150)$$

Similarly, the other energy potentials yield

$$dH = (\mathbf{grad}_{S,P}\, H) \cdot d\mathbf{z}_H , \qquad (151)$$

$$dF = (\mathbf{grad}_{T,V}\, F) \cdot d\mathbf{z}_F , \qquad (152)$$

and

$$dG = (\mathbf{grad}_{T,P}\, G) \cdot d\mathbf{z}_G ; \qquad (153)$$

in which

$$\mathbf{grad}_{S,P} = (\partial/\partial S)\mathbf{i}_S + (\partial/\partial P)\mathbf{i}_P$$

$$\mathbf{grad}_{T,V} = (\partial/\partial T)\mathbf{i}_T + (\partial/\partial V)\mathbf{i}_V$$

and

$$\mathbf{grad}_{T,P} = (\partial/\partial T)\mathbf{i}_T + (\partial/\partial P)\mathbf{i}_P ,$$

while the displacement vectors are now given by

$$d\mathbf{z}_H = dS\ \mathbf{i}_S + dP\ \mathbf{i}_P \qquad (154)$$

$$d\mathbf{z}_F = dT\ \mathbf{i}_T + dV\ \mathbf{i}_V \qquad (155)$$

and

$$d\mathbf{z}_G = dT\ \mathbf{i}_T + dP\ \mathbf{i}_P . \qquad (156)$$

The resultant effect vectors conjugate to each of the displacements correspond to the actual slope in the specific direction for the given energy potential surface considered. In addition to the T-P vector given above for the Energy, the following complete the set:

$$\begin{aligned}\mathbf{grad}_{S,P}\ H &= (\partial E/\partial S)\mathbf{i}_S + (\partial E/\partial P)\mathbf{i}_P \\ &= T\ \mathbf{i}_S + V\ \mathbf{i}_P ,\end{aligned} \qquad (157)$$

$$\begin{aligned}\mathbf{grad}_{T,V}\ F &= (\partial F/\partial T)\mathbf{i}_T + (\partial F/\partial V)\mathbf{i}_V \\ &= S\ \mathbf{i}_T + P\ \mathbf{i}_V\end{aligned} \qquad (158)$$

and

$$\begin{aligned}\mathbf{grad}_{T,P}\ G &= (\partial G/\partial T)\mathbf{i}_T + (\partial G/\partial P)\mathbf{i}_P \\ &= S\ \mathbf{i}_T + V\ \mathbf{i}_P .\end{aligned} \qquad (159)$$

In the network "space" the gradient directions are specified by the independent variable. In particular, if the independent port variable is a force ("voltage"), the direction at the port consists of parallel branches, if it is a current it consists of series branches. The complete gradient is formed by considering all the ports with their respective independent variables.

2.27   CONNECTIVITY AND THE CHEMICAL POTENTIAL

The issue of topological connectivity, as defined here in network terms, is very relevant in the definition of a proper chemical potential which is consistent with the rest of thermodynamic theory. The usual approach consists in asking what is the force that is conjugate to the flow of mass at constant temperature and pressure. Given that the Gibbs free energy G is the proper thermodynamic energy potential when the independent variables are T and P the chemical potential μ is defined as

$$\mu = \left(\partial G/\partial n\right)_{T,P} . \qquad (160)$$

There is a very serious lack of generality in this definition which arises by building an energy function (G in this case) by using only two independent variables (T and P) and then attempting to extend the application of the energy function to a quantity which was not used in the original construction of the manifold. In fact, it is more consistent to assume that chemical energy is on the same footing with thermal and mechanical energies and that it, too, contributes to the definition of the manifold by adding a separate degree of freedom. Restrictions imposed by the phase rule can then be incorporated latter.

Rather than begin with G, then, we ask for a function of state f such that

$$f = f(S,V,n) . \qquad (161)$$

Using the graphical network approach this requirement translates into a three port machine in which, say, the extensive quantities S, V and n are represented by currents, while the respective intensive variables T, P and μ represent voltages. Unlike the two port situation, in which there are only four possible networks consistent with cause and effect, there are now eight networks depending on which port quantities are considered to be the independent variables. In turn, and by analogy with the two port Legendre transformations, each of these networks will now define a (three port) energy function. The problem is to find out how the three dimensional energy functions relate to the familiar two dimensional energy potentials E (or U), G, F (or A) and H.

The possible three port networks are shown in FIGURE 15, in which the following economical notation has been utilized: i) each unlabelled branch represents a resistance (conductance), ii) each branch labelled with a "+ -" pair represents a controlled force ("voltage") source regulated by non conjugate ports, iii) each branch labelled with an arrow represents a flow source controlled

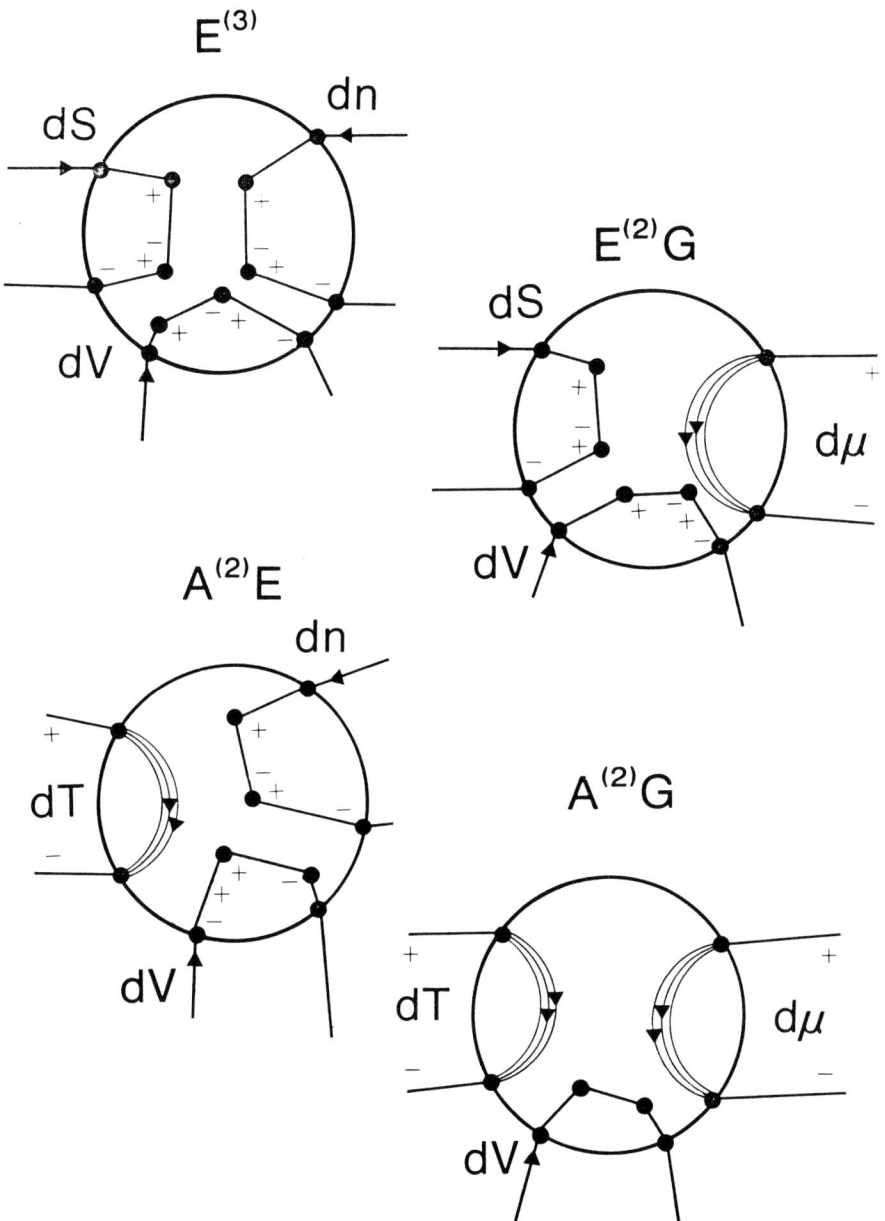

Fig. 2.15 Definition of chemical potential based on disconnected three ports. If reciprocal, these lead to generalized energy functions.

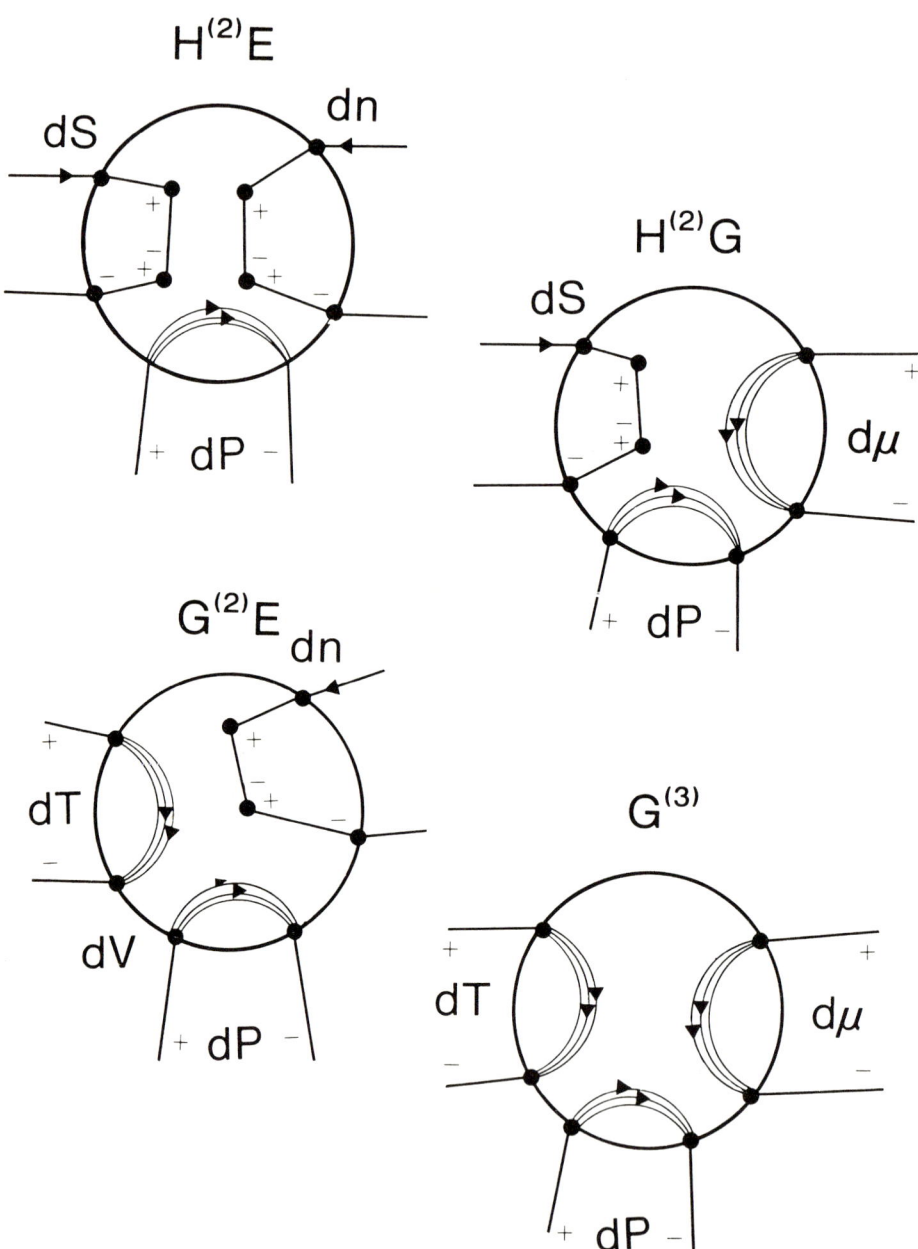

Fig. 2.15 (continued).

by a non conjugate port, iv) in a port in which the branches are in series the current (extensive variable) is the independent variable and v) in a port in which all the branches are in parallel the intensive variable (" voltage") is the independent variable.

In order to retain the ideas of the two dimensional representation as closely as possible, it makes sense to label the network in which all the extensive variables are independent as E(3) and the network in which all the intensive variables are independent by G(3). The labels E(2), G(2), F(2) and H(2) are used to represent the two dimensional energy potentials. Note that in the same way that G(2) is the graphical dual of E(2), E(3) is the dual of G(3). It can also be seen that in two of the networks the elimination of the chemical degree of freedom reduces the networks to H(2), so that we label these as H(2)E or H(2)G, depending on whether the additional independent variable is n or μ. This process can be carried out for the remaining networks. Note that G(3) is also G(2)G and that E(3) is also E(2)E, thus showing the consistency of the labelling procedure chosen.

With these definitions, we can proceed to define the chemical potentials in terms of the newly constructed (three variable) energy functions:

$$\mu = (\partial E(2)E / \partial n)_{S,V} \quad (162)$$

$$\mu = (\partial A(2)E / \partial n)_{T,V} \quad (163)$$

$$\mu = (\partial H(2)E / \partial n)_{S,P} \quad (164)$$

$$\mu = (\partial G(2)E / \partial n)_{T,P} \quad (165)$$

$$n = (\partial E(2)G / \partial \mu)_{S,V} \quad (166)$$

$$n = (\partial A(2)G / \partial \mu)_{T,V} \quad (167)$$

$$n = (\partial H(2)G / \partial \mu)_{S,P} \quad (168)$$

and

$$n = (\partial G(2)G / \partial \mu)_{T,P} \quad (169)$$

Note that Eq. (165) agrees with the commmon thermodynamic

definition of the chemical potential if E is interpreted to mean E(3). However, in that case G must be interpreted to mean G(3), in which case it must be assumed that the definition from the free energy is also taken using G(3), but this is incorrect at constant T and P, as the three dimensional energy potential which leads to the chemical potential by keeping T and P constant is (Eq 165) G(2)E, not G(3).

The consistent topological definition of the chemical potential then requires to either begin from E(3) or to use G(2)E in all formulas which heretofore used G; otherwise, no potential function has been defined.

These potentials also serve to define an extended General Reciprocity in the form

$$dS' \, dT + dV' \, dP + dn' \, d\mu = dS \, dT' + dV \, dP' + dn \, d\mu' . \quad (170)$$

In fact, any one of the above potentials imply all the other potentials and the reciprocity itself.

Appendix 1

A.2.1 REPRESENTATION IN H(u)

The network model proposed above can be put in correspondence with an approach familiar in Functional Analysis. Mathematicians have shown that every Hilbert space, including the euclidean space R is isomorphic to $H = L^2(u)$, the space of all square, integrable functions on a measure space (X, M, u).

The inner product of two vectors is given by

$$\langle x, y \rangle = \int x \, y \, du \quad (A.2.1)$$

which is equivalent to $\langle x, y \rangle = \vec{x} \cdot \vec{y}$ in R^3. Altough the theorems of Functional Analysis can expedite the transformations, we retain the geometrical analogy, in the spirit of the present network presentation. To establish the correspondence between R and L(u), we postulate the existence of some three dimensional manifold bounded by a surface. Let

1. x be a vector field inside the sphere given by x = -grad ϕ,

in which ϕ is a potential function specified at each point inside the volume,

2. y be a (flow) like vector field proportional to the field x at every point-- i.e., y = g(v) x , in which g(v) is a scalar which depends on the position in the volume.

The dot product of the vector fields x and y integrated inside the volume is given by

$$\int x \cdot y \, dV = \int g(v) \, x \cdot x \, dv = \int g \, \|x\|^2 \, dv ,$$ (A.2.2)

which is an integral of a square function and has the same form as a Hilbert space. Moreover, the integral can also be changed by means of the identity

$$\text{div} (\phi \, \vec{y}) = \vec{y} \, \text{grad} \, \phi + \phi \, \text{div} \, \vec{y}$$ (A.2.3)

and Green's theorem to yield

$$\int x \, y \, dv = \int \phi \, y \, dA - \int \phi \, \text{div} \, y \, dv.$$ (A.2.4)

If one of the surface potentials is used as a reference value and arbitrarily set to zero, the port forces

$$x(i) = \phi(i) - \phi(0)$$ (A.2.5)

can be defined at the surface. If, moreover, the field y inside the volume acts as an incompressible fluid with no sources, it follows that div y = 0 anywhere inside the volume (note that this condition is also consistent at the surface where the sum of all flows entering at all points $\phi(i)$ at the surface equals the flow leaving at $\phi(0)$) and

$$\langle x(i) \, y(i) \rangle = \int \vec{x} \cdot \vec{y} \, dv = \int g \, \|\vec{x}\|^2 \, dv ,$$ (A.2.6)

in agreement with the statement of Functional theory quoted above. This indicates the general relationship between the topology of thermodynamics, the conservation of "distance"--i.e., some physical invariant such as energy and the view that there is an incompressible "liquid" flowing in some imaginary volume with no sources.

Although this is only a conceptual device, it agrees fully with thermostatic results and theory. One would then expect that some physical insight could be gained by inquiring further about the nature of this "fluid". In the case of the energy networks considered above, the candidate for the conserved fluid must be an extensive quantity which must be recognizable by all ports. The only such variable is Entropy itself. By defining a generalized entropy S(i) flowing into each port i in terms of the corresponding extensive variable at that port,

$$S(i) = y(i) / k(i) , \qquad (A.2.7)$$

in which k is a calibration constant whose numerical value may be set equal to 1, and defining the generalized temperature at a port by

$$T(i) = X(i) k(i), \qquad (A.2.8)$$

the whole construct becomes consistent and one can avoid the puzzling " mixing" of different extensive variables --e.g., volume and entropy-- at the nodes of a network.

These results suggest the natural generalization for the use of K.V.L. and K.C.L. in the case of irreversible processes: entropy will have irreversible sources which account for the extra fluid which appears during an irreversible process. K.V.L. and all other results remain invariant, however. By contrast, geometrical thermodynamics does not have a formalism which can incorporate such irreversible transitions. (The use of irreversible sources of entropy is of course familiar from Onsager thermodynamics). These ideas will be developed further in Chapter 4.

A.2.2 INTEGRATION IN L^2 REDUCES TO A LUMPED PARAMETER NETWORK

In order to demonstrate the correspondence between the continuous parameter or flow geometry given above and the lumped parameter network, or geometrical representation, we proceed to integrate the continuous equations. We assume that there are a finite number of potentials specified at the surface enclosing the

L^2 space.

In complete analogy with field theory or hydrodynamics, one can use the superposition principle (ref. 72). Chose a streamline $s_k(v)$ going between potential k and the reference potential, both located at the surface. Assuming linear superposition the flow along $s_k(v)$ will be given by

$$j_s = d\phi_k \int g_s \, ds_k \qquad (A.2.9)$$

The total flow contribution to the output extensive quantity dy_i is obtained by integrating this expression along all paths that go between i and k, which can be done by varying the parameter v, as shown in FIGURE 15. The net result is

$$dm_i = \sum d\phi_k \, g_s \iint ds_k \, dv_i \qquad (A.2.10)$$

which has the desired form

$$dy_i = \sum g_{ik} \, d\phi_k. \qquad (A.2.11)$$

Thus, the continuous representation can also be replaced, as desired, by a discrete representaiton in which the metric is appropriately integrated to obtain lumped parameters.

In principle, the integration process could be very complex; in practice, however, it is a simple matter to find the network, as any network found for a given dimensionality will represent all possible systems of that dimension as far as the metric is concerned.

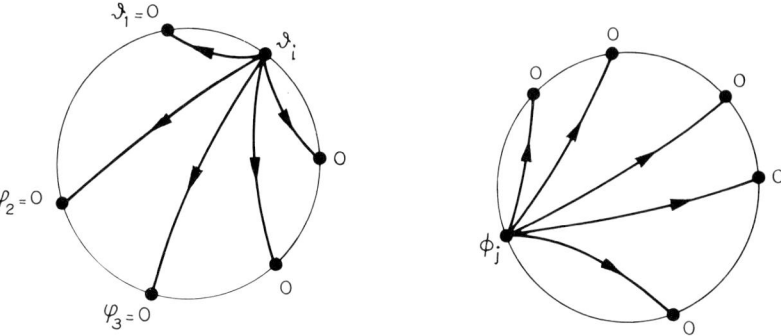

Fig. 1.A.1 Flow lines and principle of superposition used to integrate in the thermodynamic manifold.

Chapter 3

ON THE CONSISTENT DEFINITION OF KINETIC POTENTIALS IN DIFFUSION-REACTION-CONVECTION COUPLING AND RELATED TRANSPORT PROBLEMS

3.1 BACKGROUND

Linear chemical dynamics with stoichiometry has been thoroughly described in terms of algebra (ref. 50), topology (140-142) and differential geometry terms (ref. 175, 176). In addition to the well known work of Clarke , who has derived a complete set of steady states for the general dynamical problem, special computational (ref.83) and analytical tools (ref. 21) have been introduced to analyize reactions consisting of sequences of elementary unimolecular steps in the quasi steady state approximation. Nonlinearities, especially as they apply to oscillations, have been considered by Higgins (ref. 94),Tyson and Light(ref. 252) and others. In addition, the study of oscillatory behavior has seen considerable development in the past few years, especially in the chaotic range . The somewhat more intricate problem of reaction-diffusion coupling has been treated by analytical and algebraic techniques by Wei(ref.257) , Aris (refs.7-10) , Mikulecky (ref.147) , DeSimone and Caplan (ref 62) , Bunow (ref. 36) and others , including the present author (refs.133, 182,184). Parallel series non reactive systems have been further analyzed by Kedem and Katchalsky (refs. 112, 113) , by Moore (ref. 161) and by Weinstein and associates (ref. 261). In the context of reaction diffusion coupling, graphical analysis has some obvious advantages over the lengthy and sometimes ambiguous treatment of analytical kinetics, which lacks a precise definition of forces as well as a clear topological interpretation (refs. 181, 176) ; in particular, the graphical techniques introduced to solve complex linear chemical kinetic problems by King and Altman (ref 118) , and by T. Hill (ref 96-98) and extended by Kuo-Chen and Min (ref. 127) and others (refs. 182,211) have found a wide range of application in biological transport problems. A network approach has also been introduced by the author to study chemical reaction coupled to diffusion (refs. 182,184).

3.2 THE MEMBRANE INTEGRATION PROBLEM

One of the central goals of Membrane Science is to find port descriptions in which flows of matter, electric charge, etc. through a membrane or slab can be associated with differences in quantitites measured at the boundaries--e.g., solute concentrations, electrical potential difference ,etc.

These global descriptions involve integrating local partial differential equations --which express associations between molecular or ionic motion and local forces and conservation laws. The integration process is routinely carried out and its validity rarely questioned . There is little doubt that the procedure works relatively well in simple cases such as diffusion, although specific gradients inside the membrane may need to be assumed.

But a more serious problem arises as soon as one attempts to incorporate sources in the form of reactive layers or unidirectional biases such as convective flows or electrical forces.

In a typical situation, one writes separate sets of equations describing the individual processes considered and then couple these through simple physical constraints such as the conservation of mass or charge .

But, unfortunately, there is no obvious way of telling from the equations whether the imbedding of a chemical reactions in a complex network may not alter the continuity of the very potentials one is attempting to integrate.

This can be seen by looking at a basic property of a linear one port having a through flow j and (global) force f -- I use this nomenclature to avoid the immediate identification with Onsager forces and flows. If the linear law has the form $f = R j$, one can define the dissipation

$$\sigma = (1/2) R j^2, \qquad (1)$$

which has the property that its derivative with respect to j retrives the force f:

$$(d\sigma/dj) = Rj = f. \qquad (2)$$

and that, conversely, its derivative with respect to the force f retrieves the flow j,

$$d\sigma/df = d[(1/2) R j^2]/df = d[(1/2) f^2/R]/df$$
$$= f/R = j. \qquad (3)$$

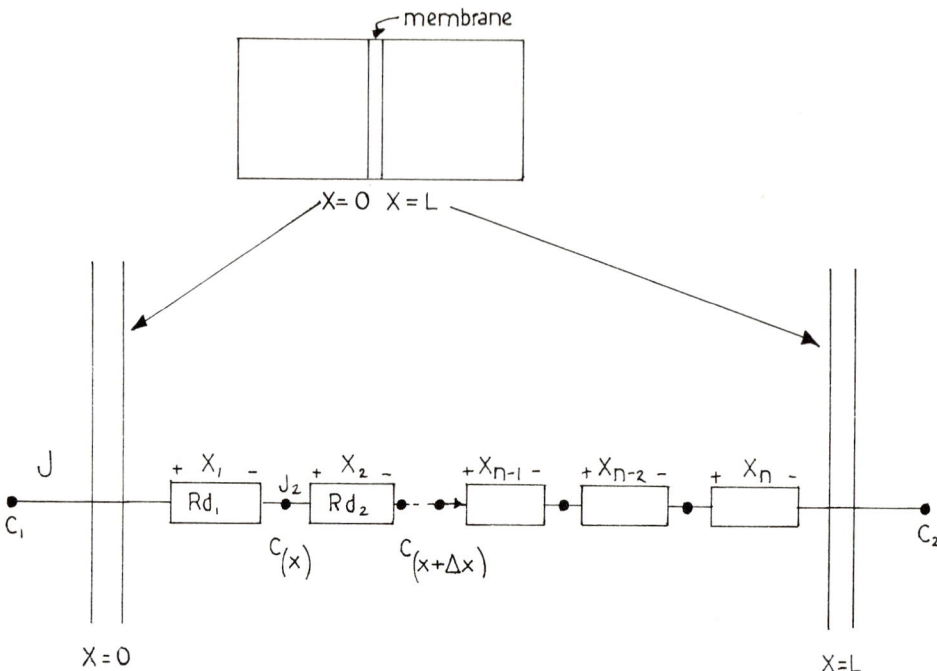

FIGURE 3.1 The diffusion of a single solute in a membrane may be represented by a string of resistances, as discussed in the text.

Thus, the dissipation function σ acts as a potential surface whose slope in the direction of j yields the force. This gives a consistent, or "proper" force-flow model, because σ becomes a scalar invariant, or the physical quantity which is invariant when frames of reference change.

This is certainly true of Fick's diffusion law, for example,
$$J_s = P \Delta c, \qquad (4)$$
with $J_s = j$, $\Delta c = f$ and $P = (1/R)$. It is also true of an electrical resistance with $i = j$, $E = f$ and $R = R$. In fact, Fick's diffusion and Ohm's law are isomorphic, so that -- if conservation of mass is invoked --, the diffusion of a single solute in a membrane may be represented by a string of resistances, as shown in FIGURE 1.

On the other hand, if one considers the simple chemical reaction
$$A \rightleftarrows B \qquad (5)$$

with linear first order rate given by

$$J_{AB} = k_{BA} [A] - k_{AB} [B], \qquad (6)$$

the situation is different. The reason is that if we assume the force is given by the difference in concentrations [A] and [B],

$$f = [A] - [B] \qquad (7)$$

the resultant "dissipation function"

$$\sigma = (1/2) \{k_{BA} [A] - k_{AB} [B]\} \{[A] - [B]\} \qquad (8)$$
$$= (1/2) \{k_{BA} [A]^2 + k_{AB} [B]^2 - [k_{BA} + k_{AB}] [B] [A]$$

does not yield the force when differentiated with respect to the rates or the rate when differentiated with respect to the global force. It turns out, however, that the dissipation

$$\sigma' = fj$$

does yield, the <u>unidirectional</u> rates

$$J_{AB} = \partial \sigma'/\partial [A] = k_{AB} [B]$$
and
$$J_{BA} = \partial \sigma'/\partial [B] = k_{BA} [A] \quad .$$

Given that Fick's diffusion and reaction are based on two completely different potentials, then, there is no reason to assume the two views can be merged directly in transport models, as it is often done.

The problem then becomes: **given a dynamical system consisting of dissimilar processes--e.g., chemical reaction, diffusion and compartmental storage--,match the internal coordinates so that the potentials seen by each subprocess is the same--or, alternatively, balance the forces exactly, if the system allows.** In addition, a consistent physical model requires that each internal force be superimposed with a proper displacement like,or contravariant variable.

In a potential theory based on thermodynamics, this situation does not arise because points in both diffusion and reaction space are <u>assigned</u> the same potentials. (In fact chemical reactions are forced to look like transport processes in the thermodynamics of "irreversible" processes.)

We cannot proceed in that direction , however, if it is required that the potential be defined from kinetics and not from thermodynamics except close to equilibrium, in which case the two should coincide). This is because Chemistry presents the curious situation of having two well developed non converging physicomathematical models: kinetics and thermodynamics. The first can predict rates without making reference to forces and the second has well specified forces but can make no mechanistic prediction of rates (except through phenomenology).

In order to give a proper covariant model --i.e., one in which each force is associated with a single conjugate rate and in which there is an energy like invariant scalar-- it is necessary to look at the similarities and differences between potential theories and kinetic theories .

From a structural point of view, the two differ in a fundamental way. The geometrical theatre for a potential theory is the energy surface or hypersurface from which forces are found in the direction of steepest descent (ref.140-2). Take two arbitrary points on this surface, say A and B having potentials ϕ_A and ϕ_B, respectively, which will be determined by external constraints and measure the resultant difference in height between those points in order to approximate the gradient --i.e., to obtain the force. Now reverse the role of the external physical constraints so that A becomes B and viceversa.

The measured force will have the same magnitude and opposite sign from the one measured in the previous experiment. This is not surprising, as it follows from the conservative nature of the paths on a potential surface--i.e., the curl **F** =0 statement, or K.V.L.

Consider now a rate theory. What is the primitive geometrical structure here? It is simply the arrows with the associated rate constants; for example,

$$\underset{k_2}{\overset{k_1}{\rightleftarrows}}$$

This can be viewed as a machine with two slots, in this example: one on the right and one of the left. If a certain concentration of A is placed on the left slot and a certain concentration of B is placed on the right slot we measure a rate. But if the situation is reversed, the rate is not simply the negative of the previously measured rate. Of course, one can also exchange the rates, but this simply shows that the geometrical structure of the problem is not uniform everywhere, unlike the case of a potential theory.

3.3 DIFFUSION AS A DEGENERATE EXAMPLE

To understand further the transition between the two models one can take a problem in which there are conservative forces and attempt to represent it by means of rates. This is a well known situation which has been studied by many workers, notably T. Hill for the representation of transport problems using rates (ref. 96-8).

Thus, consider as an example a diffusional problem in which a slab of permeability P_A separates two regions held at concentrations c_1 and c_2 of some solute which can diffuse freely through the slab. The familiar expression

$$J_S = P_A \{ [c_1] - [c_2] \} \qquad (9)$$

is clearly "reversible", in the sense that the roles of 1 and 2 can be exchanged. However, the permeability may be viewed as a rate multiplier. If the diffusion law is rewritten as

$$J_S = P_A [c_1] - P_A [c_2] \} \qquad (10)$$
$$= k_1 [c_1] - k_1 [c_2] \}$$

--i.e., if the permeability P_A is equated to both k_1 and k_2 --, it follows that **the potential theory problem has been transformed into a rate problem having the characteristic that the forward and reverse rates are equal.** This indicates that the chemical reaction given above is not a potential theory because its forward and backward rates are not the same.

As simple as these ideas are, they point into the direction of what is required in order to transform a rate problem into a potential theory problem: find a representation in which the forward and reverse rates are the same. This search is not simply a game; a potential theory problem (or an equivalent network problem) yields equations much more readily than a rate problem of the same level of complexity. Moreover, there are many physical chemistry problems with the same type of unidirectional charactersitics built in --e.g., convection -diffusion, ionic transport,etc .-- and these can benefit from the same treatment.

Although it is clear that what is missing from these descriptions is an appropriate potential which yields flows from gradients, the question remains as to what space these gradients define.

Network Theory provides a solution to this problem, because if it is possible to find some set of potentials which are consistent with the known rates and the known phenomenology, these potentials must automatically define proper forces and flows in a covariant formalism.

One would correctly suspect that there may be several possible manifolds, so the problem becomes finding at least one in which reaction and/or reaction diffusion spaces are woven so that there is continuity of both flows and potentials. An idea of how this may be achieved comes from thermodynamic work of Kramers. (ref. 60). In this technique one defines internal coordinates $x,y,z,..$etc. at which variables $v(x)$, $v(y)$, $v(z)$, etc and flows $J(x)$, $J(y)$, $J(z)$... can be defined. These flows account for the rate of change of internal quantities such as density and essentially establish a continuity equation.

3.4 FIRST ORDER REACTION DIFFUSION SYSTEM

Let the monomolecular reaction

$$A \underset{k_2}{\overset{k_1}{\rightleftarrows}} B \qquad (11)$$

have first order kinetics

$$d[A]/dt = k_1 [A] - k_2 [B] = - d[A]/dt \qquad (12)$$

in which k_1 and k_2 are the forward and reverse kinetic constants As pointed out above, there is no equivalent "permeability" or conductance which gives the appropriate flow if one insists that the potential difference (force) be equal to $[A] - [B]$.

One readily discovers, however, that a chemical resistance of value 1 placed across the potential difference
$k_1 [A] - k_2 [B]$
does the job as does a conductance --i.e., inverse resistance or permeability-- of value k_1 placed across a potential
$[A] - k_2 [B] / k_1$
or a conductance of value k_2 placed across a kinetic force
$k_1 [A] / k_2 - [B]$.
A chemical kinetic dissipation can now be constructed such that it yields proper forces and flows, in the sense of the derivatives

taken along the potential surface. Thus, for example, the dissipation

$$\sigma = (1/2) \{ k_1 [A] - k_2 [B] \} \{ k_1 [A] - k_2 [B] \} \qquad (13)$$
$$= (1/2) \{ k_1 [A] - k_2 [B] \}^2$$

yields the correct rate, when differentiated relative to the force $\{ k_1 [A] - k_2 [B] \}$.

3.5 CONSISTENCY BETWEEN KINETICS AND THERMODYNAMICS

If the forces given as proper kinetic forces are thermodynamically correct, these should be reducible to proper thermodynamic forces using topological transformations. This is somewhat more subtler than it appears on the surface: it is trivial to transform thermodynamic forces into the present kinetic forces by using analytical transformations, the question is: can the individual potentials be transformed back and forth between the chemical potential form and the kinetic form used here? The answer is that, in general, this is not possible--i.e., it is impossible to find a general kinetic potential consistent with thermodynamics. In the special case of near equilibrium, however, this is a feasible step.

Consider the kinetic conductance $1/R = 1$ and subtract $k_{forward}[A]^{equilibrium}$ from the A potential and $k_{back}[B]^{equil}$ from the B potential. (This is possible as the equilibrium condition requires

$$k_{forward}[A]^{equil} = k_{back}[B]^{equil} . \qquad (14)$$

The new potential remains unchanged if the A node is multiplied and divided by $[A]^{equil}$ and the B node is multiplied and divided by $[B]^{equil}$. Moreover, the net flow rate remains invariant if the potentials are divided by $k_{forward} [A]^{equil}$ and the conductance is multiplied by the same factor. This yields potentials

$dA/[A]_{equil}$

and

$dB/[B]_{equil}$,

in which the differential sign has been used to represent

(close to equilibrium) concentration increments, for convenience. If the potentials are now multiplied by the factor RT and the conductance is divided by the same factor and the near equilibrium approximations

$dA/[A]_{equil} = RT \ln [dA/[A]_{equil} + 1]$
and
$dB/[B]_{equil} = RT \ln [dB/[B]_{equil} + 1]$

the potentials reduce to the chemical potentials corresponding to small displacements from equilibrium (I am indebted to D.C. Mikulecky for this calculation) Again one should stress that this equality between the kinetic and the thermodynamic forces is only valid close to equilibrium.

3.6 GLOBAL REACTION DIFFUSION EQUATIONS: REACTION IMBEDDING IN DIFFUSION SPACE

Further insight on the interaction between transport and chemical reaction can be obtained by studying the possible ways of imbedding the chemical reaction in a diffusion space, or viceversa. A puzzling topological and algebraic restriction is the so called Curie principle of Thermodyanmics which forbids the local coupling between processes of different tensorial order--such as diffusion (a vectorial process) and chemical reaction (a scalar process) in isotropic systems. The question arises as to the conditions, if any, under which such coupling can occur. DeSimone and Caplan (refs. 61, 62) looked into the asymmetries that needed to be introduces in order to achieve net flow of matter when a chemical reaction took place in the center of a membrane. Using Onsager thermodynamics , they extended the thermodynamic interpretation of reaction- diffusion systems and stated a global Curie theorem: in symmetric, reactive membranes the net global chemical reaction -- defined as the sum of boundary flow going into the membrane--does not couple to osmotic forces. Concurrently, Bunow, who was performing a parallel line of work by pursuing the problem in terms of a kinetic analysis, found out that the global equivalent of Curie´s principle held in the kinetic analysis as well. However, the resultant global equations were not reciprocal, as one would have expected if the thermodynamic treatment transformed linearly into the kinetic form (ref. 35, 36).

The interesting part of this result is not as much the discrepancy between kinetics and thermodynamics as the observation that one would expect any linear, covariant theory--i.e.,having forces and conjugate displacements or, alternatively, some bilinear invariant defined -- to be reciprocal. Given that chemical kinetics, even in the first order case is not reciprocal, it cannot lead to a covariant theory of forces and flows directly. Moreover, we have indicated that the source of this reciprocity is, from a topological point of view, the mathematical continuity of the physical space being described. This continuity, we have seen,is directly related to the possibility of defining a potential at each point in the manifold ("Kirchhoff´s voltage law") and the absence of sources of divergence ("Kirchhoff´s current law"). Clearly, the second requirement is automatically obeyed by chemical systems through the principle of mass conservation, but is missing from these descriptions so that an appropriate potential which yields flows from gradients cannot be given directly.

By contrast, I suggested a rather brute force approach in which the appropriate continuity of potentials is <u>imposed</u> on the coordinates. This leads to results which are both self consistent and have physical meaning (ref. 182, 183, 184). In addition, it leads to interesting insights in the analysis of passive vs. active transport and it allows the extension to convection- diffusion and convection and reaction systems. The possibility of deriving some obtuse, but verified results as well as the possibility of extending the models to the non-linear convection-diffusion, convection reaction diffusion and oscillatory domains strongly suggest this is a theoretical approach worth considering.

3.7 FIRST ORDER REACTION DIFFUSION SYSTEM

The simple reaction diffusion system used by Caplan, DeSimone, Bunow and by the author consists of a membrane placed between two baths at which two different moieties, A and B, which can diffuse freely into and out of the membrane are held at different boundary concentrations -- $[A_o]$ and $[B_o]$ at the left hand side of the membrane (x= 0) and $[A_L]$ and $[B_L]$ at the right hand side of the membrane (x= L). Moreover, it is also assumed that a reaction

$$A \underset{k_2}{\overset{k_1}{\rightleftarrows}} B \qquad (15)$$

takes place in the center of the membrane following linear, first order kinetics, so that

$$d[A]/dt = k_1 [A] - k_2 [B] = -d[A]/dt \qquad (16)$$

in which k_1 and k_2 are the forward and reverse kinetic constants A and B are assumed to diffuse freely in and out of the membrane with diffusion constants or permeabilities defined as a function of position. If the membrane is not symmetric there will be, in the simplest case, four permeabilities P_A, P_B, Q_A and Q_B corresponding to A and B on each side of the membrane. This is indicated in FIGURE 2.

The concentrations of A and B in the left hand side, center and right are indicated by (A_o, A, A_L) and (B_o, B, B_L), respectively. With this notation, due to Bunow, the linear diffusion laws can be written as

$$J_A(0) = P_A ([A_o] - [A]) \qquad (17)$$
$$J_A(L) = Q_A ([A_L] - [A]) \qquad (18)$$
$$J_B(0) = P_B ([B_o] - [B]) \qquad (19)$$
and
$$J_B(L) = Q_B ([B_L] - [B]). \qquad (20)$$

in which J_o and J_L are the flows of A and B going _into_ the membrane at x=0 and x=L, respectively.

3.8 NETWORK REPRESENTATION

Section 3.3 considered the possible driving potentials in a first order chemical reaction. The assignment of diffusional potentials is simple when one notices that whatever the form of the potential is on one part of the particular diffusional path it should be the same for any point in the path. Thus if A sees a potential [A] in the center of the membrane, according to one of the assignments given above, it must see potentials $[A_o]$ and $[A_L]$ at the right and left of center, respectively. On the other hand, if the potential assigned at the center is $k_1 [A]$, it will of necessity be $k_1 [A_o]$ and $k_1 [A_L]$ at those points, for Fick's law to hold. All of this has little to do with networks as such, but only with the requirements imposed by conservation, the continuity of potentials from point to point and linear laws. Once

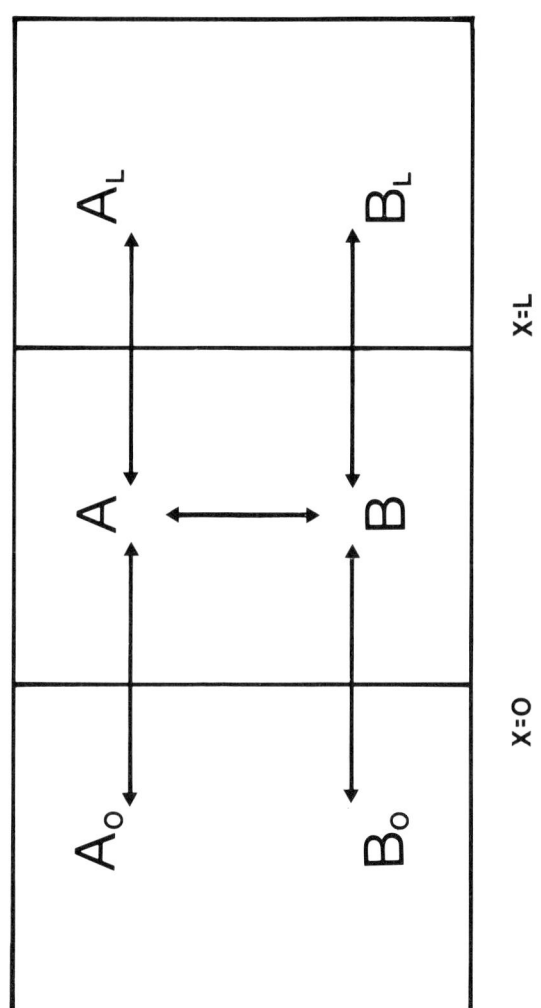

Fig. 3.2 The simple transport system described by Bunow. There are two diffusing substances (A and B) which couple through a first order chemical reaction at the center of the membrane.

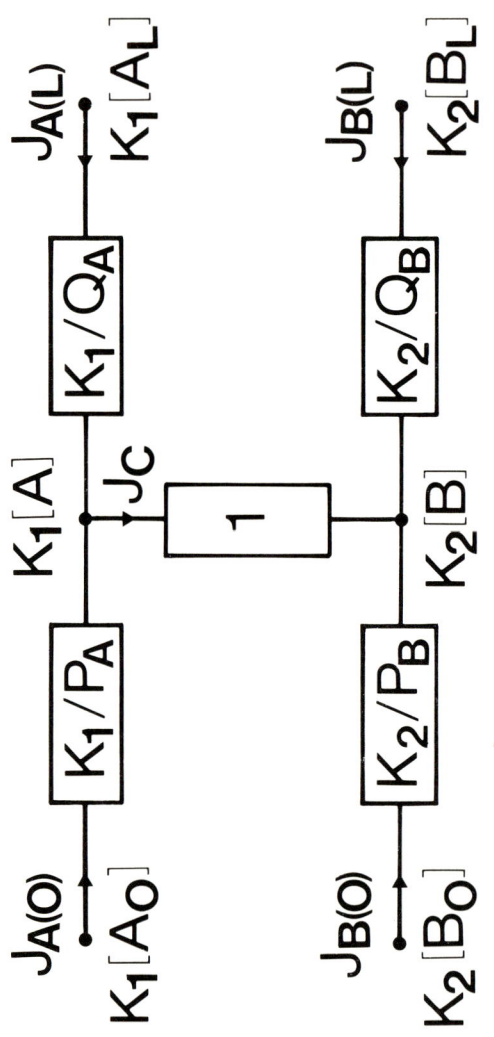

Fig. 3.3 Network representation of the transport system in Fig. 3.2 .

these are imposed it follows that the system can be represented by the resistive network as shown in FIGURE 3 , for example.

The symmetric potential arbitrarily chosen at the center of the membrane -from other consistent potentials- is $k_1[A] - k_2[B]$. This determines, then, that all potentials for the diffusional path A have the form $k_1[A(x)]$ while all potentials for the path B have the form $k_2[B(x)]$.

The connected network allows the use of Circuit Theory to write proper equations, which are reciprocal by definition; this excercise is done in the following section. However, there is another observation which is much more relevant: once the connected, resistive network is given, a metric can be obtained as a " power" invariant and this metric corresponds to the bilinear invariant which is conserved in the specific physical situation considered. This power dissipated corresponds to the Raleigh function which serves to define a velocity dependent potential in Lagrangian formulations. One could have arbitrarily assumed, as has been done by some, that the metric can be derived by considering only regular kinetics, as given from the flows, and completely ignore potentials. This is incorrect as it leads to a disconnected metric in which there is no dissipation like analog which serves as "glue" or potential in the manifold.

3.9 NETWORK EQUATIONS

Using Kirchhoff's voltage and current laws and the constitutive resistive equations it is now possible to set up the global equations for the system. In the case of the simple chemical reaction -diffusion system being considered there are four boundary flows and (given K.C.L.) only three independent equations. Several linear combinations of potentials and flows are possible, as shown in FIGURE 4.

The main requirements for a consistent terminal port description are:
(1) the chosen ports must constitute a linearly independent set of flows or forces,
(2) the force at each port be defined as a difference between a "high" potential (+) and a low potential (-).
and
(3) the conjugate flow which goes into the + terminal comes out of the - terminal unchanged in magnitude. This requires use of Kirchhoff's laws and consistency with the topology

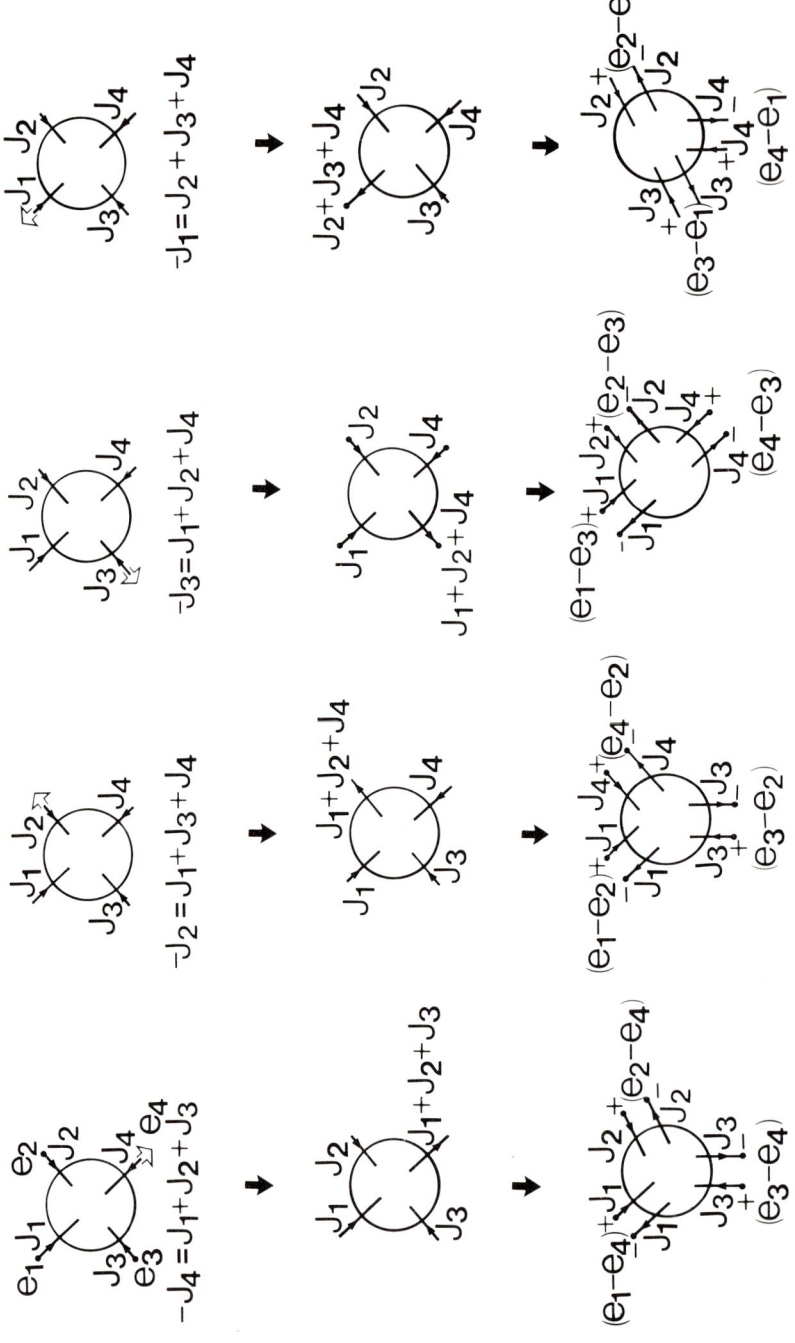

Fig. 3.4 Some possible choices of input-output ports in a network with four flows which obey K.C.L.

of the system.

We choose the following two sets of flows and forces:

$$\text{SET A}$$

	Force	Conjugate flow	
Port 1	$k_1 \Delta A$	$J_A(0)$	
Port 2	$k_2 \Delta B$	$J_B(0)$	(21)
Port 3	$a_L = k_1 A_L - k_2 B_L$	$J_C = J_A(0) + J_A(L)$	

and

$$\text{SET B}$$

	Force	Conjugate flow	
Port 1	$k_1 \Delta A$	$-J_A(L)$	
Port 2	$k_2 \Delta B$	$-J_B(L)$	(22)
Port 3	$a_O = k_1 A_O - k_2 B_O$	$J_C = -J_B(0) - J_B(L)$	

in which a_L is the boundary affinity at x=L, a_O is the boundary affinity at x=0 and J_C is the chemical flow which according to K.C.L. is either equal to

$$J_C = J_A(0) + J_A(L) \tag{23}$$

or equal to

$$J_C = -J_B(0) - J_B(L). \tag{24}$$

The networks corresponding to descriptions A and B are given in FIGURES 6 and 7, respectively.

3.10 SOLUTIONS TO THE NETWORK EQUATIONS

While sets A and B specify the chosen ports, they do not determine which are the independent and which are the dependent variables. For convenience, we choose a hybrid force-flow description in which the dependent variables are $J_A(0)$, $J_B(0)$ and a_L, while the independent variables are $k_1[A]$, $k_2[B]$ and J_C.

The resultant equations can now be obtained by superposition, setting two of the independent variables equal to zero and observing the effect of the remaining independent variable on the

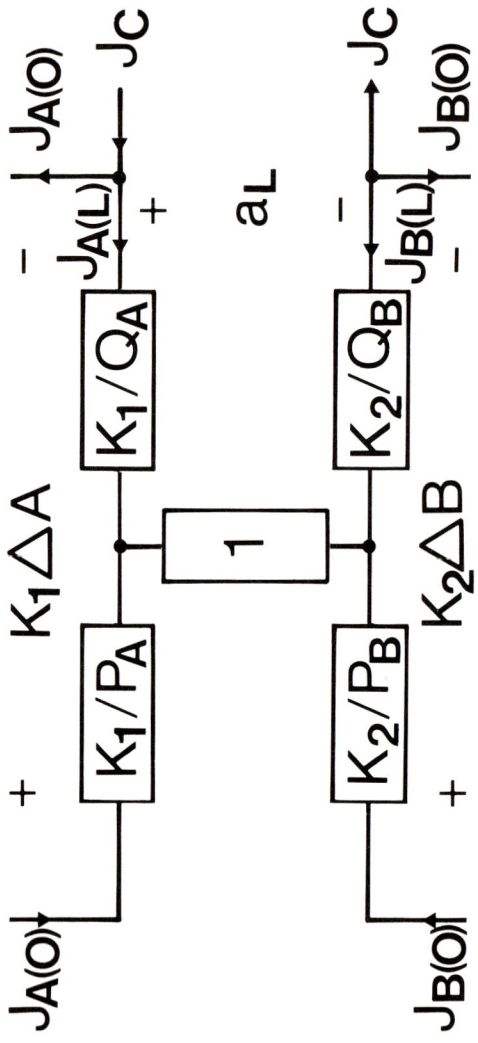

Fig. 3.5 Definition of ports for Set A.

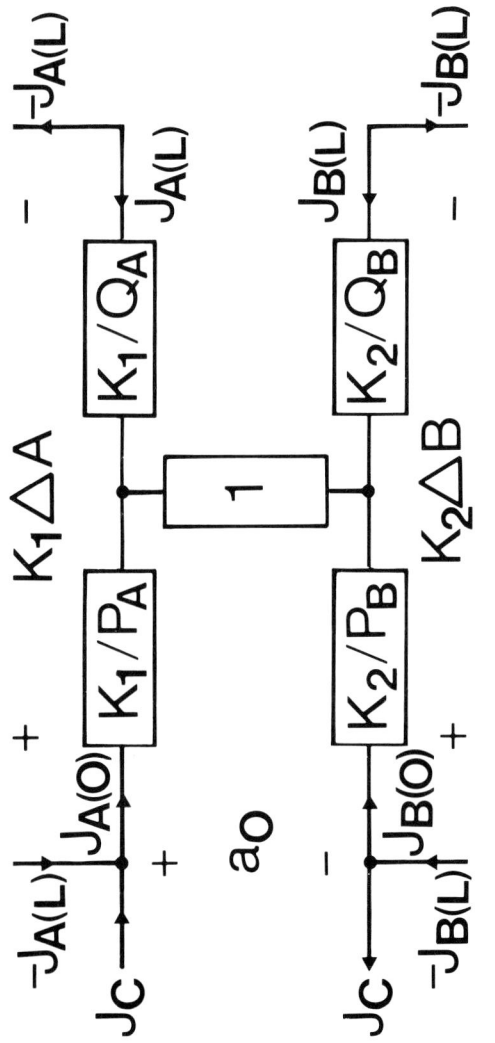

Fig. 3.6 Definition of ports for set B.

other two dependent quantities. For example, if only $k_1[A]$ is non zero, while $k_2[B]$ and J_C are set to zero, the flow $J_A(0)$ will be given by the force $k_1[A]$ placed across the sum of resistances k_1/P_A and k_1/P_A, flow $J_B(0)$ will be zero and the force or "voltage" a_L will be the "voltage" across resistance k_1/Q_A --i.e., it will be given by the "current" or flow $J_A(0)$ multiplied by that resistance (refer to FIGURE 8).

Similar network arguments yield all the entries of the following equation (25)

$$\begin{bmatrix} J_A(0) \\ J_B(0) \\ a_L \end{bmatrix} = \begin{bmatrix} (1/k_1)[(1/P_A)+(1/Q_A)] & 0 & \rho_A \\ 0 & (1/k_2)[(1/P_B)+(1/Q_B)] & -\rho_B \\ -\rho_A & \rho_B & (1+b) \end{bmatrix} \begin{bmatrix} k_1 \Delta A \\ k_2 \Delta B \\ J_C \end{bmatrix}$$

in which ρ_A and ρ_B have been defined as

$$\rho_A = P_A / (P_A + Q_A) \quad (26)$$

$$\rho_B = P_B / (P_B + Q_B) \quad (27)$$

while the coefficient b is defined by

$$b = [k_1/(P_A + Q_A)] + [k_2/(P_B + Q_B)] \quad (28)$$

is the Thiele modulus for this problem. Note that this hybrid equation is antisymmetric. In network terms, the Thiele modulus acquires a specific meaning. Referring to FIGURE 9, in which the osmotic forces $k_1[A]$ and $k_2[B]$ have been set up to zero, it can be seen that the resistance seen by the total chemical path under these conditions,

$$R_{chemical/zero\ osmotic\ forces} = a_L / J_C \quad (29)$$

is simply $1+b$. Both $J_A(0)$ and $J_B(0)$ under these conditions follow from simple current (flow) divider considerations or from Kirchhoff's laws directly.

By defining

$$\rho_A^* = Q_A / (P_A + Q_A) \quad (30)$$

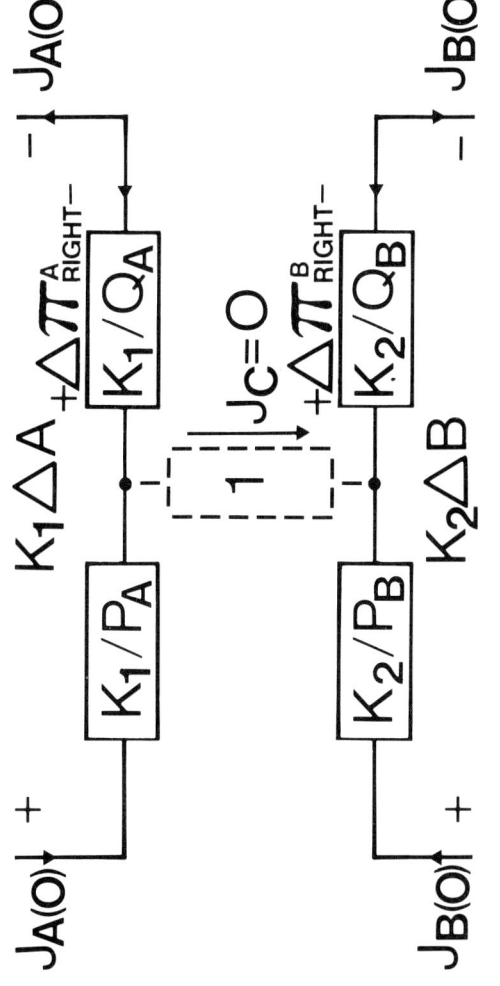

Fig. 3.7 Network giving the forces and flows when the chemical flow J_C is zero.

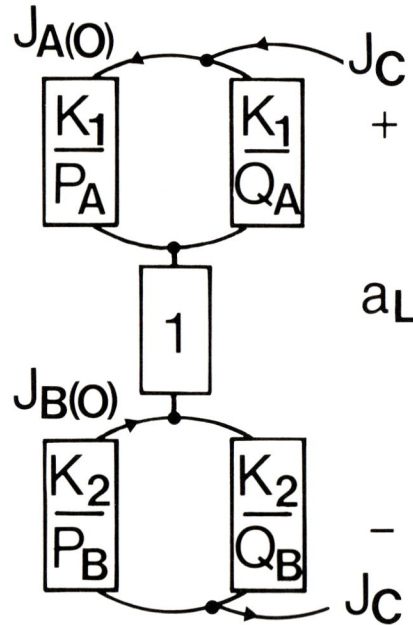

Fig. 3.8   Physical interpretation of the Thiele modulus as the resistance seen by the chemical path when the osmotic forces are zero.

$$\rho_A^* = Q_B / (P_B + Q_B) \qquad (31)$$

and using the ports given by set B above, we obtain the matrix equations (32)

$$\begin{bmatrix} -J_A(L) \\ -J_B(L) \\ a_o \end{bmatrix} = \begin{bmatrix} 1/k_1[(1/P_A)+(1/Q_A)] & 0 & -\rho_A^* \\ 0 & (1/k_1[(1/P_B)+(1/Q_B)]) & -\rho_B^* \\ \rho_A^* & -\rho_B^* & (1+b) \end{bmatrix} \begin{bmatrix} K_1 \Delta_A \\ K_2 \Delta_B \\ J_C \end{bmatrix}$$

which is analogous to the matrix equation given above for the A ports.

The entries for matrix B can be obtained directly by calculation or by noticing through symmetry arguments that the following exchanges can be made between then A and B descriptions

$$k_1 \begin{bmatrix} A \end{bmatrix} \longrightarrow -k_1 \begin{bmatrix} A \end{bmatrix}$$
$$k_2 \begin{bmatrix} B \end{bmatrix} \longrightarrow -k_2 \begin{bmatrix} B \end{bmatrix}$$
(33)

$$J_C \longrightarrow J_C$$
$$a_L \longrightarrow a_0$$
(34)

$$P_i \longrightarrow Q_i$$
$$Q_i \longrightarrow P_i$$
(35)

$$J_A(0) \longrightarrow J_A(L)$$
$$J_B(0) \longrightarrow J_B(L)$$
(36)

$$\wp_A \longrightarrow \wp_A^*$$
$$\wp_B \longrightarrow \wp_B^*$$
(37)

Either equations (25) or (32) are appropriate to describe this system. However, in order to obtain a more symmetric description, it is useful to follow Bunow in defining an average of both descriptions, as follows:

(38)

$$\begin{bmatrix} J_A^D \\ J_B^D \\ a \end{bmatrix} = \begin{bmatrix} 2 P_A Q_A / k_1 (P_A + Q_A) & 0 & \Delta \wp_A \\ 0 & 2 P_B Q_B / k_1 (P_B + Q_B) & \Delta \wp_B \\ -\Delta \wp_A & \Delta \wp_B & 2(1+b) \end{bmatrix} \begin{bmatrix} k_1 \Delta A \\ k_2 \Delta B \\ J_{C'} \end{bmatrix}$$

in which

J_A^D = net diffusional flow for species A = $J_A(0) - J_A(L)$

J_B^D = net diffusional flow for species B = $J_B(0) - J_B(L)$

a = total boundary afinity = $a_0 + a_L$

$\Delta \wp_A = \wp_A - \wp_A^*$ = measure of asymmetry for the membrane, as seen by A.

$\Delta \rho_B = \rho_B - \rho_B^* =$ measure of asymmetry for the membrane, as seen by B.

Finally, matrix equation (33) may be rearranged to yield the conductive form in which the independent variables are the forces $k_1 \Delta_A$, $k_2 \Delta_B$ and a:

(39)

$$\begin{bmatrix} J_A^D \\ J_B^D \\ J_c \end{bmatrix} = 1/2(1+b) \times$$

$$\times \begin{bmatrix} \Delta\rho_A^2 + [4(b+1)P_AQ_A/k_1(P_A+Q_A)] & -\Delta\rho_A\Delta\rho_B & \Delta\rho_A \\ -\Delta\rho_A\Delta\rho_B & \Delta\rho_B^2 + [4(b+1)P_BQ_B/k_2(P_B+Q_B)] & -\Delta\rho_B \\ \Delta\rho_A & -\Delta\rho_B & 1 \end{bmatrix} \begin{bmatrix} k_1\Delta_A \\ k_2\Delta_B \\ a \end{bmatrix}$$

3.11 DISCUSSION OF THE SOLUTIONS

The solutions presented above give a complete description of the system and, unlike those which are obtained without considering the continuity of potentials, they are reciprocal--i.e., represented by symmetric matrices. The issue of reciprocity, which had caused considerable argumentation when the analytical calculus approach was used reduces to simple considerations when network theory is introduced. From a physical point of view, the conclusion to be reached is simply that the continuity of the process must precede the definition of a metric. Thus Fick's potentials can no longer be measured in terms of concentrations but in terms of the proper modulated concentrations which will allow to define a continuous process from point to point in the membrane. This continuity is, of course, also implicit in the principle of mass conservation (K.C.L.). Once it is recognized that such a system must obey both K.C.L. and K.V.L. and linearity is introduced reciprocity follows from Tellegen's theorem .

In this view, then, the topology and dynamics of the system

determine the yardsticks to use, so that the dynamical process imposes restrictions on the allowable geometries. Geometry in turn modulates the dynamics.

The expression obtained for the chemical flow,

$$J_s = [1/2(1+b)][\Delta \rho_A k_1 \Delta A - \Delta \rho_B k_2 \Delta B + a]$$

can also be given as

$$J_s \Big|_{\substack{A=0 \\ B=0}} = \{(\Delta \pi^A_{right} - \Delta \pi^A_{left})_{J^{c=0}} + (\Delta \pi^B_{left} - \Delta \pi^B_{right})_{J^{c=0}} + a\}/2(1+b) \quad (40)$$

which expresses the total chemical flow as osmotic and "chemical" components acting on the net resistance (1+b) seen by the system when the osmotic forces are zero. Note that the chemical flow increases with increasing osmotic A gradient and it decrease with increasing osmotic B gradient, which agrees with the direction of flows in the network.

The results obtained give a precise indication of what characteristics of a "real" or artificial membrane should have depending on what its primary transport function is. For example, a membrane whose main purpose is to transform A into B would ideally have P_A and Q_B large and Q_A and P_B small, leading to a large $\Delta \rho_A$ and small $\Delta \rho_B$. In the special case of a symmetric membrane, in which

$$\rho_A = \rho_A^* \quad (41)$$

and

$$\rho_A = \rho_A^* \quad (42)$$

the phenomenological equations reduce to the canonical form

$$\begin{bmatrix} J_A^D \\ J_B^D \\ J_c \end{bmatrix} = \begin{bmatrix} P_A/k_1 & 0 & 0 \\ 0 & P_B/k_2 & 0 \\ 0 & 0 & 1/2(1+b) \end{bmatrix} \begin{bmatrix} k_1 & A \\ k_2 & B \\ & a \end{bmatrix} \quad (43)$$

The significance of this diagonal form has been considered in more detail elsewhere in terms of symmetric and antysymmetric components (ref. 182) . Note, however, that in the case of a symmetric membrane the main statement of these equations is that reaction and diffusion will not couple in this system, in agreement with the macroscopic statement of the Curie principle proposed by Caplan, Bunow and DeSimone.

The few differences with Bunow's equations should be mentioned here. The major difference in nomenclature is his use of average affinities and the use of the term $[\rho - (1/2)]$ instead of $\Delta\rho$. It is interesting to point out that although Bunow's problem does not represent a connected topology --i.e., they obey K.C.L. but not K.V.L.-- and although the forces and flows are not related at ports, he was able to identify the Thiele modulus for this system through very deep physical insight, as it does not appear in the analytical solutions as easily as in the network formalism (where it is an equivalent resistance).

3.12 FACILITATED TRANSPORT

An interesting, though straightforward application of the network formalism described above arises in the description of facilitated transport, a problem which had also been inspired by Bunow's analytical study. In the presence of facilitated transport a substance diffuses passively down an osmotic gradient and its flow is enhanced by a concurrent chemical reaction taking place in the membrane even though there may not be a net chemical reaction involving the diffusing substance. This is illustrated in FIGURE 10 (a) in which two reactions take place at the boundaries in which diffusing component A is converted to a component B which can diffuse freely inside the membrane but cannot escape.

The chemical and diffusional paths are shown in the drawing. The reason for the enhanced characteristic of the facilitated transport process becomes obvious in terms of the resistive network: a two positive resistances in parallel have a net resistance which is smaller than either of the component resistances. An elementary calculation involving addition of resistances in parallel or series yields an equivalent diffusional resistance (FIGURE 10 b)

$$R_{equiv} = R_A k_1 (R_{chemical(0)} + R_{chemical(L)} + R_B k_2) / (R_A k_1 + R_{chemical(0)} + R_{chemical(L)} + R_B k_2). \quad (44)$$

If $R_{chem(0)} = R_{chem(L)} = 1$, the above expression reduces

$$R_{equiv} = R_A k_1 (2 + R_B k_2)/(R_A k_1 + R_B k_2 + 2),$$

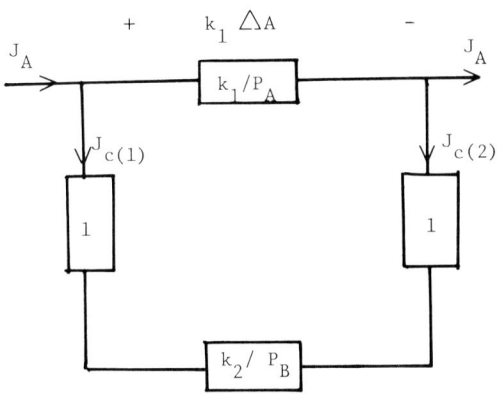

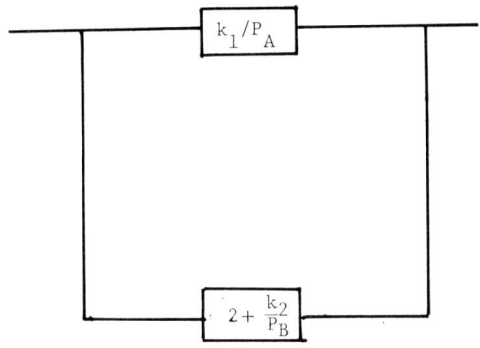

Fig. 3.9   Network interpretation of a facilitated transport example proposed by Bunow to show how the presence of a chemical reaction path can lead to enhanced flow without an additional --e.g., metabolic -- supply of energy. In the present example the flow is larger than the flow that would result from having simple diffusion. In network terms, the reason is that two resistances in parallel yield a smaller resistance than either of the constituent resistances.

which is less than $R_A \kappa_1$, if the resistances are positive--i.e., if positive forces give rise to positive flows, as required by thermodynamics. Thus, the presence of a chemical reaction can lead to an enhanced flow, as predicted by Bunow, even though there is no net use of substance A.

Note that no external source of energy is required beyond the osmotic gradient of A, so that transport of A is "passive".

3.13 NEED FOR CONNECTIVITY IN PASSIVE PROCESSES

If the original diffusion reaction example treated here is "passive" --i.e., if all molecular motions inside the membrane can be accounted for in terms of boundary concentrations-- it should be possible to define a chemical dissipation at the boundaries which matches the amounts of the same quantity dissipated inside. Tellegen's theorem provides a unique result for a given topology : the sums of products of forces and flows at the boundary will equal the internal analog, taken over all branches, provided that the network obeys K.V.L. and K.C.L.

The network given has these properties, so that the bilinear dissipation function will be conserved. We can show this by direct calculation. Consider the Set A equations. The bilinear form calculated at the ports is

$$\Phi = k_1 J_A(0) \, dA + k_2 J_B(0) \, dB + J_C(0) \, a_L \qquad (45)$$

in which I have denoted the difference in A or B across the membrane by the differential symbol, for convenience. Denote the difference in concentration between the left hand side and the middle either by dA_{left} or dB_{left} and the differences to the right of center by dA_{right} and dB_{right}. With this notation we can rewrite the port summation by

$$\begin{aligned}
& J_A(0) \, k_1 \, dA_{left} + J_A(0) \, k_1 \, dA_{right} \\
& \qquad + \\
& \qquad J_B(0) k_2 \, dB_{left} + J_B(0) k_2 \, dB_{right} \qquad (46) \\
& \qquad + \\
& - J_C k_1 \, dA_{right} + J_C \{k_1[A] - k_2[B]\} + J_C k_2 \, dB_{right}
\end{aligned}$$

Grouping terms together, the same expression can be reorganized to give

$J_A(0) \, k_1 \, dA_{left}$ (dissipation in left A resistance)

+

$J_B(0) k_2 \, dB_{left}$ (dissipation in left B resistance)

+ (47)

$\{J_A(0) - J_c\} \, k_1 \, dA_{right}$ (dissipation in right A resistance)

+

$\{J_B(0) + J_c\} \, k_2 \, dB_{right}$ (dissipation in right B resistance)

+

$J_c(k_1[A] - k_2[B])$ (dissipation in chemical resistance).

Thus, the bilinear form defined for the networks can be split into sums of individual dissipations conceivably corresponding to real physical processes if the model has been adequately set up.

I leave to the reader the excercise of attempting the same match on forces based on concentrations and concentration differences. By trial and error one is inexorably forced to the conclusion that the modulated concentration models are the only ones which can conserve--i.e. leave invariant or account for-- a bilinear form **if the chemical reaction is treated as a one port**.

If, on the contrary, one leaves the reaction space disconnected, as a two force-two unidirectional rates process, there are many models which will do the job. Their practical limitation is : since they do not conserve a dissipation like quantity, they cannot give information as to whether there are internal sources of energy --e.g., in "active" transport-- or not.

3.14 NON-LINEAR CONVECTION- DIFFUSION EQUATIONS

The same idea of continuity used above to derive a potential in the case of reaction diffusion systems can be extended to apply to convection and then convection-reaction-diffusion systems. The approach leads to discrete non linear equations which reduce to the known limits for small convective flows. In addition, the obvious advantage that will surface through the use of K.V.L. and K.C.L. is that the topology of several types of problems considered is the same (determined by the structural and functional aspects of the problem); this fact allows to write equations practically by inspection once the Kirchhovian laws are identified.

Manegold and Solf (ref. 129,130) derived an equation that gives the coupling between solute and volume flows in a slab or membrane. By starting with the <u>apparently</u> linear local equation

$$J_S = c_S(x)(1 - \sigma)J_V + P(x)(\partial c_S / \partial x) \qquad (48)$$

in which $P(x)$ is the local permeability and σ is the local reflection coefficient, a nonlinear global solution of the form

$$J_S = \left[\frac{c_0}{[1 - \exp(-aJ_V)]} + \frac{c_L}{[1 - \exp(aJ_V)]} \right] h J_V \qquad (49)$$

is obtained. This expression gives the steady state flow of solute, J_S, for a membrane of thickness L kept between concentrations c_0 and c_L. The parameters h and a are given by

$$h = (1 - \sigma) \qquad (50)$$

and

$$a = h L / P(x), \qquad (51)$$

respectively. This equation has been used many times, including in Teorell's analysis of his own membrane oscillator (refs. 232-241). Mikulecky (ref. 144) points out that this equation is periodically rediscovered in many different contexts and reviews close to a dozen references where the equation has appeared. Moreover, he shows that the point $J_V=0$ is a bifurcation point which is smeared by the Kedem-Katchalsky linearization (ref.111-113), thus leading to discrepancies in terms of reciprocity asssumptions.

The present goal is to derive proper equations which apply to this problem both in the absence and the prescence of a chemical reaction using the previously introduced methods of network analysis. We first note that the Equation (48) considered as an operator which acts on J_V is non-linear, because it is of the form

$$J_S = k_1 J_V + k_2, \qquad (52)$$

so that doubling the volume flow does not double J_S, even when the concentration remains constant.

3.15 DISCRETE REPRESENTATION OF THE CONVECTION-DIFFUSION PROBLEM

We begin the analysis by subdividing the slab or membrane into N segments so that the local equations can be represented by the difference equations

$$J_S = c_i h J_V + [c(i) - c(i_+)] P_A , \qquad (53)$$

in which $h = (1 - \sigma)$, P_A is the local permeability of the subregion, and we have assumed that the flow originates at i and moves towards i+ (the plus x direction). Equation (53) cannot be represented directly by means of a one port resistor because it is not a linear operator --the reason is that the term J_V introduces a bias level which is modulated by c_i but not related to c_{I+}. However, if the above equation is rearranged in the form

$$J_S = P_A [1 + h J_V/P_A] [c(i) - c(i_+)]/[1 + h J_V/ P_A], \qquad (54)$$

it can be modelled by means of a linear one port resistance of value

$$R_d = (1/P_A) / [1 + h J_V/ P_A] \qquad (55)$$

placed between two potentials set at c_i and $c_{i+}/(1 + hJ_V/P_A)$, respectively, as indicated in FIGURE 13. Note that the values of the resistance and the potentials are functions both of position and J_V and that, moreover, the potentials are not simple

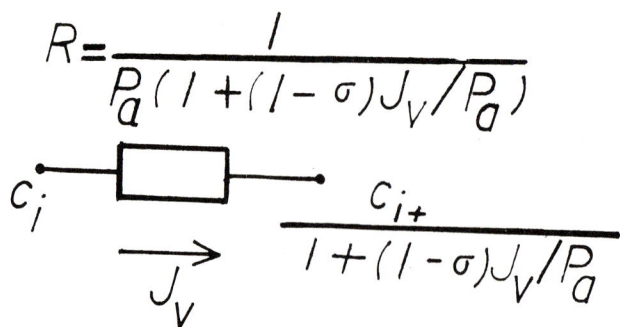

Fig. 3. 10 Representation of the convection diffusion problem as a difference in potentials across a resistance with a through convection-diffusion flow.

concentrations, but modulated quantities.

The above process may be continued on the whole slab by taking small resistive segments and matching the potential at the end of one resistance to the next potential, to obtain a continuous function of position. In the steady state in which the flow is assumed to have value J_s a string of resistances given by the series

$$R_d^i = [(P_A)(1 + h J_v)]^{-i} \qquad (56)$$

is obtained, while the concentration profile fits a series described by

$$c_o , [c_i/(1 + h J_v/P_A)^i] , [c_n/(1 + h J_v/P_A)^n] \qquad (57)$$

This is indicated in FIGURE 14. The total convection-diffusion resistance may be found by adding individual series components. If we define the term

$$(1/F) = 1 / (1 + h J_v/P_A)^n] \qquad (58)$$

the total resistance becomes

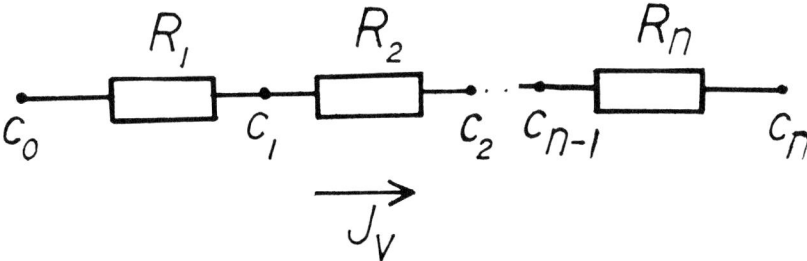

Fig. 3.11 Resistance model of the convection-diffusion system which yields a continuous potential along the x direction--i.e., the potential is the same at any region x whether it is seen from the point x+ dx or from x-dx.

$$R_d{}^{total} = (1/P_A) [F^{-1} + F^{-2} + \ldots + F^{-n}) \qquad (59)$$

a geometric progression which reduces to

$$R_d{}^{total} = (1/hJ_v) [1 - 1/(1 + h J_v/P_A)^n] . \qquad (60)$$

The potential difference --i.e., the force -- between the end points is given, according to Eq (54) by

$$X = [c_o - c_n / (1 + h J_v/ P_A)^n] \qquad (61)$$

so that the resultant steady convection-diffusion flow is

$$J_s = X / R_d{}^{total} = (c_o F^n - c_n) h J_v / (F^n - 1) \qquad (62)$$

in which $F = (1 + h J_v/ P_A)]$. In the case in which the convective flow goes from right to left, the potentials must be redefined, starting from the right hand side. The overall force X across the slab is then given by

$$X' = [c_o/ (1 + h J_v/ P_A)^n] - c_n \qquad (63)$$
$$= (c_o / F^n) - c_n ,$$

while the total resistance has the value previously found for the convection diffusion flow in the opposite direction. It then follows that the convection diffusion flow from right to left is

$$J_s = X^i / R_d = (c_o - F^n c_n) / (1/ hJ_v) (F^n - 1). \tag{64}$$

Note that Eqs (62) and (64) represent two distinct solutions, thus showing the bifurcation at $J_v = 0$. This bifurcation disappears in the continuous limit of small variables considered by Manegold and Solf, which is considered below.

3.16 LIMIT OF DOMINANT DIFFUSION RANGE: MANEGOLD AND SOLF EQUATIONS

Introducing the approximations

$$(1 + x)^n = 1 + Nx + N(N-1)(x^2/2!) + N(N-1)(N-2)(x^3/3!) + \ldots \tag{65}$$

for $(X \ll 1)$ and

$$\exp(Nx) \sim 1 + Nx + N^2 x^2 / 2! + N^3 x^3 / 3! + \ldots \tag{66}$$

it follows that, for large N and small x

$$\exp(Nx) \sim (1 + x)^n \tag{67}$$

so that in the approximation in which $J_v/P_A \ll 1$, when diffusion is dominant,

$$(1 + h J_v / P_A)^n] \sim \exp(h J_v N/P_A). \tag{68}$$

Moreover, by introducing the normalization condition $P(x)/P_A = L/N$, Eq (60) simplifies to

$$R_d^{total} = (1 - \exp -aJ_v) / h J_v, \tag{69}$$

while the force X is given by

$$X = c_o - c_L \exp(-aJ_v), \tag{70}$$

in which c_L is the concentration at x=L (the right hand side of the slab). it then follows that the steady flow is given by

$$J_s = (c_o - c_L \exp(-aJ_v)) h J_v / (1 - \exp(-aJ_v))$$

and further rearrangement leads to the expression

$$J_s = \left[\frac{c_o}{[1 - \exp(-aJ_v)]} + \frac{c_L}{[1 - \exp(aJ_v)]} \right] h J_v . \qquad (71)$$

When the volume flow is in the negative direction the potential difference becomes

$$[c_o / (1 + h J_v / P_A)^n] - c_n = c_o \exp(aJ_v) - c_L \qquad (72)$$

while the resisistance is still given by Eq (69), with negative J_v:

$$R_d^{total} = (1 - \exp(aJ_v) / (-hJ_v) . \qquad (73)$$

The convection-diffusion flow for negative J_v is

$$J_s = X/R_d = \left[\frac{c_o \exp(aJ_v) - c_L}{1 - \exp(aJ_v)} \right] (-hJ_v) \qquad (74)$$

which leads, after further rearrangement to

$$J_s = \left[\frac{c_o}{[1 - \exp(-aJ_v)]} + \frac{c_L}{[1 - \exp(aJ_v)]} \right] h J_v . \qquad (75)$$

Although equations (71) and (75) give the same expression for positive and negative volume flow, in this approximation, it should be clear that they correspond, in the general case, to two distinct sets of equations corresponding to the bifurcations $J_v > 0$ and $J_v < 0$.

3.17 CONCENTRATION PROFILES

The network given can also be utilized to find the concentration profile in the M.S. approximation. We first note that the node potentials are not concentrations, but are given instead by

$$o_i = c_i / (1 + hJ_v/P_A)^i) = c_i \exp(-a\, J_v\, x). \qquad (76)$$

To find the potential at any point the "voltage divider" relation can be used to yield

$$\frac{\phi_o - \phi_i}{\phi_o - \phi_n} = R_i / R_{total} = \frac{[1 - \exp(-aJ_v x)]}{[1 - \exp(-aJ_v)]}. \qquad (77)$$

This leads to

$$[c_o - c(x) \exp(-aJ_v x)] = [c_o - c(x) \exp(-aJ_v)] \frac{1 - \exp(-aJ_v x)]}{1 - \exp(-aJ_v)]} \qquad (78)$$

Rearranging and solving for $c(x)$, this expression can be written as

$$c_x = c_o \exp(aJ_v x)] - \frac{(\exp(-aJ_v x) - 1]}{(\exp(aJ_v) - 1]}(c_o \exp(aJ_v) - c_L) \qquad (79)$$

Further rearrangement leads to

$$c_x = c_o - (c_o - c_L)\frac{1 - \exp(aJ_v x)}{1 - \exp(aJ_v)} \qquad (80)$$

in agreement with previous results.

3.18 THE CONVECTION REACTION DIFFUSION PROBLEM

The presence of a chemical reaction, in addition to convection and diffusion adds a new level of complexity when a covariant description is desired.

Consider, again, the simple system in which a reaction $A \rightleftharpoons B$ takes place at the center of a membrane which, for the sake of generality, will be assumed to be asymmetric. Moreover, let the permeabilities of the membrane be those given before the application of the biasing flow J_V and let the membrane have reflection coefficients σ_A, σ_A', σ_B and σ_B', in which the primed coefficients correspond to the right hand side of the membrane and the unprimed coefficients correspond to the left hand side. The basic global form of the equations, for small volume flows is, in the light of the above discussion,

$$J_A(0) = A_O J_V (1 - \sigma_A) + A_O P_A - A P_A, \quad (81)$$

for the flow of A from the left hand side to the center and

$$-J_A(L) = A J_V (1 - \sigma_A') - A_L Q_A + A Q_A \quad (82)$$

for the flow of A from the center to the right hand side. These expressions may be rearranged to yield

$$J_A(0) = A_O [J_V (1 - \sigma_A) + P_A] - A P_A \quad (83)$$

and

$$-J_A(L) = A [J_V (1 - \sigma_A')] + Q_A - A_L Q_A. \quad (84)$$

Moreover, denoting the bracketed quantities by single coefficients K_A and K_A', we can further simplify these to

$$J_A(0) = A_O K_A - P_A A \quad (85)$$

and

$$-J_A(L) = A K_A' - Q_A A_L. \quad (86)$$

Clearly, when $J_V = 0$, K_A reduces to P_A and K_A' reduces to Q_A. Moreover, if the reflection coefficients for A on both sides of the membrane are the same it follows that $K_A = K_A'$, but this need not be true in general.

When the chemical reaction is introduced, it imposes the requirement that the potential for A becomes k_1 [A] rather than [A], as discussed above. The convection diffusion equations can be rewritten so that the potentials are compatible both from the convection -diffusion and the reaction points of view to yield

$$J_A(0) = (P_A/K_1) [(K_1 \ A_0 \ K_A/P_A) - A \ K_1] \quad (87)$$

and

$$J_A(L) = (K_A'/K_1) [(K_1 \ A_L \ Q_A/K_A') - A \ K_1]. \quad (88)$$

Note the asymmetry between the two equations, reflecting the bias introduced by the convective flow.

Equations (89) and (90) may now be interpreted to represent a connected network in which two resistances of value P_A/K_1 and K_A'/K_1 are joined at the center of the membrane and have a common node held at potential $k_1 [A]$. The potential adjacent to the left compartment is $k_1 \ A_0 \ K_A/P_A$ and the potential on the right is $k_1 \ A_L \ Q_A/K_A$. Clearly, the system is asymmetric, as expected. Following the above line of reasoning, we can write equations for the flows $J_{B(0)}$ and $J_{B(L)}$ in the form

$$J_B(0) = (P_B/K_2) [(K_2 \ B_0 \ K_B/P_B) - B \ K_2] \quad (89)$$

and

$$J_B(L) = (K_B'/K_2) [(K_2 \ B_L \ Q_B/K_B') - B \ K_2], \quad (90)$$

in which

$$K_B = J_V (1 - o_B) + P_B \quad (91)$$

and

$$K_B' = J_V (1 - o_B') + Q_B. \quad (92)$$

Again, the two expressions for the B path--constructed to agree with the potential for the B potential at the center of the membrane are asymmetric.

The complete reaction-diffusion-convection network is that shown in FIGURE 12, in which the potentials for the chemical and diffusional paths and the relevant resistances are indicated. It should be pointed out that the topology of the network is the same as in the original diffusion-reaction network, but the resistances and potentials are different. This is a practical

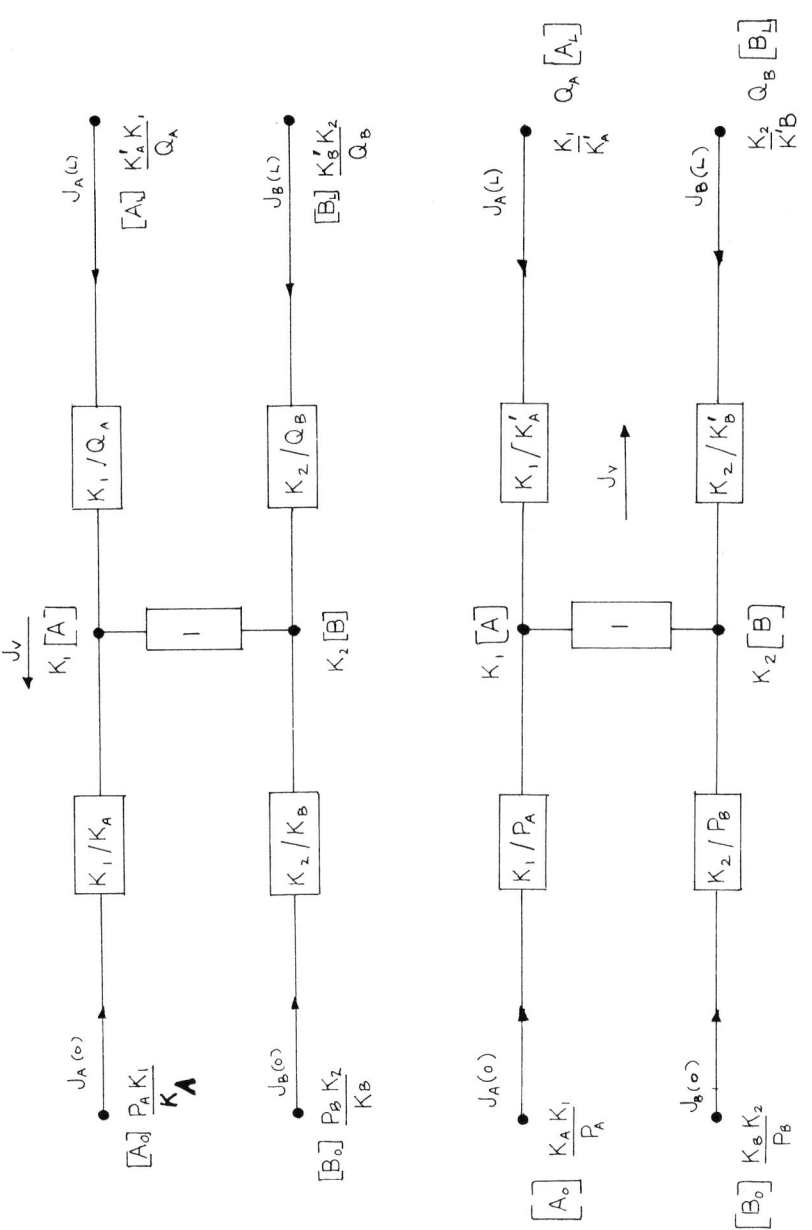

Fig. 3.12 Connected resistive networks used to analyze diffusion-convection flows coupled to a reactive layer.

TABLE I

Correspondence between variables in the diffusion reaction and convection reaction problems

Original Network	$\vec{J}_V$ network	$\vec{J}_V$ network
P_A	P_A	$P_A + J_V(1 - \sigma_A)$
Q_A	$Q_A + J_V(1 - \rho_A')$	Q_A
P_B	P_B	$P_B + J_V(1 - \sigma_B)$
Q_B	$Q_B + J_V(1 - \sigma_B')$	Q_B
A_o	$(A_o/P_A)(P_A + J_V(1-\sigma_A))$	$A_o P_A / (P_A + J_V(1-\sigma_A))$
B_o	$B_o((P_B + J_V(1-\sigma_B))/P_B)$	$P_B B_o / (P_B + J_V(1-\sigma_B))$
A_L	$Q_A A_L / ((Q_A + J_V(1 - \sigma_A'))$	$A_L(Q_A + J_V(1 - \sigma_A'))/Q_A$
B_L	$Q_B B_L / ((Q_B + J_V(1 - \sigma_B'))$	$B_L(Q_B + J_V(1-\sigma_B'))/Q_B$
$\beta = \dfrac{k_1}{P_A + Q_A} + \dfrac{k_2}{P_B + Q_B}$	$\dfrac{k_1}{P_A + Q_A + J_V(1-\sigma_A')} + \dfrac{k_2}{P_B + Q_B + J_V(1-\sigma_B')}$	$\dfrac{k_1}{P_A + Q_A + J_V(1-\sigma_A')} + \dfrac{k_2}{P_B + Q_B + J_V(1-\sigma_B')}$

advantage, because the resultant global equations can be obtained by inspection, simply by identifying corresponding variables in both drawings. In the same way, the network for convective flow from right to left can be represented by the same topology and different resistances. TABLE 1 gives the corresponding variables in the reaction diffusion and the reaction-diffusion-convection with right to left and left to right biases.

The relevant variables and physical interpretations are:

$A' = (K^A \ A^O \ K^1 / P^A) - (K^1 \ A^L \ Q^A / K^{A'}) =$ difference in concentration of A species modulated by J^V (left to right).

$B' = (K^B \ B^O \ K^2 / P^B) - (K^2 \ B^L \ Q^B / K^{B'}) =$ difference in concentration of B species modulated by J^V (left to right).

$a^{O'} = (K^A \ A^O \ K^1 / P^A) - (K^2 \ B^O \ K^B / P^B) =$ chemical affinity at x=0, in the presence of convection from left to right.

$a^{L'} = (K^1 \ A^L \ Q^A / K^{A'}) - (K^2 \ B^L \ Q^B / K^{B'}) =$ chemical affinity at x=L, in the presence of convection from left to right.

$A'' = (P^A \ A^O \ K^1 / K^A) - (K^1 \ A^L \ K^{A'} / Q^A) =$ difference in concentration of A species modulated by J^V (right to left).

$B'' = (P^B \ B^O \ K^2 / K^B) - (K^2 \ B^L \ K^{B'} / Q^B) =$ difference in concentration of B species modulated by J^V (right to left).

$a^{O''} = (P^A \ A^O \ K^1 / K^A) - (K^2 \ B^O \ P^B / K^B) =$ chemical affinity at x=0, in the presence of convection from right to left.

$a^{L''} = (K^{A'} \ A^L \ K^1 / Q^A) - (K^2 \ B^L \ K^{B'} / Q^B) =$ chemical affinity at x=L, in the presence of convection from right to left.

3.19 DISCUSSION OF SOLUTIONS

There are a few interesting points about the equations which are obtained by replacing the appropriate variables in Equations (38). First, the equations will be <u>reciprocal</u> even though the phenomenological coefficients are now functions of the forces (this shows the difference between a linear operator and a non linear equation, discussed in Chapter 1). Second, the convective flow turns a symmetric membrane into an asymmetric membrane, for functional purposes, as has been pointed out by Bunow . Third, when the convective flow is zero, the equations reduce to those of the reaction diffusion problem, Eq. (38), as expected.

Appendix

A.3.1 NON-ADDITIVE TERMS

The above ideas also have relevance in practical considerations regarding the attachment of two or more membranes in series or parallel.

When two or more Onsager converters attach one would expect them to do so simply following the rules of matrix addition or multiplication, depending on the variables chosen. For example, if two membranes are placed in parallel (the same boundary pressures and concentrations and the total flow equal to the sum of the individual flows through the membranes), it would seem that the individual hydraulic conductivities would add. Instead, it is found that the addition follows the rule

$$(1/L_{p(total)}) = (1/L_{p(1)}) + (1/L_{p(2)}) + \text{"non additive term"}. \tag{93}$$

The appearance of the "non additive term" is rather puzzling if one looks at the problem within the framework of linear operations and nothing else. On the other hand, it starts to make some sense when one requires that the potentials seen by both membranes be the same. We shall analyze the case of two membranes *in series*.

To fix ideas, consider only the J_v path to see how the matching of the potential works. Given two membranes in series, we shall represent the first membrane by unprimed quantities and the second membrane by primed quantities. The volume flow equations --in linear terms-- are given by

$$J_v = L_p (\Delta P - \sigma \Delta \pi) \tag{94}$$

and

$$J_v' = L_p' (\Delta P' - \sigma' \Delta \pi') \tag{95}$$

for the first and second membranes, respectively. We denote the relevant regions considered as a, b, and c, as shown in FIGURE . At each of these regions we have measurable hydrostatic and osmotic pressures ΔP and $\Delta \pi$. Equations (94) and (95) can be rewritten as

$$J_v = L_p [(P_a - P_b) - \sigma(\pi_a - \pi_b)] \tag{96}$$

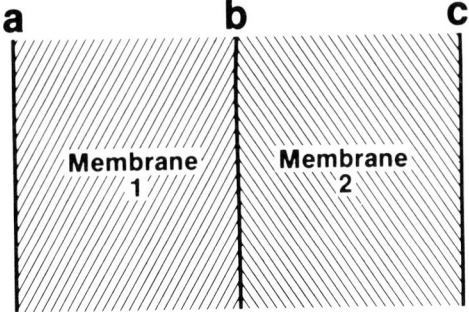

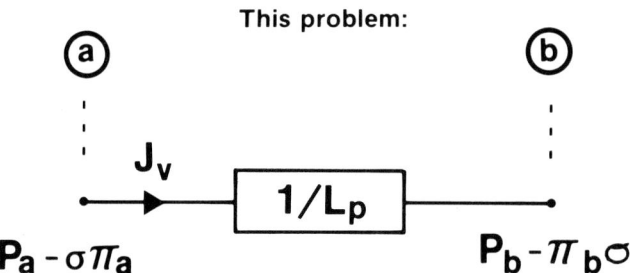

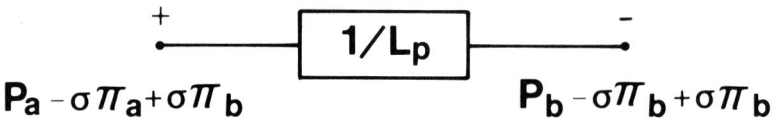

Fig. A.3.1 The upper drawing shows two membranes in series. The lower drawing indicates that the physical problem remains invariant if one adds (subtracts) the same pressure--i.e., potential for bulk flow motion-- to both boundaries of a given membrane.

and

$$J_v' = L_p' [(P_b - P_c) - \sigma' (\pi_b - \pi_c)], \quad (97)$$

respectively. When the two membranes are placed in series, it is clear that $J_v = J_v'$. If one simply assumes that the force differences add up for the series composite membrane, the overall membrane would have a volume flow given by

$$J_v = L_p \text{(total)} [(\Delta P_{total} - \sigma \pi_{ab} - \sigma' \pi_{bc})] \quad (98)$$

in which

$$\Delta P_{total} = P_a - P_c \quad (99)$$

and

$$(1/L_p\text{(total)}) = (1/L_p) + (1/L_p') . \quad (100)$$

In taking this step, however, we have assumed that there are definable potentials at a, b, and c which have the form

$$P_i - \sigma \pi_i . \quad (101)$$

If this were actually the case there would be four such potentials: $(P_a - \sigma \pi_a)$ and $(P_b - \sigma \pi_b)$ --for the first membrane -- and $(P_b - \sigma' \pi_b)$ and $(P_c - \sigma' \pi_b)$ for the second membrane. But the requirement of the operation performed is that the potentials match at the b interface, so that

$$(P_b - \sigma \pi_b) = (P_b' - \sigma' \pi_b'). \quad (102)$$

However, the osmotic and hydrostatic pressures themselves are only a function of the region b, not a function of the membrane; therefore, this equation would state that the reflection coeffcients of two arbitrary membranes are the same ! This is clearly an unreasonable result.

The only possible solution to the riddle consists, as in the case of the diffusion/ reaction problem considered before, in looking for potentials which meet the continuity requirements at

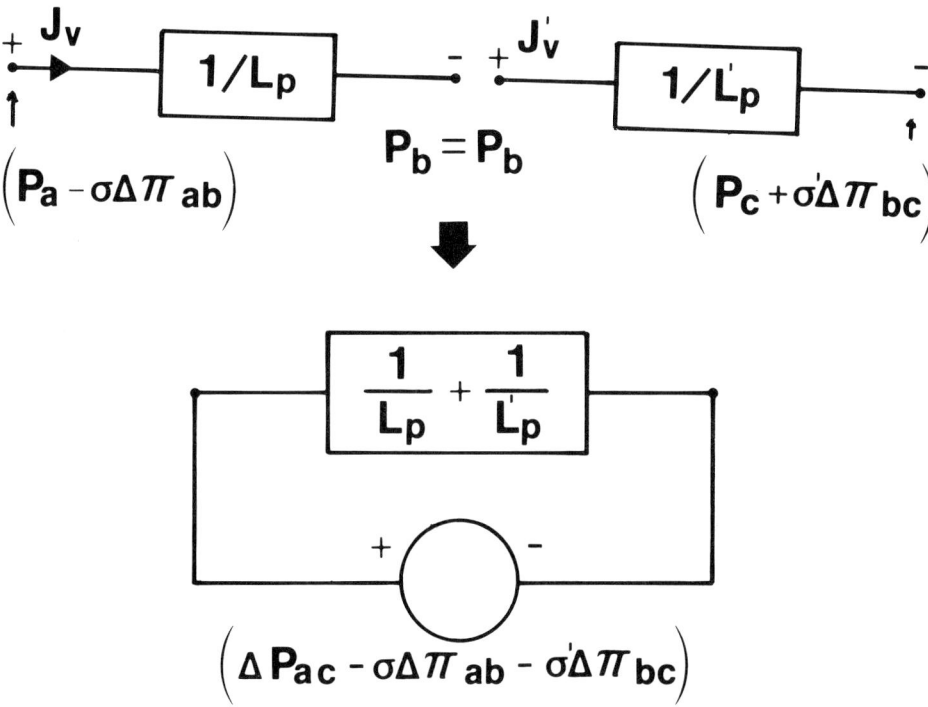

Fig. A.3.2 The principle illustrated in Fig. A3.1 is now applied to two membranes in series having hydraulic conductivities L_p and L_p'. The pressure P_b is matched in the right and left membranes by subtracting the pressures $\sigma \pi_b$ from a and b and the pressure $\sigma' \pi_b$ from b and c in the second membrane. Note that $\pi_{ab} = -\pi_{ba}$.

the interfaces. Consider again the potential difference across the first membrane,

$$(P_a - \sigma \pi_a) - (P_b - \sigma \pi_b). \qquad (103)$$

All potentials are only defined within an arbitrary additive constant, so that we can add the term $\sigma \pi_b$ to both ends of the membrane without changing the value of the resultant flow. The resultant pressure difference then looks like

$$P_a - \sigma(\pi_a - \pi_b). \qquad (104)$$

on the left hand side and a pressure P_b on the right hand side (this is just a rearrangement of terms, nothing has changed physically). A similar operation can be performed on the pressure difference for the second membrane by adding the term $\sigma' \pi_b$ to both sides of the membrane. As a result, the two potentials across the membrane are

$$P_b \qquad \text{(left)} \qquad (105)$$

and

$$P_c - \sigma' \pi_c + \sigma' \pi_b \quad \text{(right)}. \qquad (106)$$

In order to match the two membranes we introduce the physical assumption that, at the interface, there is a single pressure P_b. The complete membrane looks as shown in FIGURE 2. The pressure, or potential at a is given by

$$P_a - \sigma \pi_a + \sigma \pi_b = P_a + \sigma \pi_{ab}, \qquad (107)$$

while the pressure or potential at c is given by

$$P_c - \sigma' \pi_c + \sigma' \pi_b = P_c - \sigma' \pi_{bc}, \qquad (108)$$

while the overall potential difference is now given by

$$(P_a - P_c) - \sigma(\pi_a - \pi_b) - \sigma'(\pi_b - \pi_c) =$$

$$= P_{ac} - \sigma \pi_{ab} - \sigma' \pi_{bc'} \qquad (109)$$

which is the expected formal result, although it has been obtained by requiring continuity of the physical potentials defining forces.

Consider now the complete model which includes a solute path. The solute flow is given by

$$J_s = c(1-\sigma)J_v + K_f \pi_a - K_b \pi_b \qquad (110)$$

in the first membrane and it can be represented by the network shown in FIGURE 3. The source $c(1-\sigma)$ is a "proper" source, but the source ($K_f \pi_a - K_b \pi_b$) is not a linear element--this is the same situation encountered in the case of the first order chemical reaction-- as there is no resistance which gives the correct flow when placed across the proper force $\pi_a - \pi_b$. At this point, we could either call the system nonlinear or attempt to obtain a continuous potential representation. For example, the force $[(K_f/K_b)\pi_a - \pi_b]$ placed across a conductance K_b, as shown in FIGURE 3 gives the rate $K_f \pi_a - K_b \pi_b$, in agreement with the original equation. The potential across the second membrane could not have the same form, however, because the interface values would not match. We can set the force across the second membrane equal to

$$\pi_b - (K_b'/K_f')\pi_{c'} \qquad (111)$$

provided the conjugate conductance is K_f' (FIGURE 3).

We can now convert to the proper series representation, the **R** form of the equations.

The resultant **R** operators can now be added term by term to obtain the global parameters:

$$R_{11}\text{ total} = R_{11} + R_{11}' = (1/L_p) + (1/L_p') \qquad (112)$$
$$R_{12}\text{ total} = R_{12} + R_{12}' = 0 + 0 \qquad (113)$$
$$R_{21}\text{ total} = R_{21} + R_{21}' = -[c(1-\sigma)/K_b + c'(1-\sigma')/K_f'] \qquad (114)$$

and $\qquad (115)$

$$R_{22}\text{ total} = R_{22} + R_{22}' = -[c(1-\sigma)/K_b] - [c'(1-\sigma')/K_f'].$$

Note that, although the resultant matrix is not symmetric the coefficients add without introducing "non additive" terms.

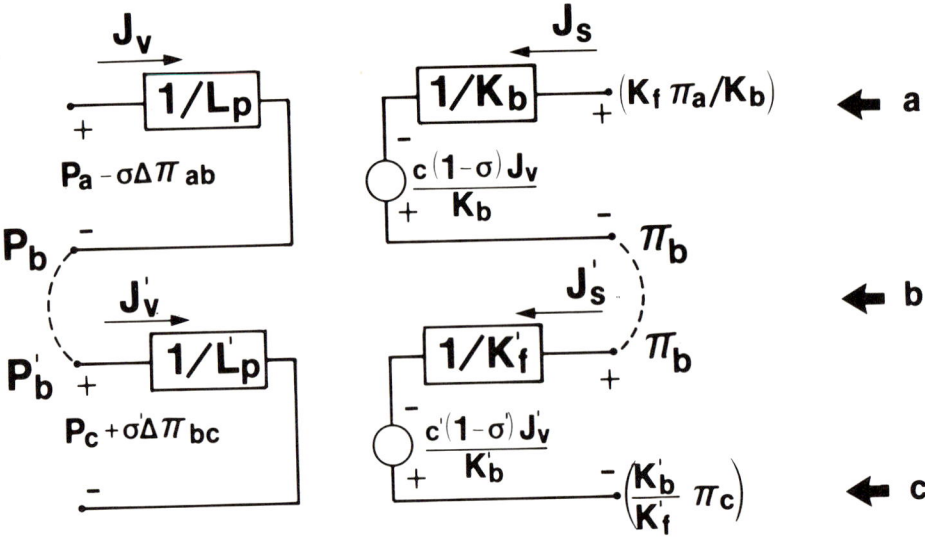

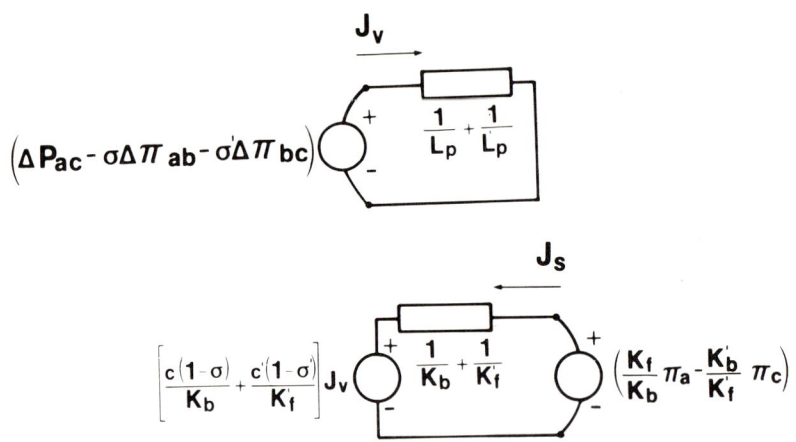

Fig. A.3.3 Use of the potential matching technique described in the text in order to obtain complete osmotic and hydraulic interactions. The final two port is not reciprocal in this version. If reciprocity is desired, the approach described for the single membrane can be used.

Chapter 4

MICROSCOPIC REVERSIBILITY, ONSAGER THERMODYNAMICS AND KINETIC NETWORKS

4.1 INTRODUCTION. RECIPROCITY IN ONSAGER THERMODYNAMICS

The celebrated theory of near equilibrium processes proposed by Onsager (refs. 168-170) deals with the phenomenology of interacting processes of various kinds--e.g., electrical, transport, chemical, etc. Given the thermodynamic forces X_i and their conjugate flows J_i which appear in the bilinear dissipation function

$$\Phi = T\, d_i S/dt = X_1 J_1 + X_2 J_2 + X_3 J_3 + \ldots + X_n J_n \qquad (1)$$

the theory establishes that

1. Irreversible coupling between various processes can only occur spontaneously when
$\Phi \geq 0$ and

2. if the forces and flows are related by linear equations of the form

$$X_i = \sum R_{ik} J_k \qquad (k=1,2,\ldots,n) \qquad (2)$$

the coefficient matrix R is symmetric --i.e.,

$$R_{ik} = R_{ki}\,. \qquad (3)$$

In addition to the proof of reciprocity, the original theory proposed by Onsager dealt with the extremum properties of energy, entropy and the dissipation function. Proponents of the model have exhibited notable creativity. Thus, Machlup and Onsager (ref. 128) extended the original treatment to include time decay of fluctuations and Casimir (ref. 47) extended the theory to include fluctuations which depend on odd functions of the velocities -- e.g., in the case of the magnetic field -- and, furthermore, he pointed out that the reciprocities required that the decay of fluctuations was linear. Properties of Onsager frames of reference, in particular the effects of reference velocities as well as further general extensions and criticisms have been given by several authors (e.g., refs. 58, 59 , 60, 65, 76, 86, 87, 101-103, 105, 106, 110, 119, 136).

Various geometrical treatments have also been given,either to show that the Onsager model is Euclidean (ref.163) or to extend the fluctuation treatment to the non-linear domain (refs. 17, 209). Network relations, extensions and isomorphisms have been given by Meixner (ref. 138), Oster, Perelson and Katchalsky (ref. 178),van Kampen, the author and others (refs.255, 182, 144).

In spite of all this work, however, the general concensus appears to be that Onsager's theory presents a very restrictive analysis in which all systems are seen as being linear, passive, realizable, causal in a world of continuous potential functions and continuous derivatives.

This narrow view of physical phenomena has basically divided the thermodynamic community into two groups: the diehards-- who look for reciprocities even in situations in which the theory clearly excludes them-- and those who believe the Onsager model will not lead anywhere and a new approach must be taken if a valid extension of thermostatics is to be given for the non-equilibrium irreversible and non-linear ranges. Some of these "new " routes include statistical mechanics , stochastic theory, non-linear programming , geometric analyses ,etc.

Although it is always useful to present new physical models and views, many of these approaches underestimate Onsager's physical insight: rather than throw the baby with the dirty water, it proves more profitable to dissect Onsager's model and strip it of some of its unnecessary metric charactersitics. In so doing, the basic logical/topological structure of Onsager's theory is retained and the model can be naturally extended to irreversibility, non reciprocity, non-equilibrium and non-linear situations.

Before one can proceed in this direction , however, it is mandatory to get to the root of the problem. Onsager's theory is ostensibly based on Einstein's elegant analysis of statistical fluctuations , but there is a central paradox which
surfaces in the attempts to verify the reciprocities in actual physical systems (ref. 159) :

When the reciprocities expressed by Eq. (2) hold, they are valid for a range way beyond the "near equilibrium " requirements

of fluctuation theory

while

There are many known situations in which the reciprocities do not hold, even in the equilibrium range.

Thus, the theory is a bad physical model, even if it gives the correct final result from time to time, because the range in which the phenomena are predicted is not the same as the range in which the phenomena are observed. By contrast, Einstein's fluctuation predictions in the theory of Brownian motion, for example, agrees closely with experimental ranges of observation.

Inconsistencies of this sort suggest that the property of a system being reciprocal or not has little to do with being close or far from equilibrium.

In order for Onsager's fluctuation analysis to be valid, then, fluctuations must be stripped of their most characteristic feature--i.e.,their small size ! On the other hand, a further problem arises because there is no known macroscopic way to prove Onsager's reciprocities and because,in addition , it is the fact that the fluctuations are small which allows the application of microscopic reversibility.

With this background, there are two types of general questions that arise:

i) Is there is any physical assumption, in addition to macroscopic reversibility -- which leads to macroscopic reciprocities?

ii) Can a theory of irreversible , non equilibrium, possibly non-linear processes be generated such that it can apply to the range of observed phenomena?

The purpose of this chapter is : i)to show the basic Kirchhoffian structure of Onsager's theory and ii) to show the close relation between reversibility and Kirchhoff's voltage law --i.e., the possibility of defining a potential. In following chapters a direct application of these results will be made to specific examples involving non equilibrium, irreversibilities and non-linearities.

4.2 MICROSCOPIC FOUNDATION OF ONSAGER THERMODYNAMICS

We first consider some of the basic ideas involved in the thermodynamic model proposed by Onsager in his seminal papers (refs. 168, 169). These involve only alpha fluctuations which are even functions of the velocity and, moreover, they exclude time decays.

The most fundamental assumptions made by Onsager is what Fitts (ref. 77) refers to as "postulate I" which is the idea that for a state that is close to equilibrium there are functions of state which can be calculated by extrapolation from their values at the equilibrium state. In particular, the entropy can be calculated as a near equilibrium entropy. Thus, the internal thermodynamic space is seen as consisting of continuous paths that go between the various macroscopic processes (ports) where n-coordinates or variables $a_1...a_n$ can be measured by ordinary means. These displace the system from its equilibrium state S_o, as shown in FIGURE 1.

In the Onsager model the displacements are identified with (small) fluctuations from the equilibrium state A^o, so that

$$a_i = A_i - A_i^o. \qquad (4)$$

This physical identification is restrictive, however, as it will

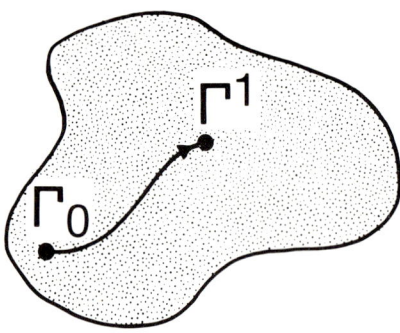

FIGURE 4.1 Displacement in the Onsager manifold.

be indicated that a self consistent theory does not require that the "fluctuations" be small--i.e., it does not require displacements to be fluctuations. Topologically, the Onsager assumption corresponds to drawing tangent planes close to equilibrium , thus changing a point set into an open set.

A second basic assumption is that we deal with an **aged system**-- i.e., with a system which has been isolated for a length of time. Such a system, it is claimed, will go through all possible states $\Gamma, \Gamma', \Gamma''....$, etc. consistent with the conditions of isolation and the length of time spent in each state (t, t´, t",...) will depend on the number of microstates in the specific configuration (ergodic assumption).

The question then arises as to how the thermodynamic state can be specified precisely. The simplification of macroscopic thermodynamics is that, on the average of many observations, every time a set of fluctuations $a_1...a_n$ occurs, the same thermodynamic state $\Gamma(a_1, a_2, ..., a_n)$ is observed.

This is the state that maximizes the entropy given all possible arrangements and interactions of those fluctuations and can be uniquely specified by the entropy change from the reference value S_o:

$$\Delta S = S(a_1, a_2,..., a_n) - S_o \qquad (5)$$

Thus, a given set of fluctuations always takes the system to a <u>precise</u> thermodynamic state uniquely and unequivocally determined by the set of fluctuations. As Boltzmann (ref. 25) pointed out and then Einstein (ref. 70), Onsager (ref. 169) and others stressed, this observation (the second law) is only valid in the sense that --on the average taken for many measurements-- a most likely value of entropy will be obtained. Statistically, this follows because, although all microstates are equally likely the overwhelming number of microstates are those of equilibrium.

Given a different set of fluctuations

$a_1',.....a_n'$

a new thermodynamic state

$\Gamma'(a_1',.....,a_n')$

can be specified through the same functional dependence

$\Gamma(a_1', \ldots a_n')$

and, in general, for an arbitrary set of fluctuations a^r, the thermodynamic state Γ^r observed would be given by $\Gamma(a^r)$
The implied assumption is, clearly, that every time a set of fluctuations is given, a specific state follows from the macroscopic point of view. Moreover, the entropy of this state is given by Boltzmann's equation

$S = k \ln W + \text{constant}$,

which gives the probability of any state relative to the probability of the equilibrium state.

Clearly, a non equilibrium theory must also include a distribution of states away from equilibrium. In Onsager's model, it is assumed that the Boltzmann expression holds not only at equilibrium, but also in all the intermediate states traveled in the decay pathway from the non equilibrium displacement point to the initial equilibrium state of maximum entropy. This entropy is expressed, as usual, as a Taylor approximation around an equilibrium state

$$\Delta S = S - S_o = (1/2) \sum g_{ij} a_i a_j \, , \tag{6}$$

in which higher order terms are ignored. The coefficients g_{ij} are the partial derivatives

$$g_{ij} = \partial^2 \Delta S / \partial a_i \partial a_j . \tag{7}$$

Given that S is a function of state, it follows that $g_{ij} = g_{ji}$ (these are not Onsager reciprocities, of course !).

If the forces which drive the system to equilibrium are found from

$$X_i = - \partial \Delta S / \partial a_i$$

it follows that the forces are linear combinations of the fluctuations

$$X_i = \sum_j g_{ij} a_j. \qquad (8)$$

By identifying the averages of the time variations of fluctuations in their approach to equilibrium,

$$(da_i/dt) = \iiint_{-\infty}^{+\infty}\cdots P(a_1, a_2, \ldots a_n)(da_i/dt)\, da_1\, da_2\cdots da_n \qquad (9)$$

--in which P is a probability distribution function -- with the macroscopic flows J_i and then introducing microscopic reversibility, Onsager showed that the macroscopic phenomenological equations are reciprocal.

4.3 INTEGRATION OF MICROSCOPIC VARIABLES

The microscopic reciprocity given by equation (7),

$$g_{ik} = g_{ki} \qquad (10)$$

--which is clearly <u>not</u> Onsager's reciprocity-- allows to synthesize a network in which, in agreement with the tenets of Onsager's theory the forces (voltages) are given as

$$X_i = \partial \Delta s / \partial a_i = \sum g_{ik} a_k$$

provided that the dissipation is

$$\Delta s = (1/2) \sum g_{ij} a_i a_j$$

and the fluctuations are identified with port currents conjugate to the forces (voltages) X_i.

The resultant resistive network has internal fluctuations which obey K.C.L., by construction of the resistive network. Thus, at any network node the following holds

$$\sum_{\text{all } i} a_{ik} = 0. \qquad (11)$$

The internal fluctuations are clearly linear functions of the independent fluctuations at the ports. This expression may be integrated as follows

$$\iiint_{-\infty}^{\infty} \cdots \sum a_{ik}(a_1, a_2, \ldots, a_n) \; P(a_1, a_2 \ldots a_n) \; da_1 \; da_2 \ldots da_n = 0 \quad (12)$$

in which P is the distribution function giving the probability of finding joint fluctuations within a certain "probability volume " range. If the order of summation and integration is exchanged, we obtain

$$\sum \iiint_{-\infty}^{\infty} a_{ik}(a_1, a_2, \ldots, a_n) \; P(a_1, a_2 \ldots a_n) \; da_1 \; da_2 \ldots da_n = 0 \quad (13)$$

Denote the integral inside the summation by a_{ik} -- these are not the average quantities which define the measurable flows at the inputs, but averages of fictitious fluctuations inside the connected network ! It then follows that these averages also obey K.C.L.,

$$\sum_i \bar{a}_{ik} = 0 \quad (14)$$

in the same branches and nodes as the original internal fluctuations. It then follows that the input averages will belong to the same network, for any P chosen. Moreover the averaging process may inlcude dividing by a time increment, t and that the internal fluctuations divided by an arbitrary t will also obey K.C.L. The macroscopic assumption is that the phenomenological equations are linear. These phenomenological equations will now arise from the relationships between the internal average fluctuations and the local internal (tree) forces. These will depend on the characteristics of P. But, whatever they are, the same topology and Kirchhoff's laws will again lead to reciprocity.

4.4 MACROSCOPIC EQUATIONS AND TOPOLOGICAL AVERAGING OF THE MICROSCOPIC NETWORK

The fundamental axiom of Onsager's thermodynamics is, as pointed out above, that the "non equilibrium" states we deal with are so close to equilibrium that a Taylor expansion of any of the functions of state at the nearby equilibrium point yields the corresponding function of state at the non equilibrium point. (Clearly, this is a debatable point, but we proceed from Onsager's own axioms for consistency). A second assumption --which is never explicitly stated-- is the following: all the non equilibrium points under observation are in the neighborhood of the same equilibrium state Γ_o.

The first axiom allows us to construct a microscopic network, which is connected and resistive (with the reciprocity $g_{ik} = g_{ki}$). The second axiom says that, for the range of measured fluctuations these g's are constant; we can therefore calculate all variables from the same microscopic network in the case of Onsager's alpha variable theory.

The macroscopic theory also leads to a connected resistive network. The question is how the coefficients in the two are related--or, in geometric terms, how the microscopic and macroscopic metrics relate.

The statistical weight of the fluctuations can either be given in terms of a probability distribution P or in terms of the relative time periods T^i a given fluctuation set

$$a^i = [a_1, a_2, ..., a_n]^i$$

lasts on the average. Moreover, when the set of fluctuations a^i is plugged into the network machine a set of forces

$$x^i$$

will be measured which correspond to those attempting to bring the system to equilibrium in Onsager's model. Each of the set of forces will exist--again, only in the model that excludes inertial decays-- for the time T^i.

The total force measured in a period

$$T = \sum_i T^i$$

will then be given by the average force measured during the time T. This is simply found by a weighted average of the network forces which yields

$$X_{average} = \sum_i x^i \, T^i / T$$

$$= \sum_i g \, a^i \, T^i / T .$$

Given the constancy of **g**, it follows that

$$X_{average} = \mathbf{g}\, a^i{}_{average}$$

$$= \mathbf{g}\, J^i$$

Therefore, the microscopic model of Onsager alpha variables leads to the (undesirable) result that the microscopic and the macroscopic coefficients are the same. One way to avoid this result would be to assume that the **g**'s measured are not constant; however, this would also be inconsistent as it would mean that two sets of fluctuations $[a]^i$ and $[a]^{i'}$ found in the same steady state would be expanded around two different equilibrium points, say Γ_o and $\Gamma_o{}'$. In this scheme, the probability of a path joining the two fluctuation sets in time is higher than the probability of either of these decaying to their (now distinct) nearby equilibrium points. This contradicts the main postulate.

4.5 EUCLIDEAN VECTOR SPACES AND ONSAGER THERMODYNAMICS

Before a procedure for quantitating entropy changes is given, the space of Onsager thermodynamics is devoid of shape or size, it is simply a collection of states specified by n coordinates or variables --i.e., an n dimensonal manifold. Thus, each collection of points or state Γ is specified in terms of n fluctuations $a_1, a_2, \ldots a_n$. We can ask questions regarding topological properties of the manifold--e.g., how each state is connected to neighboring states, whether mappings between a set of points and a second set of points can be achieved through continuous bijective transformations (homeomorphisms) or what the topological properties which remain invariant during these homeomorphic transformations are.

The questions which are explicitly excluded are, however, those regarding measurement of distances between points. The introduction of a metric allows to measure the distance between two arbitrary states and transforms the manifold into a metrical manifold. It then becomes possible to look at geometrical properties --e.g, distances and angles. In particular, the precise definition of the entropy change when going from an equilibirum state Γ_o to a non equilibrium state Γ requires the introduction of a metric. The entropy change

$$\Delta S = S'(a_1, a_2, \ldots a_n) - S_o(0, 0, \ldots, 0) \qquad (15)$$

-- in which we have identified the equilbrium state S_o with the origin of the manifold is graphically represented in FIGURE 2.

Given that all the information about Onsager's thermodynamics is contained in this metric information, we can refer to the Onsager manifold as a metric generated topology. In particular, the implicit assumption that the coefficients g_{ik} are constant for

a given problem indicates that the model describes an euclidean (flat) space as opposed to a non-euclidean (non linear, or folded) space. This follows because Eq. (6) is a generalization of Pythagoras theorem, as discussed before. The metric manifold of Onsager processes may be viewed as a family of n dimensional

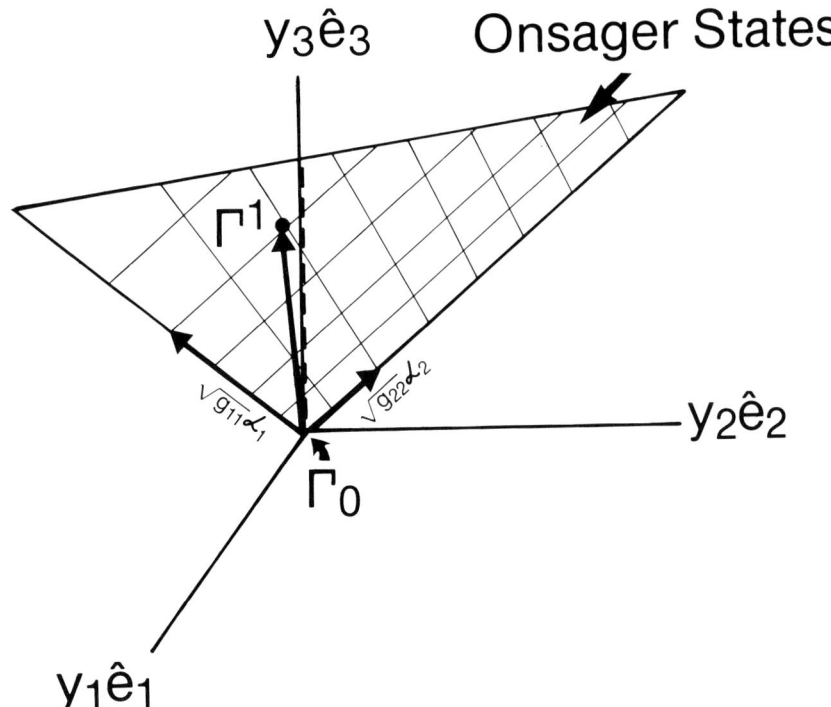

FIGURE 4. 2   Geometric displacements in the Onsager Manifold.

hyperprisms having sides $a_1 \sqrt{g_{11}}$, $a_2 \sqrt{g_{22}}$, $a_i \sqrt{g_{ii}}$
The distance (change in entropy) can again be specified in terms of the self dot product $||\mathbf{d} \cdot \mathbf{d}||$ of a vector $\mathbf{d}$ and all the previous requirements for a vector space are met for the fluctuations and the forces.

The physical components of the vectors in the directions of the parametric coordinates are, for the case of two dimensions,

$$d_1 = a_1 \sqrt{g_{11}}\, \mathbf{i}_1 \qquad (16)$$

and

$$d_2 = a_2 \sqrt{g_{22}}\, \mathbf{i}_2 . \qquad (17)$$

These form a basis for the vector space of two dimensions and all other vectors are linear combinations of these. The resultant vector $\mathbf{d}$ which reaches a non-equilibrium state from the origin S_o is given by

$$\mathbf{d} = \mathbf{d}_1 + \mathbf{d}_2 = a_1 \sqrt{g_{11}}\, \mathbf{i}_1 + a_2 \sqrt{g_{22}}\, \mathbf{i}_2 \qquad (18)$$

and the squared length (norm) of this vector is given by the dot product

$$||\mathbf{d}^2|| = \mathbf{d} \cdot \mathbf{d} = \mathbf{d}_1 \cdot \mathbf{d}_1 + 2\, \mathbf{d}_1 \cdot \mathbf{d}_2 + \mathbf{d}_2 \cdot \mathbf{d}_2 \qquad (19)$$

$$= |\mathbf{d}_1|^2 + 2\, |\mathbf{d}_1|\, |\mathbf{d}_2| \cos\theta + |\mathbf{d}_2|^2 .$$

in which $\cos\theta_{12} = g_{12} / \sqrt{g_{11}\, g_{22}}$. The distance is then proportional to the change in entropy and may be expressed as

$$\Delta S = ||\mathbf{d} \cdot \mathbf{d}||^2 / 2. \qquad (20)$$

Clearly, the cosine can have only values between 0 and 1. The function of state requirement for the entropy, $g_{12} = g_{21}$ means that there is only one angle $\theta = \theta_{12} = \theta_{21}$, as opposed to two different angles, say θ_{12} and θ_{21}. This fact allows to construct

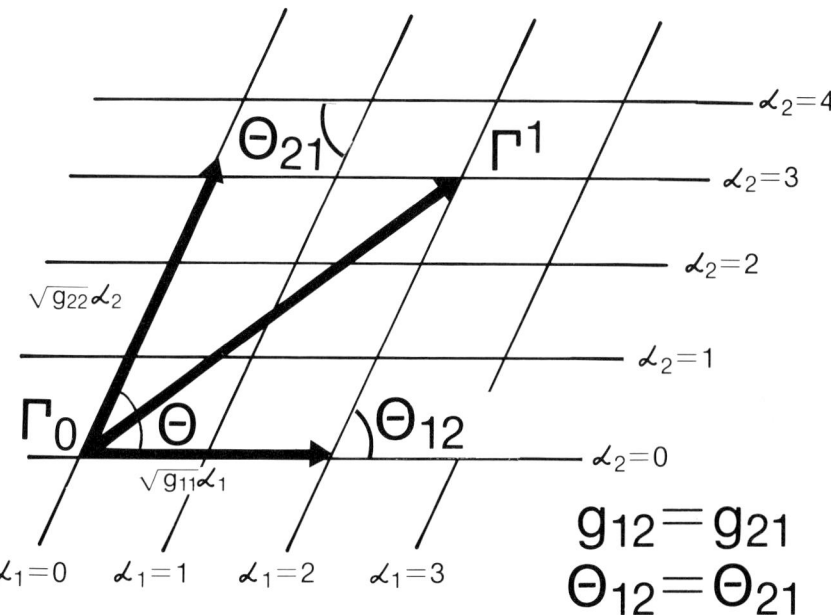

FIGURE 4.3 The existence of a reciprocal metric implies that the parallelogram interstate parallelogram can be constructed (or a parallelopiped in the case of n dimensions).

the parallelogram shown in FIGURE 3, which implies the same change in entropy is measured, in absolute value, regardless of whether the process goes forward or backwards--i.e., the process is reversible and entropy is a function of state.

At this point it is instructive to ask what is the physical meaning of θ, which is not evident from the dot product interpretation. Geometrically, this is the projection of one coordinate onto another, physically this is related to the efficiency of transfer of energy between coordinates, as will be considered in the following chapter.

Thus, the main characteristic of Onsager thermodynamics is that it establishes a calibration or common yardstick to go from one state to another. This problem has also been considered, in a slightly different manner by Nathason and Sinanoğlu (ref. 163).

4.6 MICROSCOPIC PROOF OF ONSAGER RECIPROCITIES

It will now be shown that Onsager's reciprocities can be proven, within the framework of the theory, without invoking microscopic reversibility. The point of this is not that Onsager's equations are valid, but that microscopic reversibilities constitute a redundant assumption.

Consider an arbitrary Onsager system described in terms of its component forces, X_i, and conjugate fluctuations, a_i, at n boundary ports. We can also assume, without loss of generality, that the internal workings of the system are enclosed inside a volume bounded by a surface A and that the entropy generated inside the volume by some irreversible process is measured by

$$\Delta s = \sum g_{ik} a_i a_k \qquad (21)$$

at the ports and that this corresponds to the integrated value of a local entropy taken inside the volume,

$$\Delta S = \int_V s(v)\, dv, \qquad (22)$$

in which $s(v)$ is the entropy generated per unit volume, which is a function of the position in v.

In order to have internal consistency with Onsager's theory, the internal processes must be a subset of the macroscopic processes--i.e., we require that at each point $P(v)$ the entropy generated is a function of the _local_ fluctuations,

$$\int s(v)\, dv = \int g(v)\, a(v)\, a(v)\, dv = \int g(v)\, ||a(v)||^2\, dv \qquad (23)$$

while the local force $X(v)$ can be found from the local entropy production by

$$\mathbf{X}(v) = -\mathbf{grad}_a\, s(v) = -g(v)\,\mathbf{a} \qquad (24)$$

Equation (24) simply states that the force can be found from a potential which is uniquely defined at each point inside the volume within an arbitray constant, so that $\mathbf{curl\ X(v)} = 0$, while the first equation follows directly from the assumptions.

THEOREM. The field of internal fluctuations behaves like an incompressible, constant density fluid in potential motion.

Proof. Identify individual the port forces and conjugate fluctuations a_i with vectors which pierce the boundary surface so that

$$\Delta S = \int_A \mathbf{a} \cdot \phi \, dA \qquad (25)$$

in which ϕ is a local potential defining the force at the ports. From (22) and (23) it follows that ΔS inside the volume is given by

$$\Delta S = \int_V \mathbf{X}(v) \, \mathbf{a}(v) \, dv. \qquad (26)$$

Thus, from (23) and (25) we obtain

$$\int_V \mathbf{X}(v) \, \mathbf{a}(v) \, dv = \int_A \mathbf{a} \cdot \phi \, dA. \qquad (27)$$

Introducing the identity

$$\text{div } A \, \mathbf{B} = A \text{ div } \mathbf{B} + \mathbf{B} \text{ grad } A \qquad (28)$$

and the divergence theorem, it follows that the only general way for the potentials to be continuous at the ports and inside the Onsager volume is

$$\text{div } \mathbf{a} = 0, \qquad (29)$$

in which this divergence is taken locally at each point inside the volume. This proves that there are no sources of fluctuation inside the volume. Note that this is a continuous corollary of Tellegen's theorem.

One may question the physical significance of this result. In the case of ports consisting of the same type of flux--e.g., heat flows, diffusional flows of a single substance, etc.-- there is no problem and this type of integration inside a volume is well known. However, the present example considers some generalized fluctuation "flow" which can exchange with all other fluctuation forms. This should not present a problem, however, as the process is a thought experiment which is consistent with the logic of the Onsager construct. We can simply say that <u>the fluctuation space of Onsager Thermodynamics behaves as if there existed a fluctuation which can exchange with all other fluctuating quantities</u>..

We now proceed to prove the following

THEOREM II. **The collection of macroscopic flows J_i considered as the average**

$$J_i = da_i/dt$$

also behave as an incompressible fluid in the same space.

PROOF. It is clear that for any two sets of port (vector) fluctuations **a** and **a'** corresponding to two different possible values, their difference

$$\Delta a = a - a' \qquad (30)$$

also lies in the same topological space, so that div $\Delta a = 0$ at any point inside the volume. Moreover, any averaging process which looks at N occurrences of a fluctuation in a time Δt, such as

$$\Delta a_{av} = (N_1 a + N_2 a' + N_3 a'' + \ldots)/N \qquad (31)$$

will also lie in the same vector space. (One can visualize the fluctuation vector as pointing in a certain direction, all the differences which can be constructed with values lying in the same direction will point in that direction as well.) We can then write

$$\text{div}(\Delta a_{av}/\Delta t) = 0. \qquad (32)$$

The resultant vector can be formally interpreted as the internal average flow

$$J(v) = \Delta a_{av}/\Delta t, \qquad (33)$$

corresponding to the situation in which the macroscopic flows are given by

$$J_i = da_i/dt. \tag{34}$$

The <u>corollary</u> which follows is clearly : **if the macroscopic equations (2) are obeyed, the local constitutive equations inside v have the linear form**

$$\mathbf{X}(v) = R(v) \mathbf{J}(v). \tag{35}$$

This follows by using the local force definition and the principle of superposition, as shown in Chapter 2 for the thermostatic case.

We can now prove the central

<u>THEOREM</u>. **The macroscopic equations (2) must be reciprocal.**

Consider two sets of measurements, $(\mathbf{X}, \mathbf{J})$ and $(\mathbf{X}', \mathbf{J}')$ corresponding to the same macroscopic set of equations. Given that

$$\mathbf{X}(v) = -\text{grad}_a\, s(v) = -g(v)\, \mathbf{a} \tag{36a}$$

and

$$\text{div}(\Delta\, \mathbf{a}_{av}/\Delta\, t\,) = 0, \tag{36b}$$

it follows that

$$\mathbf{X}(v) = -\text{grad}\, \phi, \tag{37}$$

$$\mathbf{X}(v)' = -\text{grad}\, \phi', \tag{38}$$

$$\text{div}\, \mathbf{J}(v) = 0 \tag{39}$$

$$\text{div}\, \mathbf{J}'(v) = 0. \tag{40}$$

Moreover,

$$\mathbf{X}(v) = R(v)\, \mathbf{J}(v) \tag{41}$$

and

$$\mathbf{X}'(v) = R(v) \mathbf{J}'(v) \qquad (42)$$

because the same topology applies to the primed and unprimed states. We consider, again, an integration of the form of equation (1.37) with either the potential or the flow being the primed state so that

$$\int \mathbf{J}' \cdot \phi d\mathbf{A} = \int \text{div}(\phi \mathbf{J}') \, dv = \int \mathbf{J}' \, \text{grad } \phi \, dv + \oint \phi \, \text{div} \mathbf{J}' \, dv = \int \mathbf{J}' \cdot \mathbf{X} \, dv \qquad (43)$$

and, similarly,

$$\int \mathbf{J} \cdot \phi' d\mathbf{A} = \int \text{div}(\phi' \mathbf{J}) \, dv = \int \mathbf{J} \, \text{grad } \phi' dv + \oint \phi' \, \text{div} \mathbf{J} \, dv = \int \mathbf{J} \cdot \mathbf{X}' \, dv. \qquad (44)$$

However, given that

$$\mathbf{J}(v) \, \mathbf{X}'(v) = \mathbf{J}(v) \, R(v) \, \mathbf{J}'(v) = \mathbf{J}(v)' \, \mathbf{X}(v) \qquad (45)$$

it follows that

$$\int \mathbf{J}' \cdot \phi \, d\mathbf{A} = \int \mathbf{J} \cdot \phi' d\mathbf{A} \qquad (46)$$

or, equivalently (recall that $\Delta s = \int \mathbf{a} \cdot \phi \, d\Lambda$),

$$\sum_i \mathbf{J}'_i \, \mathbf{X}_i = \sum_i \mathbf{J}_i \, \mathbf{X}'_i, \qquad (47)$$

which states the reciprocity of Onsager equations.

4.7 PROBABILITIES OF JOINT NON-EQUILIBRIUM STATES

Given the availability of a procedure to perform entropy calculations inside the dissipating box and the possibility to relate these calculations to Boltzmann's probabilistic assumptions, we can find relative probabilities for an occupancy level of non-equilibrium states in paths inside the Onsager manifold. In general, these will correspond to joint probabilities of various activated states.

To fix ideas, consider the following two simple examples. Given a two force, two fluctuations system represented either by the

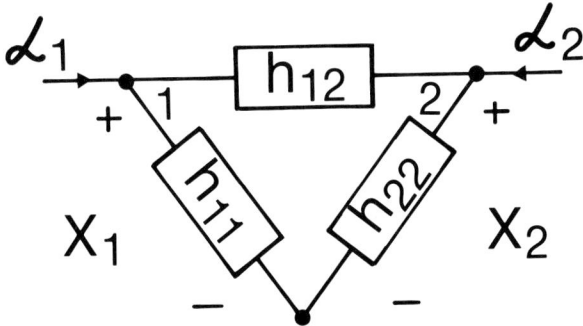

FIGURE 4.4 The two fluctuation Pi network.

equations

$$X_i = \sum g_{ik} \, a_k \qquad (48)$$

or by the inverse set

$$a_i = \sum h_{ik} \, X_k, \qquad (49)$$

the simplest connected resistive networks which will represent this system are, as before:

i) The triangular, Pi network in which the two non equilibrium states are joined to each other--the direction of decay will depend on the relative levels of the non-equilibrium states--and the equilibrium state (FIGURE 4) and

ii) The "T" network in which each of the two non-equilibirum states first decays to a common intermediate state which then decays to equilibrium (FIGURE 5).

The relative probabilites for these transitions can be calculated

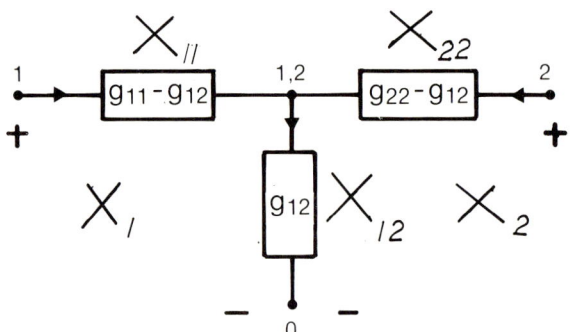

FIGURE 4.5 The T network for microscopic fluctuations.

by equating the entropy dissipated on a given jump with Boltzmann's formula. For example, the transition $S_1 \longrightarrow S_{12}$ in the T network has an associated entropy change

$$a_1^2 \, (g_{11} - g_{12}), \qquad (50)$$

in which

$$g_{11} - g_{12} = X_{11} / a_1 \qquad (\text{"Ohm's" law}). \qquad (51)$$

It then follows that the ratio of probabilities between states S_{12} and S_1 must be given by

$$(p_{12}/ p_1) = \exp [\, (g_{11} - g_{12}) \, a_1^2/k \,] \qquad (52)$$

Similarly, the entropy change in branch a_{12} of the triangular, Pi network is given by

$$\Delta s = (X_1 - X_2)^2 \, h_{12} \qquad (53)$$

so that the ratio of occupancies between the two states 1,2 is given by

$$(p_1 /p_2) = \exp [\, (X_1 - X_2)^2 \, h_{12}/ k \,]. \qquad (54)$$

By expressing h_{12} in terms of g parameters, we finally obtain

$$(p_1/p_2) = \exp[(X_1 - X_2)^2 \, g_{12}/(\det g)k]. \qquad (55)$$

(Note that K.C.L. must be obeyed at every node).

4.8 MICROSCOPIC REVERSIBILITY

The question arises as to the role and relationship of microscopic reversibility to the network ideas.

The view of microscopic reversibility, central to the derivation of reciprocity, is that if an event **a** --which depends only on molecular and atomic configurations or on squares of velocities-- is followed t seconds later by an event **b**, **a** will occur in a reversible system as often as the event **b** followed t seconds later by the event **a**. This can be expressed in terms of averages of reversible microscopic variables a_i and a_j by

$$A_{ji}(t) = a_j(t) \, a_i(t + t_o) = a_i(t) a_j(t + t_o) = A_{ij}(t). \qquad (56)$$

Microscopic reversibility is a bothersome notion, because it has no obvious counterpart or equivalent meaning in the macroscopic domain, except for Onsager's insight into the nature of kinetic reversibility.

It has been shown above that there is form of _macroscopic_ reversibility included in the assumptions of the theory. As a result, microscopic reversibility is a redundant notion. At first the validity of such statement may be put in question and raise similar criticisms to those (rather unfairly) made against Thomson's quasi irreversible theory of non equilibrium processes. Thomson pointed out that in abstract dynamics the instantaneous reversal of motion of every moving particle of a system causes the system to move backwards. This would also be true in macroscopic Physics, he noted, except for the observation that in physical, dynamics this simple and perfect reversibility fails, on account of forces depending on friction of solids, the imperfect fluidity of fluids, the imperfect elasticity of solids, the solution of solids influids, the generation of heat by electric currents induced by motion and the like. The fact that such processes do exhibit reciprocity, then, shows that irreversibility and reciprocity are not necessarily inconsistent. A relevant example is the reciprocity shown by resistive networks, which prompted me to postulate that these and Onsager systems were isomorphic, even though I had no particular idea of how Kirchhoff's voltage law would have entered in Onsager's original proof (Kirchhoff's current law was more obvious, at least in the case of the diffusion of a single substance or heat).

An important clue into the relationship between reversibility and K.V.L. is provided by the three phase monomolecular reaction considered by Onsager to motivate the correlation between microscopic reversibility and the kinetic concept of detailed balance, which has already been discussed in Chapter 3.

The main result which emerged from that discussion was that reversibility, which is primarily a displacement concept, is closely associated with the definition of a conservative force-- i.e, one which can be found from a potential. Included in those considerations is the dissipative form (Raleigh) of the dissipation function in which the potential arises from forces which are functions of the velocities.

We now consider the sources of these continuous potentials from another point of view.

4.9   KINETIC REVERSIBILITY AND DETAILED BALANCE: THE TRIANGULAR CHEMICAL REACTION

We can now extend the idea to a reaction network consisting of linear rates. Of particular interest is the case of a hypothetical equilibrium scheme in the case of a homogeneous substance which can coexist in three different phases, as represented by the triangle shown in FIGURE 6.

This scheme was used by Onsager to motivate an analogy with the concept of microscopic reversibility in equilibrium situations.

Onsager was able to show the similarity between this molecular equilibrium reversibility and the idea of reciprocity by looking at backward and forward time transitions in Maxwell's velocity distribution. By contrast, he dismissed on a heuristic argument the possibility of having a unidirectional, truly irreversible, spontaneous reaction such as that shown in FIGURE 7, except for the case of permanent magnets (or other magnetic fields) and Coriolis forces. Although these are restrictive conditions which

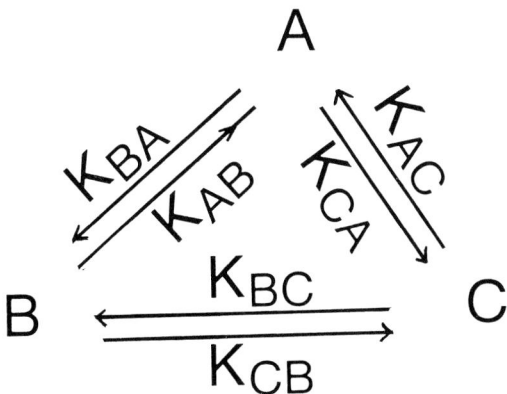

Figure 4.6 The triangular chemical reaction used by Onsager to illustrate reversibility.

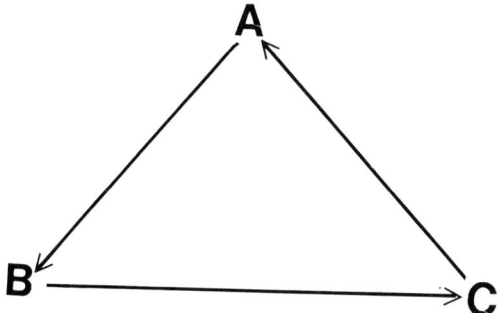

Fig. 4.7 A triangular reaction with unidirectional flows.

will be shown not to be necessary , we retain Onsager's model for purposes of discussion.

If one assumes that a simple mass action law can be postulated to explain the equilibrium transitions, the exchange of mass between any two phases of the molecular reaction can be found by multiplying the concentration at the phase where the motion originates by an appropriate rate factor, so that, for example, the arrow going from A to B in FIGURE 4.6 is associated with the rate of mass transfer $[A] k_{BA}$, while the total mass that appears in a given phase must be obtained by adding the arrows entering a phase and subtracting the arrows leaving the phase. Referring to FIGURE 4.6 the rate at which molecules enter phase B would then be given by

$$d[B]/dt = k_{BA} [A] + k_{BC} [C] - (k_{AB} + k_{CB})[B]$$

Clearly, the same mass action laws could also be specified in terms of the number of moles n_i in each phase. The complete system of kinetic equations becomes, with the postulated linear mass action law ,

$$J_{AB} = k_{BA} [A] - k_{AB}[B] \qquad (57)$$
$$J_{BC} = k_{CB} [B] - k_{BC}[C] \qquad (58)$$
$$J_{CA} = k_{AC} [C] - k_{CA}[A] . \qquad (59)$$

In general, there will be external sources of A,B, and C which, according to the law of conservation of mass will then obey

$$J_A^{ext} = J_{AB} - J_{CA}$$

$$J_B^{ext} = J_{BC} - J_{AB}$$

and

$$J_C^{ext} = J_{CA} - J_{BC} .$$

Such sources allow steady states to be established. In an equilibrium situation, by contrast, all the time derivative vanish so that

$$d[A]/dt = d[B]/dt = d[C]/dt = 0. \qquad (60)$$

In this equilibrium situation it is assumed that reaction rates take place as often forward or backward (clockwise or counterclockwise in the drawing), so that the detailed balance condition

$$k_{AB} k_{BC} k_{CA} = k_{AC} k_{BA} k_{CB} \qquad (61)$$

holds. The present goal is to demonstrate that equilibrium is not a necessary condition for reversibility and, conversely, that reversibility implies a condition of conservative forces through the definition of a potential analogous to the K.V.L. condition which goes beyond the equilibrium restriction. This serves to tie in the principle of "microscopic" reversibility with a "macroscopic" principle--i.e. K.V.L. As a byproduct, this excercise in networks shows how to considerably simplify kinetic diagrams such as those invented by T. Hill .

Given the previous discussion on kinetics, it follows that one possible way of representing the above equations by means of a network--i.e., having a graph which obeys K.V.L., as well as the K.C.L. condition imposed by mass conservation-- is to assign the potential [A] to node A and then proceed by asking what is the conductance which placed across A and B will give the proper rate of flow. Clearly, once [A] is chosen as the A potential this conductance must have the value k_{BA} in order to yield the term k_{BA} [A] in Eq (57). However, once this conductance is chosen it follows that the potential at node B must be given by

(k_{AB} / k_{BA}) [B]

so that the proper backward rate k_{AB} [B] is obtained when the B end is "connected" to this conductance. Following step by step with this process we obtain the potentials

$(k_{AB} k_{BC}) / (k_{BA} k_{CB})$ [C] \qquad (62)

and

$(k_{AB} k_{BC} k_{CA} / k_{AC} k_{BA} k_{CB})$ [A] \qquad (63)

for nodes C and A (upon the return to the starting point) and the conductances

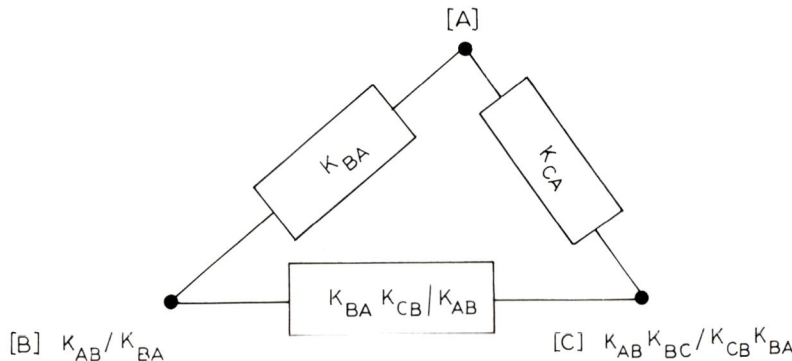

Fig. 4.8 Connected resistive network for the cyclic reaction ABC.

$$(k_{BA} K_{CB} / K^{AB}) \tag{64}$$

and

$$(k^{BA} K^{CB} K_{AC} / K_{AB} K_{BC}) \tag{65}$$

between nodes B and C and C and A, respectively (see FIGURE 4).

The two potentials found for node A must be consistent with each other, whence it follows that

$$(k_{AB} k_{BC} k_{CA} / k_{AC} k_{BA} k_{CB}) = 1 . \tag{66}$$

This last equation is simply the conditions for detailed balance. It is worth pointing out that there is no equilibrium demand on the system, as described,
because we have allowed for external sources, which may lead to a steady state, for example. The important point is that reversibility follows by requiring that a consistent potential set may be defined for a kinetic loop with linear law of mass action. The same will be found to be true for a loop of any length. This can be shown by considering one triangle at a time and then equating end potentials in a given branch to enlarge the loop.

It should be stressed that neither equilibrium nor the lack of dissipation was a necessary condition to derive reversibility, as it holds in the presence of sources and dissipative elements (the resistances). It should also be clear that each resistive branch represents both a direction and the dissipative law associating force and flow in the kinetic system. Thus, **this model associates with each rate a conjugate force**, as expected, linearity is a requirement for the reversibility found, as non linear resistors would have led to inconsistencies with reversible kinetics. But the surprising macroscopic connection is the relationship between K.V.L. and microscopic reversibility.

4.10   STEADY STATE VS. EQUILIBRIUM

A very crucial result which is obtained from the above considerations is: in a closed isolated loop consisting of only linear first order kinetic transitions the only possible steady state which can be attained is equilibrium itself. This can be shown directly by setting up equations of the form

$$J_{AB} = k_{BA} N_A - k_{AB} N_B$$
$$J_{BC} = k_{CB} N_B - k_{BC} N_C$$
$$J_{CD} = k_{DC} N_C - k_{CD} N_D \qquad (67)$$
$$\vdots \qquad \vdots$$
$$J_{ZA} = k_{AZ} N_Z - k_{ZA} N_A$$

and noting that, for the steady state,

$$J_{AB} = J_{BC} = J_{CD} = \ldots = J_{ZA} = J \qquad (68)$$

for which the only possible general solution is $J = 0$. For example, in the triangular reaction considered

$$J_{ZA} = J = k_{BA} N_A - k_{AB} N_B$$
$$J_{BC} = J = k_{CB} N_B - k_{BC} N_C \qquad (69)$$

and

$$J_{CA} = J = k_{AC} N_C - k_{CA} N_A .$$

Multiplying the first equation by k_{CA}, the last equation by k_{BA} dividing both equations by $(k_{CA} + k_{BA})$ and adding the two equations we obtain

$$J = (k_{AC} \; k_{BA} \; N_C)/2(k_{CA} + k_{BA}) - k_{CA} \; k_{AB} \; N_B)/2(k_{CA} + k_{BA}). \qquad (70)$$

By comparing this to the second equation, it is clear that all rates will be unreasonably dependent on each other when reversibility is introduced. We must then conclude that J=0 in general.

The network presents the same result from a different point of view: if there are no forces, there will be no flows; therefore, at equilibrium, when no source is present, the only solution for the flow ("current") in the resistive network is J= 0 .

4.11 NETWORK EXTENSION FOR THE PROBLEM OF MULTIPLE CHEMICAL REACTIONS WITH LINEAR MASS ACTION

The extension of the above ideas to arbitrarily large first order networks such as those considered by T. Hill is straightforward, even though it requires some graph gymnastics. The main problem consists in defining a potential for every node in the network. A straightforward, brute force approach consists in finding an algebraic transformation which would lead to one of the standard forms of the network equations, such as the conductance equations presented in Chapter 1. This can readily be done but it does not add any substantial insight. The corresponding graph approach is more enlightening. We can take advantage of the fact that once potentials are defined along any closed path they will be consistent for the rest of the network (ref. 188).

Begin by choosing an arbitrary Hamiltonian path--i.e., one which goes through every node in the network without crossing itself. Nodes in this path can be labeled 1,2,....n in increasing order and the only branches which will be present are the ordered branches $J_{12}, J_{23}....J_{n1}$, the remaining rates are ignored and are adjusted once all the potentials are found for the Hamiltonian path. The most obvious place to start is at node 1, by assigning to it some concentration (or number of moles, etc), $[N_1]$. The correct rate from 1 to the next node,2, can be obtained by placing a conductance of value k_{21} between 1 and 2, as done before. The backward rate J_{21} will only have the correct value, $k_{12}[N_2]$ if the effective concentration, or potential at 2 is given by

$$[N_2]' = (k_{12}/k_{21})[N_2] \qquad (71)$$

because multiplication of this potential by the conductance (k_{21}) will yield the desired rate. Now we turn to the next step, between 2 and 3. Given the potential just found for node 2, the forward rate

$$k_{32}[N_2] \qquad (72)$$

will only be obtained if the conductance between 2 and 3 is given by

$$G_{23} = k_{21} \, k_{32} / k_{12} \qquad (73)$$

which in turn implies that the backward rate

$$k_{23}[N_3] \qquad (74)$$

will require the potential or effective concentration at 3 to be

$$[N_3'] = (k_{12} k_{23} / k_{21} k_{32})[N_3]. \qquad (75)$$

The pattern that emerges is clear:

1. Every concentration $[N_i]$ must be changed to

$$[N_i'] = (k_{12} k_{23} \ldots k_{-p} k_{pi})/(k_{ip} k_{p-} \ldots k_{32} k_{21}) \, N_i. \qquad (76)$$

2. Every conductance $G_{i,i+1}$ is given by

$$G_{i,i+1} = k_{i+1,i} \, k_{i,i-1} \ldots k_{21} / k_{12} \, k_{23} \ldots k_{i-1,i}. \qquad (77)$$

3. The remaining conductances (in the branches removed) must then be chosen so that the end point rates agree with the required reaction rates.

A prediction of this theory, then, is that many rates in a complex chemical network are dependent. This is a topological dependence, if you will, which is related to the possibility of defining a Hamiltonian path. One can ask further: what is the physical meaning of a Hamiltonian path? First of all. I will loosely assume that all "reasonable" planar graphs have a Hamiltonian path (although at least one graph does not), so that Hamiltonian is to be considered equivalent to "planar". A planar graph, in turn is one which has a dual graph--i.e., one in which the role of covariant and contravariant variables can be exchanged. It then follows that the systems which show reversibilty in the kinetic sense and which lead to K.V.L. are those in which it is possible to describe the system either from the point of view of forces or from the point of view of conjugate flows. This is, again, a purely topological result.

In practical situations it is not necessary to always deal with a Hamiltonian graph. One can apply the same argument on short paths in a single tree, as done in the example below. The advantage of this alternate approach is that the kinetic multipliers are not as confusing as in the case of long paths (a Hamiltonian is a long path, by definition, as it goes through every node in the network.

It should be pointed out that the resultant equations will be reciprocal as they belong to a resistive network.

4.12   EXAMPLE: TRANSPORT COUPLED TO CHEMICAL REACTION

In order to illustrate some of the concepts introduced above, I now consider an example borrowed from T. Hill which involves the transport of two species, 1 and 2 --e.g. Na^+ and K^+ -- across a membrane coupled to a chemical reaction --e.g. the conversion of ATP into ADP + P -- which takes place inside the membrane at constant temperature. Hill considers eight states, as shown in FIGURE 4.9 defined by

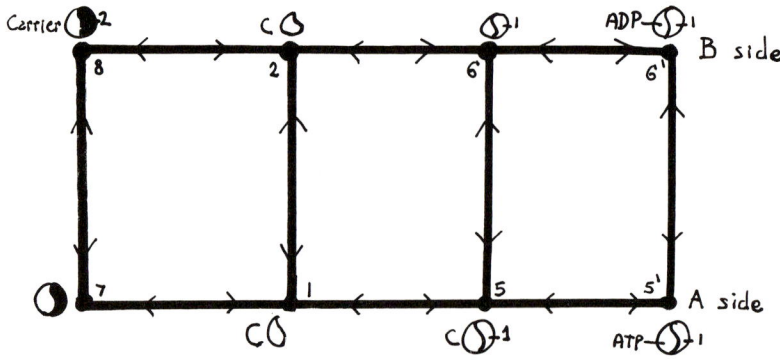

Fig. 4.9 Eight state transport system used by T. Hill (ref. 98)

State 1: carrier in the membrane side adjacent to bath A,

State 2: carrier in the membrane side adjacent to bath B,

State 5: carrier and component 1 complex on side A of the membrane

State 5′: carrier and component 1 complex activated to ATP form on side A of the membrane

State 6′: carrier and component 1 complex in ADP form on side A of the membrane

State 6: carrier and component 1 complex on side B of the membrane

State 7: carrier and component 2 complex on side A of the membrane

State 8: carrier and component 2 complex on side B of the membrane

The following unididrectional flows are first defined by Hill

J_{12}= flow of carrier from side A to side B of the membrane
J_{15}= rate of adsorption of component 1 to carrier on side A of the membrane.
J_{26}= rate of adsorption of component 1 to carrier on side B of the membrane.
J_{28}= rate of adsorption of component 2 to carrier on side B of the membrane.
J_{28}= rate of adsorption of component 2 to carrier on side B of the membrane.
J_{56}= flow of carrier-1 complex from side A to side B of the membrane
$J_{55'}$= rate of activation of carrier-1 complex on side A of the membrane
J_{56}= flow of carrier-1 complex from side A to side B of the membrane
$J_{5'6'}$= flow of carrier-1 complex from side A to side B of the membrane with concurrent chemical reaction ATP ---ADP+P

J_{71} = rate of desorption of component 2 into bath A

$J_{66'}$ = rate of desorption of ADP on side B.

Finally, the following <u>net</u> flows are defined:

J_1 = the net flow of 1 from bath A to bath B

J_2 = the net flow of 2 from bath A to bath B

J_r = the net chemical reaction rate for ATP to ADP conversion.

Note that reversal of subindices reverses the sign of the flow, in agreement with network conventions defining the physical characteristics of the problem. Moreover, the steady state assumption imposes the following restrictions on the flows:

$$J_{55'} = J_{56'} = J_{66'} \qquad (78)$$

which implies that, in the steady state the carrier-1 complex will be activated on side A of the membrane then transported to side B and the deactivated without any accumulation of any of these moieties. Similar steady state restrictions require that

$$J_{28} = J_{87} = J_{71'} \qquad (79)$$

implying that, in the steady state, component 2 will be transported from side B to side A as a complex with the carrier molecule without accumulation of either carrier or component 2 and that the rates of adsorption and desorption are equal. Furthermore,

$$J_{15} = J_{62} \qquad (80)$$

--indicating that the rate of adsorption of component 1 on side A of the membrane equals the rate of desorption of component 1 on side B of the membrane.

A possible network containing only resistances can now be given as shown in FIGURE 4.10, in which we have established the potentials using the procedure shown above on short paths, rather than on a complete Hamiltonian circuit, for convenience,

2→8→7,

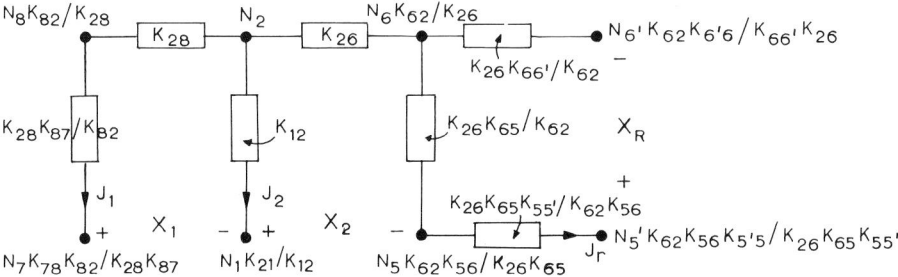

Fig. 4.10 A possible network representing the eight state transport system.

2 → 1,
2 → 6 → 5 → 5´,
and
6 → 6´.

This choice of potentials corresponds to the branches of the tree for which the independent links are the flows J_1, J_2 and J_r. Given the node potentials chosen it is clear that the corresponding conjugate forces will be

$X_1 = (k_{82}k_{78}/k_{28}k_{87}) N_7 - (k_{21}/k_{12}) N_1$

$X_2 = (k_{21}/k_{12}) N_1 + (k_{62}k_{56}/k_{26}k_{65}) N_5$

$X_r = (k_{62}k_{56}k_{55'}/k_{26}k_{56}k_{55'}) N_{5'} + (k_{62} k_{66'}/k_{26} k_{66'}) N_{6'}$.

The kinetic notation may be simplified further by defining the following resistances:

$R_1 = 1/k_{12}$
$R_2 = k_{82}/k_{28} k_{87}$
$R_3 = k_{62} / k_{26} k_{65}$

$R_4 = 1/k_{26}$

$R_5 = k_{56} k_{62} / k_{65} k_{26} k_{55}'$

$R_6 = k_{62} / k_{26} k_{66}'$

and

$R_7 = 1/k_{28}$.

The topologically proper step now consists in defining the loop currents J_1, J_2 and J_r, shown in FIGURE 4.10, as these completely specify the forces at the links and the flows in the tree branches and agree with the net flows defined by Hill. Using the principle of superposition we obtain

$X_r = (R_3 + R_5 + R_6) J_r - R_3 J_1 + 0$

$X_1 = - R_3 J_r + (R_1 + R_3 + R_4) J_1 - R_1 J_2$

$X_2 = 0 J_r - R_1 J_1 + (R_1 + R_2 + R_7) J_2$

Note that all other flows can be expressed as linear combinations of the link flows. For example,

$J_{12} = J_2 - J_1$

and

$J_{56} = J_r - J_2$.

These equations give the relation between independent flows and their conjugate forces in <u>reciprocal</u> form and yield a procedure which is much simpler than the flow graph calculations used by Hill (ref. 98) or the author(ref. 182).

Alternatively, node equations could have been set up as done in Chapter 1. Once the flows are specified in the network, however, all potentials are determined within an additive constant. The reciprocity of the phenomenological equations obtained here is clearly a direct consequence of Tellegen's theorem applied to a resistive network.

4.13 BALANCED LOOPS, TOPOLOGICAL AND METRICAL REVERSIBILITIES

In chapter 3 the connection between Weigschneider's condition (detailed balance) and Kirchhoff's voltage law was considered in the context of chemical kinetics. It is instructive to consider in some further detail the same reversibilities from the probabilistic and from the metrical points of view.

The characteristic of a balanced loop is that the probability of a forward transition is equal, at equilibrium, to the probability of a reverse transition. Given that each of these depends on the intermediate transition probabilities, we can write, for an arbitrary loop:

$$\prod p_{ij} = \prod p_{ji} \qquad (81)$$

or, rearranging,

$$\prod p_{ii}^{forward} = \prod p_{ii}^{reverse} \qquad (82)$$

Explicitly, this is given by

$$(p_{12}\, p_{23}\, \cdots\, p_{n,1}) = (p_{21}\, p_{32}\, \cdots\, p_{1,n}). \qquad (83)$$

which can be readily converted to

$$\prod \ln (p_{ik}/ p_{ki}) = 0. \qquad (84)$$

This is, in essence, a form of K.V.L., which has been studied in detail by M. Propp from the point of view of Markov processes.

This interpretation allows to extend the same topological viewpoint to imbalanced (non equilibrium) loops. If a loop contains a source branch between two nodes i and j, the resultant voltage across the nodes is given by

$$\ln (p_{ik}/ p_{ki}) = v_i - v_k . \qquad (85)$$

which is equivalent to the more familiar

$$(p_{ik}/ p_{ki}) = \exp (v_i - v_k). \qquad (86)$$

This shows the consistency of statistical theory and the introduction of sources to represent imbalanced loops.

A second reversibility which has been implicit in all the kinetic discussions above is of a metrical nature.

In the case of diffusional processes, the forward and reverse kinetic constants are the same, in the case of reactions this is not the case, but using the techniques described in Chapter 3, one basically forces the forward and reverse constants to be the same, while the potentials are redefined. In metric terms a branch resistance

$$X_b = R_b J_b \qquad (87)$$

or the (non-physical, but synthesizable) resistance

$$X_b = -R_b J_b \qquad (88)$$

are reversible, while the "resistance"

$$J_b = C_b [\exp (X_b /a) - 1] \qquad (89)$$

which appears in folding equations leading to non-linearities and constitutes a physical approximation to the ideal diode -- and will be discussed in Chapter 7 in more detail -- is not metrically reversible, as

$$C_b [\exp (-X_b /a) - 1] =/= -J_b. \qquad (90)$$

Generalization away from the steady state requires that the laws of decays of fluctuations in their approach to the steady state be given. As a first approximation, such laws may be assumed to be linear and the steady state can either be viewed as the limit in which Kirchhoff's laws are obeyed and the networks beome connected in the Onsager sense,, or they may be viewed as a linear time decay in the inertial delays and potential build ups.

4.14 MACROSCOPIC RECIPROCITY AND CONNECTIVITY

A positive definite reciprocal matrix $\mathbf{R}$ such as the Onsager phenomenological matrix can always be decomposed in the form

$$\mathbf{R} = \mathbf{Q}^T \mathbf{r} \mathbf{Q}, \qquad (91)$$

in which $\mathbf{r}$ is a diagonal matrix. In the case of the Cut set and Tie set matrices considered in Chapter 1 the matrix entries are +1, -1 and 0 and the complete phenomenological equation represents the three sets of equations that relate loop currents (flows) to link flows through K.C.L., loop branch voltages (forces) to the conjugate currents through Ohm's law (linearity) and port voltages (forces) to the loop voltages.

Thus, Onsager's reciprocal phenomenology always is isomorphic to a resistive network having the macroscopic variables as port variables and can be synthesized in this manner. As in the case

of thermostatics, the bilinear form
$$\Phi = \sum R_{ik} J_i J_k \qquad (92)$$
which represents the resultant length of n non-orthogonal axes $\sqrt{R_{ii}}\, J_i$ which make angles given by $\cos\theta_{ij} = R_{ij}/\sqrt{R_{ii}R_{jj}}$ transforms to the "Pythagorean" metric having the form
$$\Phi = \sum r_k\, j_k^2, \qquad (93)$$
in which the flows j_k are internal currents through the network resistances. This follows directly from Tellegen's theorem, as in the thermostatic case.

4.15 MACROSCOPIC ARGUMENT FOR THE RECIPROCITY OF LINEAR (ONSAGER) SYSTEMS

One can give a macroscopic argument to show that an Onsager description, if self consistent, will always lead to reciprocal equations. (This is not to imply, of course, that the system will be reciprocal, as the model has to do with the assumptions of the theory, rather than with Physics !).

First, we note that Onsager's phenomenology can be represented by black box matrix equations
$$\mathbf{X} = \mathbf{R}\,\mathbf{J}, \qquad (94)$$
for any number of ports. Let an n port system be built by means of smaller one port subsystems having the same characterisitics of the multiport system --i.e., linear phenomenology
$$X = R\,J \qquad (95)$$
and a way for the forces in the subsystems to recognize or match each other (a form of K.V.L.). Although this may not be always be possible, the model itself allows for this type of subsystem to be treated by Onsager thermodynamics. We further require that the total dissipation measured at the ports equal the sums of dissipations inside the subsystems.

Now a corollary to Tellegen's theorem requires that K.C.L. must also be obeyed --because both K.V.L. holds and the dissipation is conserved. It then follows that we are back to dealing with a resistive, connected network which obeys both K.V.L. and K.C.L. It global equations are therefore reciprocal. Note, moreover, that because of Tellegen's theorem the dissipation inside the network must again be equal to that of the macroscopic model.

One can of course argue that there are no Onsager subsystems which could possibly perform the job, because they would have to recognize all kinds of flows. This is physically correct, even though it would not affect the logical result. Moreover, this serves as an motivation for introducing a universal element which is based on fluctuations (a form of flow or displacement quantity) in the microscopic proof given by Onsager.

4.16 STATIONARY PROPERTIES OF THE ONSAGER SYSTEM:UNIQUENESS OF FLOW DISTRIBUTION IN THE STATIONARY STATE

We have indicated that an Onsager system can be represented by an infinite number of resistive, connected networks belonging to the equivalence class of systems with the same dissipation and the same set of port forces (or flows). Moreover, for the most general n-port Onsager metric there are $n(n+1)/2$ resistances in the network with least number of branches which can be constructed.

If we adopt this physical assumption that the resultant network may be considered a decomposition of an Onsager process into individual Onsager processes having a single force and conjugate flow but otherwise having the same properties of the global system--e.g., positive dissipation-- some interesting results follow. We should stress that this is a physical assumption which is not granted by the global principle of positive dissipation alone, which would only require the positive definiteness of the phenomenological matrix and, by Sylvester's law of inertia, the positive definiteness of the principal minors. (It is simple to show that these alone do not lead to positive resistances which involve coefficients of the form $R_{ii} - R_{ij}$).

Assume that a stationary state has been reached so that the solutions to the flows in the system are given by Onsager's phenomenological equations. Keep the boundary forces constant and assume there is another possible distribution of internal flows which would also describe the system. In that case we could construct the differences in forces and in flows at each branch using the two solutions obtained at the networks. Each branch i would then be assigned a dfference flow δJ_i and a difference force δX_i.

Using Tellegen's theorem we obtain

$$\sum_{\text{ports}} \delta X_o \, \delta J_o = \sum_{\text{resistances}} \delta X_i \, \delta J_i \qquad (96)$$

--i.e., the incremental variation in the dissipation function at the ports equals the incremental dissipation inside the network. If the forces (or the flows) at the ports are held constant in both measurements, the summation on the left vanishes.

If the explicit values for the forces across the internal resistances are given in the form of Ohm's law, the right hand side summation becomes

$$\sum_{\text{resistances}} \delta X_i \, \delta J_i = \sum_{\text{resistances}} R_i \, (\delta J_i)^2 = 0 \quad .. \tag{97}$$

Given the asumption that each resistance is positive and the squares of the variations must either be positive or zero, each term in the summation must be identically zero. Thus, it follows that the variations are zero and that the distribution of flows inside a given Onsager decomposition is unique and that the stationary state is stable.

We can now write the expression for the total dissipation in terms of the variations in flows and forces infinitesimally close to the steady state,

$$d\Phi = \sum X_o \, dJ_o + \sum J_o \, dX_o = d\Phi_J + d\Phi_X, \tag{98}$$

in which the summations are taken at the ports. Using Tellegen's theorem for each of the above terms, we obtain

$$d\Phi = \sum J_i \, dX_i + \sum X_i \, dJ_i, \tag{99}$$

in which the summations are now taken inside the network using two separate solutions but the same resistive equations. The constitutive laws for the resistances then yield

$$d\Phi = \sum R_i \, J_i \, dJ_i + \sum J_i \, R_i \, dJ_i \tag{100}$$

$$= \sum 2 \, R_i \, J_i \, dJ_i \, .$$

Consider a situation in which the flow through a given resistance or Onsager subsystem is positive so that the force across the resistance is also positive, by assumption. A positive increment in the flow will then make each term in the summation positive, while a negative variation will lead to negative terms. It then follows that, from this stationary state point of view., the dissipation function has a minimum in the steady state. It should be stressed that this analysis cannot justify a minimum of entropy production for the time history of the process (this question will be raised later, when storage is allowed), as it can only claim that the distribution of flows is a minimum for the possible distributions involving the same boundary forces in the steady state.

The above result can be put in more familiar form, as follows: Given the increments $d\Phi_x$ and $d\Phi_J$ we can write

$$d\Phi_x = \sum J_i \, dX_i = \sum (X_i / R_i) \, dX_i = \qquad (101)$$
$$= \sum X_i \, dJ_i = d\Phi_j ,$$

which can also be expressed in the form

$$d\Phi = 2 \, d\Phi_x = 2 \, d\Phi_J . \qquad (102)$$

This well known result then appears as a special case of Tellegen's theorem when linearity is imposed on the system.

4.17 FUNCTIONS OF STATE IN THE THERMODYNAMICS OF IRREVERSIBLE PROCESSES

Glansdorff and Prigogine (ref. 87) have shown the parallel between stationary states and thermostatics in the sense that the dissipation function behaves as a function of state from which forces and flows may be found. Transformations based on network theory can expand this view, in the case of two forces and two flows, to four "dynamic" functions of state which play the same role as the Gibbs energy potentials in thermostatics.

Given the linear equations of Onsager thermodynanmics,

$$X_i = \sum R_{ik} \, J_k , \qquad (103)$$

we can write, if the coeffficients R_{ik} are invariant, the incremental expression

$$d \, X_i = \sum R_{ik} \, dJ_k , \qquad (104)$$

in which the d's represent either finite or infinitesimal differences. In particular, in the 2x2 case, these incremental equations may be given explicitly as

$$dX_1 = R_{11} \, dJ_1 + R_{12} \, dJ_2 \tag{105}$$

and

$$dX_2 = R_{21} \, dJ_1 + R_{22} \, dJ_2. \tag{106}$$

Moreover, if the Onsager reciprocity holds, we can represent these equations by one of the two port connected resistive networks previously considered ("T" or "Pi") in which the input output variables are now incremental quantitites. The reason we can give a connected network is because of the reciprocity

$$(\partial X_1/\partial J_2)_{J_1} = (\partial X_2/\partial J_1)_{J_2}. \tag{107}$$

Application of Tellegen's theorem to the incremental T network or Pi network yields the General Reciprocity Relation

$$dX_1 \, dJ_1' + dX_2 \, dJ_2' = dX_1' \, dJ_1 + dX_2' \, dJ_2. \tag{108}$$

From this reciprocity relation we can obtain, in addition to the partial derivative equality given above in Eq (107) the following partial equalities

$$(\partial X_1/\partial X_2)_{J_1} = -(\partial J_2/\partial X_1)_{X_2} \tag{109}$$

$$(\partial J_1/\partial X_2)_{X_1} = -(\partial J_2/\partial X_1)_{X_2} \tag{110}$$

and

$$(\partial J_1/\partial J_2)_{X_1} = (\partial X_2/\partial X_1)_{J_2}. \tag{111}$$

These can be interpreted as being the reciprocities in R, L, P and H incremental networks analogous to those given earlier, or as the continuous potentials

$$\partial^2 \sigma_1 / \partial J_1 \partial J_2 = \partial^2 \sigma_1 / \partial J_2 \partial J_1 \qquad (112)$$

$$\partial^2 \sigma_2 / \partial J_1 \partial x_2 = \partial^2 \sigma_2 / \partial x_2 \partial J_1 \qquad (113)$$

$$\partial^2 \sigma_3 / \partial x_1 \partial x_2 = \partial^2 \sigma_3 / \partial x_2 \partial x_1 \qquad (114)$$

and

$$\partial^2 \sigma_4 / \partial x_1 \partial J_2 = \partial^2 \sigma_4 / \partial J_2 \partial x_1 \qquad (115)$$

with the additional equalities

$$X_1 = \left(\partial \sigma_1 / \partial J_1 \right)_{J_2} \qquad (116)$$

$$X_1 = \left(\partial \sigma_2 / \partial J_1 \right)_{X_2} \qquad (117)$$

$$J_1 = \left(\partial \sigma_3 / \partial x_1 \right)_{X_2} \qquad (118)$$

$$J_1 = \left(\partial \sigma_4 / \partial x_1 \right)_{J_2} \qquad (119)$$

$$X_2 = \left(\partial \sigma_1 / \partial J_2 \right)_{J_1} \qquad (120)$$

$$-X_2 = \left(\partial \sigma_4 / \partial J_2 \right)_{X_1} \qquad (121)$$

$$J_2 = - \left(\partial \sigma_2 / \partial x_2 \right)_{J_1} \qquad (122)$$

$$J_2 = \left(\partial \sigma_3 / \partial x_2 \right)_{X_1} . \qquad (123)$$

The set of partial differential equalities given above defines the exact differentials

$$d\sigma_1 = X_1 \, dJ_1 + X_2 \, dJ_2 \qquad (124)$$

$$d\sigma_2 = X_1 \, dJ_1 - J_2 \, dX_2 \tag{125}$$

$$d\sigma_3 = J_1 \, dX_1 + J_2 \, dX_2 \tag{126}$$

and

$$d\sigma_4 = J_1 \, dX_1 - X_2 \, dJ_2 . \tag{127}$$

These are, then, four fundamental dissipation functions which are completely analogous to the thermostatic potentials. The differentials given above may be integrated to yield

$$\sigma_1 = \sigma + J_2 X_2 \tag{128}$$

$$\sigma_2 = \sigma \tag{129}$$

$$\sigma_3 = \sigma - J_1 X_1 \tag{130}$$

and

$$\sigma_4 = \sigma - J_1 X_1 + J_2 X_2 . \tag{131}$$

I have arbitrarily chosen σ_2 as the "fundamental" dissipation σ_1, but any other could have been taken as the starting point in the derivation.

In complete analogy to the formal unification given for the thermostatic ports, we can use these results to obtain a unification of Onsager forces and flows. By reference to equations (117) and (120) we can first define a connected manifold in which o_1 is the connecting, continuous potential and the independent variables are J_1 and J_2. The force vector defined in the direction of the flow vectors is given by

$$\mathbf{X} = \partial \sigma_1 / \partial J_1 \; \mathbf{j_1} + \partial \sigma_1 / \partial J_2 \; \mathbf{j_2} , \tag{132}$$

in which $\mathbf{j_1}$ and $\mathbf{j_2}$ are unit vectors. This can be described more concisely as

$$\mathbf{X} = \mathrm{grad}(\, \mathbf{j_1} \, , \, \mathbf{j_2} \,) \; \sigma_1 , \tag{133}$$

which stresses the connection between taking K.V.L. around a loop and taking the gradient "in the direction of **J**" to obtain a force from a potential.

Clearly, this operation can be repeated for the four potentials. If we generalize the direction vector given by the independent components as C_i (for "cause i") and the resultant vector as E_i (for "effect i"), the general equation is

$$\mathbf{E} = \text{grad}_C \, \sigma, \qquad (134)$$

which represents the equations

$$E_1 = \text{grad}(C_1) \, \sigma_1, \qquad (135)$$

$$E_1 = X \qquad (136)$$

$$C_1 = J_1 \, j_1 + J_2 \, j_2 ; \qquad (137)$$

$$E_2 = \text{grad}(C_2) \, \sigma_2, \qquad (138)$$

$$E_2 = X_1 \, j_1 - J_2 \, x_2 \qquad (139)$$

$$C_2 = J_1 \, j_1 + X_2 \, x_2 \qquad (140)$$

$$E_3 = \text{grad}(C_3) \, \sigma_3, \qquad (141)$$

$$E_3 = J \qquad (142)$$

$$C_3 = X_1 \, x_1 + X_2 \, x_2; \qquad (143)$$

and

$$E_4 = \text{grad}(C_4) \, \sigma_4, \qquad (144)$$

$$E_4 = J_1 \, x_1 - X_2 \, j_2 \qquad (145)$$

$$C_4 = X_1 \, x_1 + J_2 \, j_2 \qquad (146)$$

Similar expressions can be given for problems with more than two ports, as done, for example, for the chemical potential problem.

Appendix

CHANGES IN REFERENCE FRAME WITH RESPECT TO BARYCENTRIC VELOCITY

We can utilize the dual network transformations in order to demonstrate the topological basis of a well known theorem proved by Prigogine (ref. 197), in which he demonstrated that the barycentric velocity--which serves as a reference in the definition of diffusional flows -- can be replaced by an arbitrary velocity.

The general conditions under which an Onsager describable isothermal diffusion process remains in the class of Onsager processes upon a change in the frame of reference has considerable theoretical and practical interest. The problem is relevant not only in the case of diffusion of liquids, but also in the case of diffusion across membranes. From the practical point of view, part of the problem arises in changing from the fundamental diffusion coefficients (which relate flows and chemical potential gradients) to the practical diffusion coefficients (which relate the diffusional flows to concentration gradients). A more basic problem, however, has to do with the mechanical equilibrium condition

$$-\text{grad } P + \sum_i F_i \rho_i = 0 \qquad (1)$$

which, in the case of isothermal diffusion with no reacting species, leads to the familiar dissipation expression

$$\Phi = \sum_i J_i \text{ grad } (-\mu_i^c). \qquad (2)$$

These equations are valid when referred to the center of mass and it is not immediately obvious what changes would be required if the reference frame chosen was referred to a solvent fixed frame, a volume fixed frame, etc., which are--except for the trivial case of zero concentration differences-- different from that of the velocity of the center of mass relative to the experimental cell.

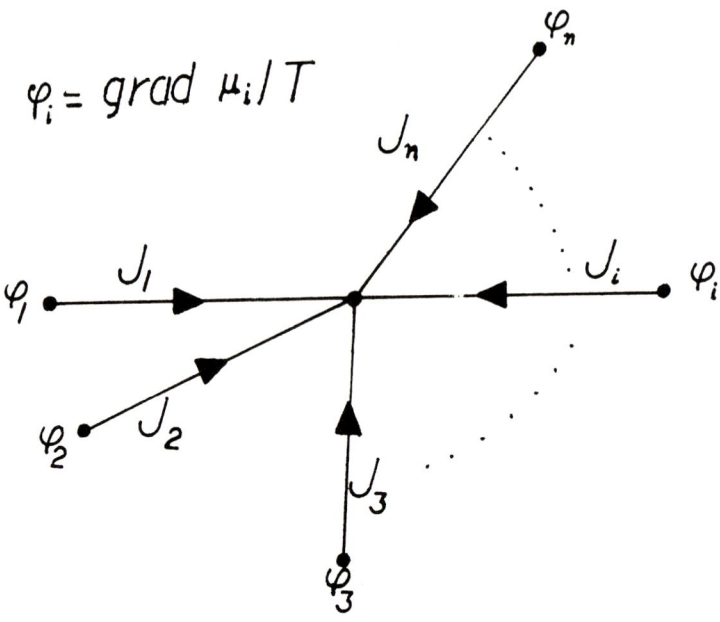

FIGURE A.4.1   Topological representation of the reference diffusional problem.(ref. 86).

Prigogine established sufficient conditions under which the changes in the frame of reference would lead to another valid Onsager process and Kirkwood et al.(ref.119), analyzed many cases of isothermal diffusion referred to various frames of reference and derived specific conditions under which changes in frame of reference would retain Onsager's reciprocities (ref.168- 9). Although this work was extremely important from a practical point of view, it failed to reveal any mathematical reasons why changes in reference velocities would leave the dissipation invariant.

The apparent assumptions required in Prigogine's theorem are: (1) mechanical equilibrium, (2) the Gibbs-Duham equality, (3) linearity and (4) Onsager's reciprocities. From a graphical/topological point of view the source of the invariant character of these transformation seems to be rooted in the duality between intensive and extensive variables, or in force flow descriptions. In the following , we stress the non-metric characteristics of the transformations by omitting resistances and reduce the networks to simple arrows connecting nodes at which potentials are defined.

The diffusional flow of component i is defined as

$$J_i = \rho_i (v_i - v) \qquad (i=1,2,\ldots,n) \qquad (3)$$

in which v_i is the velocity of component i, ρ_i is its density, and v is the barycentric velocity. The set of Eqs. (2) for an n component system may be collectively represented by the network drawing shown in FIGURE 1, in which the potentials ϕ_i at the nodes are measured relative to some arbitrary reference value $O_G = 0$ (the "ground") and are given by

$$\phi_i = (-\text{grad } \mu_i)/T \qquad (i=1,2,\ldots,n) \qquad (4)$$

The conjugate flows J_i are simply indicated by arrows in the branches of the network. The total dissipation --found from the products of flows and forces at the network branches--is the bilinear form

$$\Phi_i = \sum_i J_i (-\text{grad } \mu_i)/T \qquad (i=1,2,\ldots,n) \qquad (5)$$

The barycentric velocity is given by

$$v = \sum_{i=1}^{n} \rho_i v_i \qquad (6)$$

and, by definition,

$$\sum_{i=1}^{n} \rho_i = 1. \qquad (7)$$

At the reference node in the network, Kirchhoff's current law holds because

$$J_{ig} = \sum_{i=1}^{n} \rho_i (v_i - v) = \sum \rho_i v_i - \sum \rho_i v = v - v = 0 \qquad (8)$$

The network may be given more resolution by introducing the form of the individual flows, J_i, explicitly, as given by Eq (3). Thus, each input branch in the multiport network will now be specified in terms of the component velocities, as indicated in

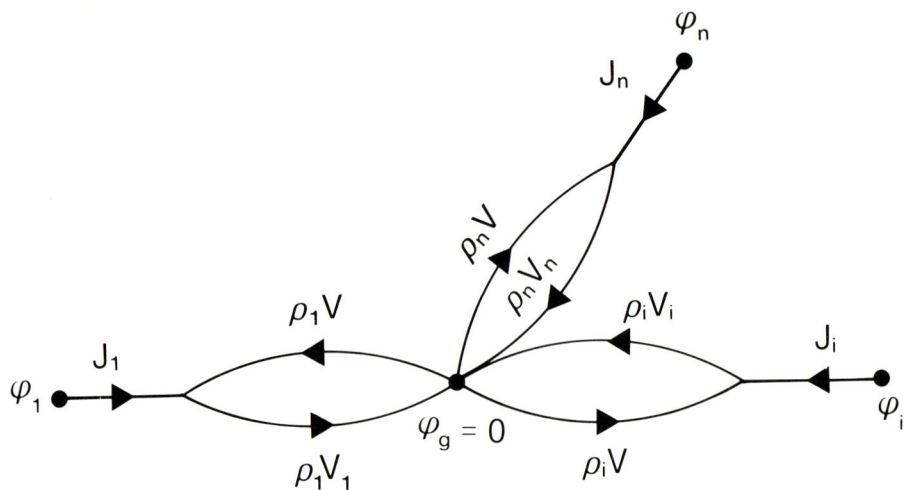

FIGURE A.4.2 Explicit representation of flows in terms of velocities.

FIGURE 2. We can now divide each flow J_i by the corresponding density ρ_i, which leaves the dissipation invariant provided that all potentials are multiplied by the corresponding ρ_i, as shown in FIGURE 3. In addition, if we assume that the entries in the original phenomenological equations are linear and have the form

$$X_i = \sum_k R_{ik} J_k \qquad (9)$$

as required by Onsager's theory, the new equations will have the form

$$\rho_i X_i = \sum_k R_{ik}' J_k' = \sum_k R_{ik}'(J_k/\rho_k). \qquad (10)$$

After substituting X_i given by Eq (9) into Eq (10) we obtain the transformation expression for R_{ik}',

$$R_{ik}' = \rho_i \rho_k R_{ik}. \qquad (11)$$

We can now construct a dual network, as shown in FIGURE 4 in which the roles of forces and flows are exchanged. The expressions

$$J_i/\rho_i = v_i - v \qquad (12)$$

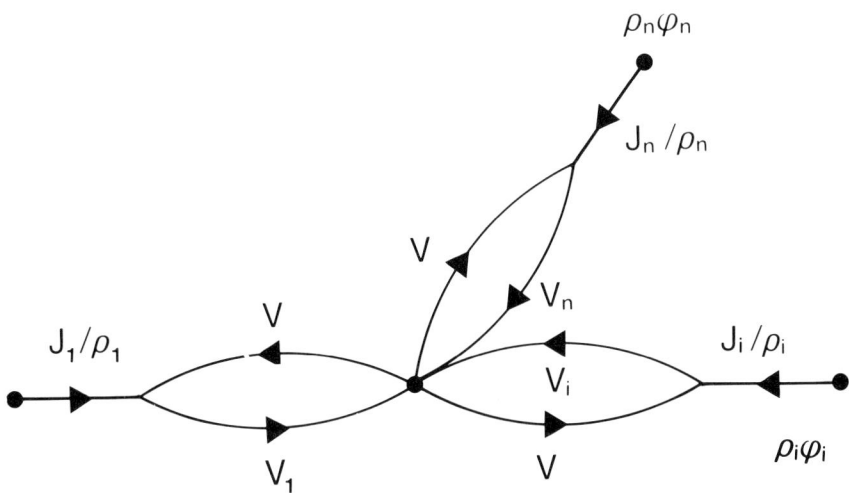

FIGURE A.4.3 Transformed diffusional network.

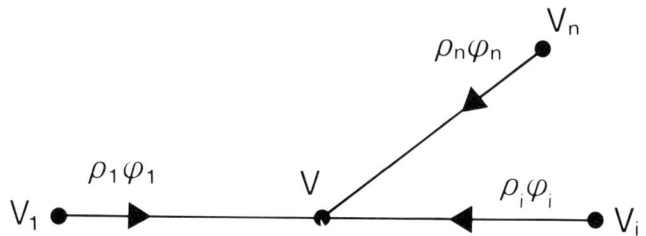

FIGURE A.4.4 Dual of the above network.

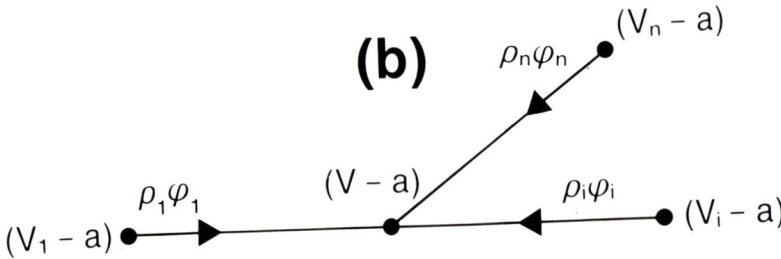

FIGURE A.4.5 Offset dual.

are now interpreted as <u>force differences</u> between potentials v_i and v in the dual network, while the expressions (ρ_i grad μ_i)/T are to be considered flows. Note that, because of the operations just performed there is a node potential ϕ_g at which all the branches converge; this potential can be adjusted arbitrarily provided that the same reference (offset) potential is either added or substracted to or from all the potentials in the network. Let the offset potential be the quantity -a and add this potential to all the nodes, as shown in FIGURE 4. Once this step is taken we can transform back to the original form of the graph in which flows are the real flows and forces are the real forces. Finally, in a reversal of the original operation, we multiply all flows by ρ_i without changing the dissipation, so that all forces and flows are restored to their original form, as shown in FIGURE 5. From a macroscopic point of view, this network is equivalent to the original system, even though now the flows seem to be different. These flow transformations--which simply offset all velocities by an arbitrary velocity -- leave the system invariant. This is then the toplogical essence of Prigogine's theorem, which is based on the indistinguishability of dual representations.

Note, in passing, that K.C.L. in FIGURE 4 is simply a form of the Gibbs-Duhem equations

$$\sum_{i=1}^{n} \rho_i \, (\text{grad } u_i)/T = 0 \qquad (13)$$

which was not assumed in this derivation, but is clearly implicit in the statement of the problem.

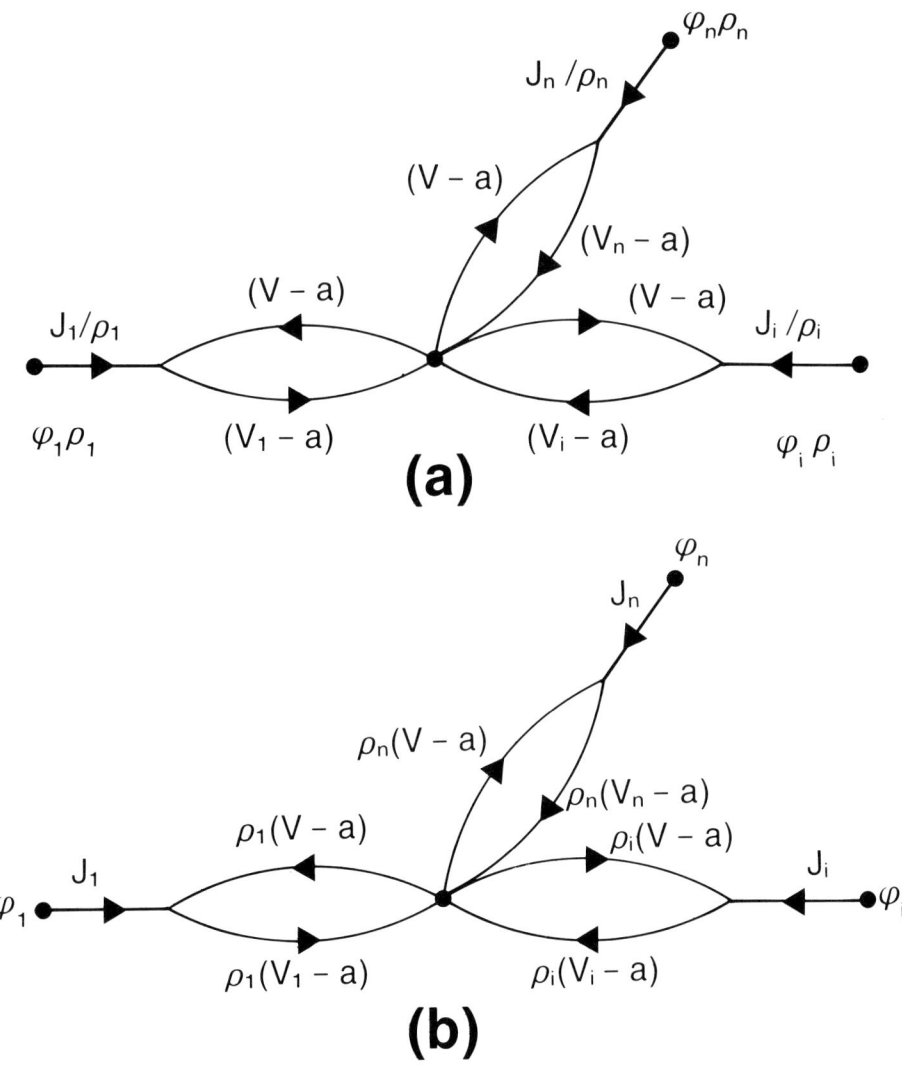

FIGURE A.4.6 (a) Dual network and (b) Offset dual returning to the original problem.

Chapter 5

MACROSCOPIC SYMMETRIC AND HYBRID NETWORKS

5.1 INTRODUCTION

The preceding chapters have stressed the relationship between connectivity and reciprocity. In particular, I have attempted to show the reciprocities which are built into classical and Onsager thermodynamics through the idea of connectivity--or, equivalently, the assumption that there is a potential which can be consistently defined along an arbitrary physical path.

This connectivity can be represented easily in the case of a system with two degrees of freedom (a two port) by means of three resistance networks. The simplest such networks are the **T** and **Pi** networks introduced in Chapter 1 in a general context. The T network follows naturally from the reciprocal **R** network, as shown in FIGURE 1.24, while the Pi network follows from the reciprocity of the L network, as shown in FIGURE 1.25.

Thus, the R and L networks --which represent the fundamental contravariant and covariant tensors, respectively -- contain information regarding the connectivity of the manifold. This connectivity can also be brought out, as in the case of thermostatic processes in terms of the Pythagorean imbedding discussed in Sections 2.7 and 4.5.

In this chapter we take a different approach. Rather than stress the descriptions which lead to connectivity, we shall look at descriptions which are consistent with Onsager thermodynamics but are antireciprocal, or antisymmetric. As wil be indicated below, these have the characteristic that the transducing step is non-energenic.

There are several reasons for going through this excercise. First, it demonstrates that there are some non Onsagerian irreversible models--e.g., Thomson´s model of the Peltier heat and Odum and Pinkerton´s model of energy conversion (ref. 165) -- which are isomorphic to the Onsager formalism. Second, this approach leads to Onsager consistent equations which are simpler and have more physical meaning than the resistive or conductive equations. Third, the antisymmetric approach allows to extend the

Onsager model to obtain both non-linear phenomenological equations and to analyze non-reciprocal systems which are otherwise thermodynamically "proper".

Following the analysis of the hybrid two ports, a review of energy conversion characteristics of both reciprocal and antireciprocal machines is undertaken following some of my thesis work (ref. 182), including the extensions to multiple ports and a possible application to thermostatic processes.

5.2 DUALITY BETWEEEN ONSAGER FORMULATIONS AND HYBRID EQUATIONS

We have indicated that the Onsager and hybrid formulations are completely equivalent, both from the point of view of input output properties and covariant relationship between forces and flows. Moreover, it has been pointed out that, in the particular case of reciprocal equations, the Onsager R and L boxes lead to connectivity. By contrast, the hybrid descriptions become anti-reciprocal in the Onsager limit ($H_{12} = -H_{21}$ and $P_{12} = -P_{21}$) and lead to "engines" in which the coupling or transduction step is nonenergenic--i.e., the power or dissipation that goes into the transduction step equals the amount that comes out of the transduction step.(Note, however, that the dissipation measured at the ports is the same as before!). As a result, the quasimetric expression which describes the H and P boxes does not contain cross terms and corresponds to a pythagorean expression which mixes a force and a flow, as in the case of thermostatics previously treated.

Do the Hybrid machines **H** and **P** have any physical role? The basic characteristic of these antisymmetric machines (tensors) is that the power, or dissipation, used in the transducing step is zero. For example, the total dissipation of the H box is given by

$$\Phi = H_{11} J_1^2 + H_{22} X_2^2 \qquad (1)$$

because the input transducer amount

$$H_{12} X_2 J_1 \qquad (2)$$

is cancelled out by the output transduction amount

$$-H_{21} X_2 J_1. \qquad (3)$$

This observation has some direct applications, as discussed below.

5.3 THOMSON ANALYSIS OF THE PELTIER HEAT

Thomson(ref. 247) proposed a theory of the Peltier heat which could be considered the first "irreversible" theory of thermodynamics even though it has been labelled pseudoirreversible by Onsagerists, perhaps to point out that it is "not as good" as Onsager's. In fact, it predicts as much as the Onsager theory --at least in the realm of 2x2 processes. This becomes more evident when the network paradigm is superimposed on it.

The central part of Thomson's model consists in postulating a reversible step for energy conversion in the actual transducing step. In this view -- and using the Onsager formalism (ref. 168)-- the dissipation function can be written as follows:

$$\Phi L_{11} = J_1^2 - L_{12} J_1 X_2 + L_{21} J_1 X_2 + \\ + X_2^2 (L_{11} L_{22} - L_{12} L_{21}), \qquad (4)$$

so that the cross terms vanish.

Although the argument is given that both the mixing of forces and flows and the "reversible" step is not as proper as Onsager's theory, it is seen that

$$(1/L_{11}) = H_{11}, \qquad (5)$$

$$(-L_{12}/L_{11}) = H_{12}, \qquad (6)$$

$$(L_{21}/L_{11}) = H_{21} \qquad (7)$$

and

$$(\det L/L_{11}) = H_{22}, \qquad (8)$$

so that the dissipation reduces to

$$\Phi = H_{11} J_1^2 + H_{12} J_1 X_2 + H_{21} J_1 X_2 + H_{22} X_2^2. \qquad (9)$$

This is the dissipation function (or power dissipated, in network terms) by the H network. Given that it is completely isomorphic, from the point of view of the outputs or ports to either the L or R networks, it follows that there is basically no difference between the Onsager´s model and the Thompson model.

In fact, the H and P machines are neither more or less reversible than the R and L machines. By introducing the conversion factors given in TABLE 1 the equivalence of all the dissipations can be demonstrated.

The more interesting impact of these representations is the simplification of the derivation of proper thermodynamic equations without resorting to the use of the dissipation function and involved Onsager calculations. To indicate this fact, some basic Onsager results will be rederived in this light. It should be stressed, however, that these results will be valid only if there is a linear phenomenology and it is not implied in any way that Onsager´s thermodynamics has a long range validity beyond this range.

The following examples --Kedem Katchalsky equations, electrical phenomena, etc.-- are not new results in terms of the final equations obtained; what is new is the possibility of deriving these from practical, common sense measurements rather than from some involved theory. In order to contrast the Onsager approach to the network engine approach, the procedure for deriving the Kedem Katchalsky equations using the Onsager theory will be reviewed first.

5.4 THE KEDEM-KATCHALSKY EQUATIONS IN ONSAGER THERMODYNAMICS

The well known transport equations proposed by Kedem and Katchalsky (ref. 112) to describe the interaction between water and solute flows in a membrane follow from the following general recipe of Onsager thermodynamics:

i) Find the dissipation function,

ii) Transform the dissipation function to include proper practical forces and flows, and

iii) Use the proper forces and flows to give macroscopic

phenomenological equations. (These will then be reciprocal in the Onsager model).

In general terms, the derivation proceeds as follows. First, the dissipation function is written in terms of products of flows and chemical potential differences across the membrane,

$$\phi = J_w \Delta u_w + J_s \Delta u_s , \qquad (10)$$

in which J_w is the water flux, $\Delta \mu_w$ is the chemical potential difference for water across the membrane, J_s is the solute flux and $\Delta \mu_s$ is the chemical potential difference for the solute across the membrane.

Introducing the explicit expressions for the chemical potentials in dilute solutions and rearranging terms, it is easy to show that the dissipation function becomes

$$\phi = (J_w \overline{V}_w) \Delta P + [(J_s/\overline{c}_s) - (J_w \overline{V}_w)] \qquad (11)$$

in which c_s is the average concentration of solute in the membrane; $\Delta \pi$ is the ideal osmotic pressure, RT Δc_s, V_w and V_s are the partial molar volumes of water and solute, respectively; and ΔP is the hydrostatic pressure difference across the membrane. According to Onsager thermodynamics, the dissipation function leads to products of proper conjugate flows and forces. Kedem and Katchalsky rewrote the dissipation function as

$$\phi = J_v \Delta P + J_D \Delta \pi \qquad (12)$$

in which

$$J_v = (J_w \overline{V}_w) + J_s \overline{V}_s \qquad (13)$$

is defined as the volume or bulk flow and

$$J_d = (J_s/\overline{c}_s) - \overline{V}_w J_w \qquad (14)$$

is the diffusional flow, which is about equal to the difference in velocities of solute and water and dilute solutions. Onsager's model then indicates that linear reciprocal equations can be set up :

$$J_V = L_P \Delta P + L_{PD} \Delta \pi \qquad (15)$$

$$J_D = L_{Dp} \Delta P + L_D \pi , \qquad (16)$$

in which L_P is the hydraulic coefficient, L_{PD} is the ultrafiltration coefficient, L_{DP} is the osmotic flow coefficient and L_D is the diffusional coefficent. Onsager theory considers only linear reciprocal situations in which $L_{PD} = L_{DP}$. This is one of its most restrictive limitations, as these reciprocities do not always hold in real systems. Morevoer, it also excludes, or, rather, it ignores practical formulations of the hybrid form. Kedem and Katchalsky therefore transformed their equations to the practical form

$$\Delta P - \Delta \pi = [\bar{c}_s (1-\sigma)^2 L_p + w] J_V/wL_p \quad - (1-\sigma)J_s / w \qquad (17)$$

$$\Delta \pi / c_s = -(1-\sigma) J_V/w \quad + J_s/w\bar{c}_s , \qquad (18)$$

in which σ is the Staverman reflection coefficient, w is the solute mobility in the membrane and c_s is the average concentration of the solute inside the membrane.

There are two interesting points that become rather obvious. First, the "practical" equations of Kedem and Katchalsky contain the physical measurements hidden in such an involved form -- especially considering the simplicity of the system-- that one would think there may be a simpler formulation of the phenomenological equations. Second, it is simple to use algebra to transform the Onsager equations to Staverman's equations, which are known from experiment and make intuitive sense,

$$J_V = L_p(\Delta P - \sigma \Delta \pi) \qquad (19)$$

$$J_s = c_s(1-\sigma) J_V + w\Delta \pi, \qquad (20)$$

but it is impossible to proceed in the reverse direction using only Onsager thermodynamics, but excluding the dissipation function. And, of course, **there is no way to proceed if the experiment happens to show that**

$$(J_V/\Delta \pi)_{\Delta P=0} \qquad (21)$$

does not equal

$$(J_D / \Delta P) \Delta \pi = 0 . \tag{22}$$

5.5 DERIVATION OF KEDEM-KATCHALSKY EQUATIONS USING NETWORK TRANSFORMATIONS

We now proceed to "derive" Kedem-Katchalsky equations, in a more usable form using only practical equations and the common network transformations. Basically, we utilize the following ideas:

1. <u>Network Representation</u> . Any set of linear equations may be represented by resistances and sources.

2. <u>Kirchhof's laws</u>.a) Kirchhoff's Voltage law allows to split a voltage (force) source into the sum of two or more voltages. In particular, a battery may be split into two or more batteries, a resistance may be split into two resistances in series with the same through flow (current),etc.

    b) Kirchhoff Current law allows to split a current source into two or more parallel current (flow) sources and a resistance into two parallel resistances with the same voltage (force) across.

3. <u>Onsager Thermodynamics</u> consists of looking for either
    a) A connected resistive network or
    b) A disconnected R or L symmetric representation or a disconnected H or P antisymmetric representation. The latter simply says that the actual transduction step delivers to the output the same "power" plugged into it.

We begin with the practical, Staverman's form of the transport equations and represent them by a disjoint network, as shown in FIGURE 1 , in which the input flow J_v and force ($\Delta P - \sigma \Delta \pi$) see an internal resistance $R = 1/L_p$ and the output flow J_s and force $\Delta \pi_s$ see a conductance w in parallel with a flow source $c_s (1-\sigma) J_v$. This is the most direct and obvious way to represent these equations by means of a two port network. The problem consists in finding an H or P representation which is antisymmetric. Given that a flow controlled flow source is given at the output in parallel with a conductance, comparison with FIGURE 1 shows that we must transform the input circuit so that

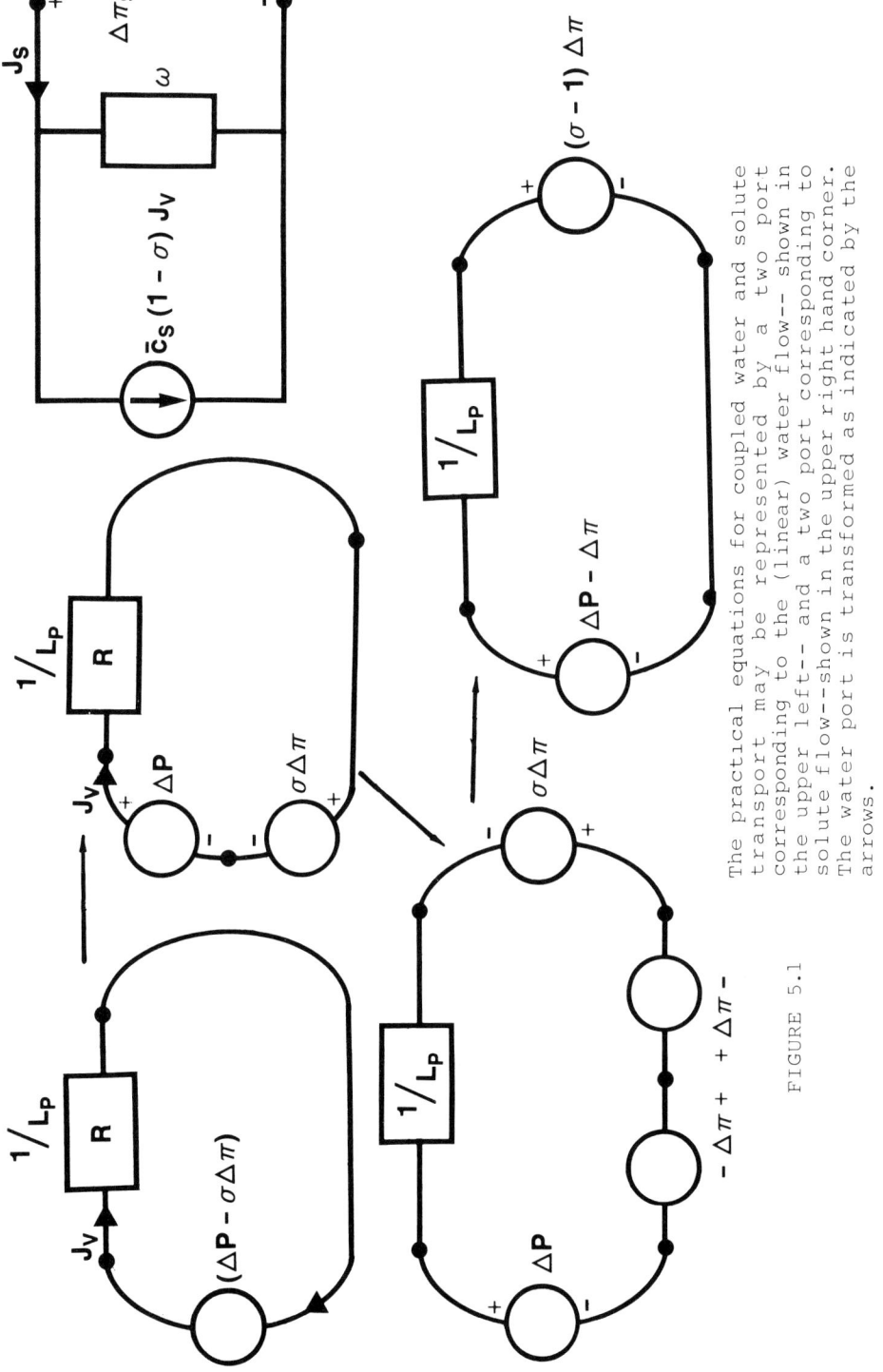

FIGURE 5.1 The practical equations for coupled water and solute transport may be represented by a two port corresponding to the (linear) water flow-- shown in the upper left-- and a two port corresponding to solute flow--shown in the upper right hand corner. The water port is transformed as indicated by the arrows.

the resistance is in series with a force controlled force source.

The first step consists in splitting the input force into a sum of batteries and assuming, as a first approximation, that the $\sigma \Delta\pi$ source is the internal controlled source we are searching (note that this is reasonable as the output force is $\Delta\pi$). When this introductory transformation which just involves application of K.V.L.-- is done following the steps shown in FIGS. 1.c and 1.d the final network is that obtained in FIGURE 1.e.

This network is not consistent with Onsager thermodaynmics, because it is not an antireciprocal H box. However, this can be readily achieved by adding and subtracting a force source $\Delta\pi$ to the input circuit, then associating one of these sources with the input and one with the internal part of the network (so that the internal source is now $(\sigma - 1)\Delta\pi$ and then multipliying and dividing this source by $\bar{c}_s$. As a result of these changes the controlling, output force seen by the input is now $\Delta\pi/c_s$. In order to have the proper force at the output it is necessary to assign this new force as the output force. Finally, in order to have the correct output flow it will be necessary to multiply the output conductance by w. The final network is shown in FIGURE 2, corresponding to the hybrid H matrix

$$H = \begin{bmatrix} 1/L_p & -(1-\sigma)\bar{c}_s \\ (1-\sigma)\bar{c}_s & w\,\bar{c}_s \end{bmatrix} \quad (23)$$

These equations are antisymmetric and therefore conform to the Onsager model. Note that this matrix--which has been derived solely from the assumptions of network theory and the practical phenomenological equations -- has as its entries the simplest physical parameters for this problem. This result, incidentally, cannot be obtained directly following Onsager's approach.

The **H** matrix leads to the following practical definitions of membrane coefficients:

$$1/L_p = (\Delta P - \Delta\pi / J_v)\Delta\pi/\bar{c}_s \quad (24)$$

$$-(1-\sigma)\bar{c}_s = (J_s/J_v)_{\Delta\pi/\bar{c}_s = 0} = -\left((\Delta P - \Delta\pi)/(\Delta\pi/c_s)\right)_{J_v=0} \quad (25)$$

and

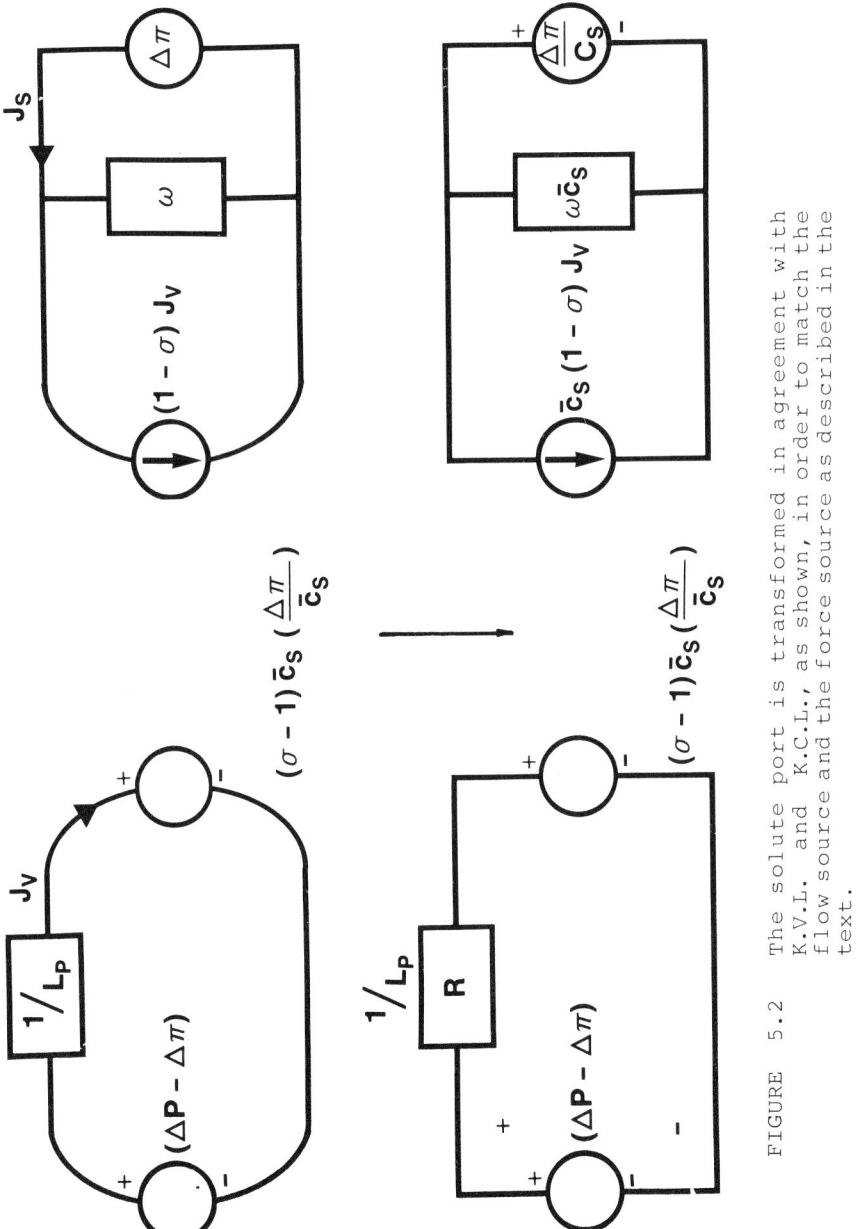

FIGURE 5.2 The solute port is transformed in agreement with K.V.L. and K.C.L., as shown, in order to match the flow source and the force source as described in the text.

$$w\, c_s = (J_s)/(\Delta\pi/\bar{c}_s)J_{v=0}. \qquad (26)$$

Although the matrix **H** corresponds to the hybrid equations

$$\begin{bmatrix} \Delta P - \Delta\pi \\ J_s \end{bmatrix} = H \begin{bmatrix} J_v \\ \Delta\pi/c_s \end{bmatrix}, \qquad (27)$$

it can readily be transformed to **L** and **R** coefficients using conversion TABLE 1.1. The corresponding matrices are

$$L = \begin{bmatrix} L_p & (1-\sigma)\bar{c}_s L_p \\ (1-\sigma)\bar{c}_s L_p & (1-\sigma)^2 \bar{c}_s^2 L_p + w\,\bar{c}_s \end{bmatrix} \qquad (28)$$

and

$$R = \begin{bmatrix} [(1-\sigma)^2 \bar{c}_s L_p + w]/L_p w & (\sigma-1)/w \\ (\sigma-1)/w & 1/wc_s \end{bmatrix} \qquad (29)$$

Clearly, the **L** and **R** matrices are more involved than those given by the simpler hybrid formalism, which leads to equations having more physical significance.

5.6 LIQUID JUNCTION POTENTIALS

In the case of motion of charged molecules in the presence of osmotic and hydrostatic gradients, the dissipation function includes additional electrical terms,

$$\Phi = J_s \Delta\pi + J_v \Delta P + I\, E. \qquad (30)$$

We next consider a network approach to deriving proper equations without the need of invoking a dissipation function.

The practical equations describing linear junction potentials for the motion of a uni-univalent ionic compound may be given by

$$J_s = w\Delta\pi + \tau I/F \qquad (31)$$

$$I = \tau K\Delta\pi/Fc_s + K E, \qquad (32)$$

in which J_S is the flow of salt through the membrane, $I = (J^+ + J^-)$ is the electrical current density; T_i is the transference number, K is the membrane conductivity, F is the Faraday constant and other variables are the same as above. The practical equations can be conveniently represented by the network shown in FIGURE 3a, which has two controlled sources. If the input circuit is changed using the Thevenin-Norton substitution theorem, as shown in FIGURE 3b, a proper hybrid, antisymmetric H representation is obtained. The resultant H matrix is

$$\begin{bmatrix} E \\ J_S \end{bmatrix} = \begin{bmatrix} 1/K & -T_i/F \\ T_i/F & w\,c_S \end{bmatrix} \begin{bmatrix} I \\ \Delta\pi/c_S \end{bmatrix}. \quad (33)$$

Again, this representation has some obvious advantages over the common L or R matrices. The main advantage resides in the possibility of spelling out what electrokinetic measurements should be performed to describe the basic physics of the system with minimal effort:

$$1/K = (E/I)_{\Delta\pi_S/\bar{c}_S = 0} \quad (34)$$

$$T_i/F = (J_S/I)_{\Delta\pi_S/\bar{c}_S} = -(E/\Delta\pi/c_S)_{I=0} \quad (35)$$

and

$$w\bar{c}_S = J_S/(\Delta\pi/\bar{c}_S)_{I=0}. \quad (36)$$

These coefficients can also be expressed in terms of R coefficients using the conversion table, to obtain

$$R_{11} = (H_{11} H_{22} - H_{12} H_{21})/H_{22} = \quad (37)$$
$$= (w\bar{c}_S/K) + (T_i^2/F^2)/wc_S \quad (38)$$
$$= (1/K) + (T_i^2/w F^2 c_S), \quad (39)$$

$$R_{12} = R_{21} = H_{12}/H_{22} = -H_{21}/H_{22} = -T_i/Fc_S w \quad (40)$$

and

$$R_{22} = 1/H_{22} = 1/w\bar{c}_S. \quad (41)$$

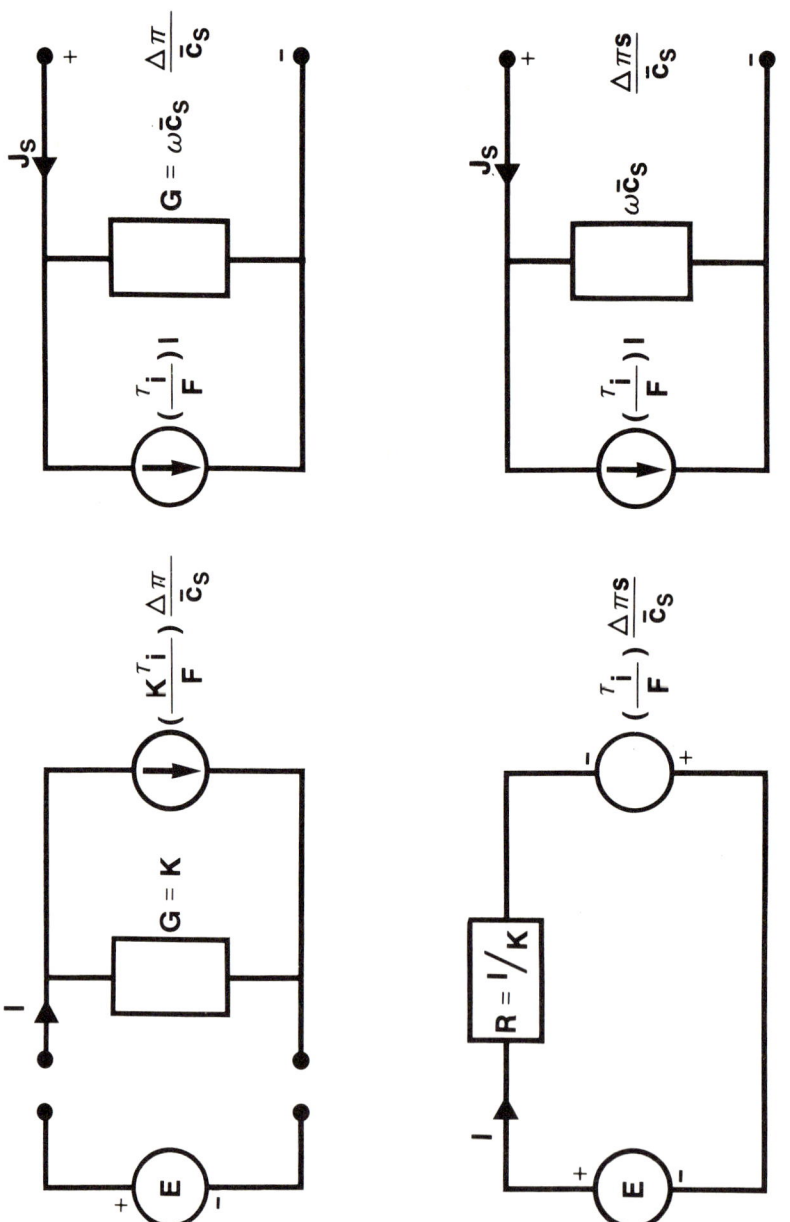

Fig. 5.3   Transformation of practical equations for the electrokinetic problem.

These coefficients correspond to the **R** matrix

$$\begin{bmatrix} E \\ \Delta\pi/c_s \end{bmatrix} = \begin{bmatrix} (1/K) + (\mathcal{T}_i^2/F^2 w c_s) & -\mathcal{T}_i/F\,\bar{c}_s w \\ -\mathcal{T}_i/F\,\bar{c}_s w & 1/\bar{c}_s w \end{bmatrix} \begin{bmatrix} I \\ J_s \end{bmatrix} \quad (42)$$

which are in agreement with the results obtained using Onsager's theory. Clearly, the R and L matrices are more involved and have less physical meaning than the hybrid formulations.

5.7 ELECTROKINETIC PHENOMENA

The volume flow caused by the passage of current through a porous membrane can be described by the practical equation

$$J_v = L_p P + \beta I \quad (43)$$

in which

$$\beta = (J_v/I)_{\Delta P = 0} \quad (44)$$

and

$$L_p = (J_v/\Delta P)_{I=0} \quad . \quad (45)$$

If we accept the premise that there is an H or P type hybrid description which accounts for the conservation of power or dissipation at the transducing step, there must be a companion equation (Fig. 4).

$$E = -\beta \Delta P + RI \quad (46)$$

--i.e., the presence of a hydrostatic pressure difference in the absence of electric current will also lead to a potential difference across the membrane. (Of course, this is known; it is only used as an example of how hybrid equations compare with the Onsager treatment). The parameters are in the P form, but we can also give an H equation

$$I = H_{21} J_v + H_{22} E \quad (47)$$

in which

256

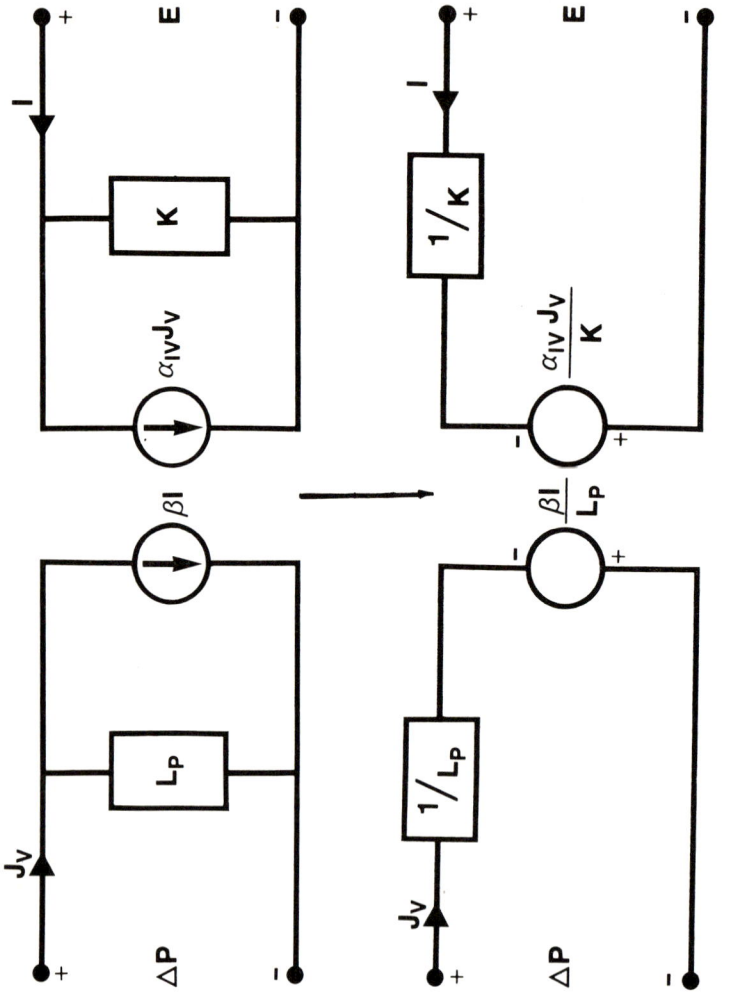

Fig. 5.4   Transformation from practical to hybrid coefficients for the electrosmosis problem.

$$H_{22} = K \quad (48)$$

is defined as the conductivity in the absence of volume flow.

From Eq (46) and conversion TABLE 1.1 we immediately obtain the R coefficients R_{11}, R_{12} and R_{22}:

$$R_{11} = 1/P_{11} = 1/L_p, \quad (49)$$

$$R_{12} = -P_{12}/P_{11} = -\beta/L_p, \quad (50)$$

and

$$R_{22} = 1/H_{22} = -1/K. \quad (51)$$

Moreover, given that

$$R_{12} = R_{21} = -H_{21}/H_{22} = -H_{21}/K \quad (52)$$

it follows that

$$H_{21} = \beta K/L_p$$

and

$$H_{12} = -\beta K/L_p$$

(by the assumption of antisymmetry). Given the complete R equatio

$$\begin{bmatrix} \Delta P \\ \\ E \end{bmatrix} = \begin{bmatrix} 1/L_p & -\beta/L_p \\ \\ -\beta/L_p & 1/K \end{bmatrix} \begin{bmatrix} J_v \\ \\ I \end{bmatrix} \quad (53)$$

the remaining H coefficient, H_{11} follows from the conversion table

$$H_{11} = \det R/R_{22} = (1/L_p) - (\beta^2 K/L_p) \quad (54)$$

and the complete H matrix is given by

$$\begin{bmatrix} \Delta P \\ I \end{bmatrix} = \begin{bmatrix} 1/L_p) - (\beta^2 K/L_p) & -\beta K/L_p \\ \beta K/L_p & K \end{bmatrix} \begin{bmatrix} J_v \\ E \end{bmatrix} \quad (55)$$

An additional change invariables leads back to the P coefficients

$$\begin{bmatrix} J_v \\ E \end{bmatrix} = \begin{bmatrix} L_p & \beta \\ -\beta & (1/K) - (\beta^2/L_p) \end{bmatrix} \begin{bmatrix} \Delta P \\ I \end{bmatrix} \quad (56)$$

The newly found P_{22} must then equal to the inverse of the conductance k' measured at zero current flow

$$K' = 1/R = 1/(1/K) - (\beta^2/L_p). \quad (57)$$

Given that

$$\beta^2/L_p = R_{12} R_{21}/R_{11} = R_{12}^2/R_{11} \quad (58)$$

it follows that
$$(1/K) - (1/K') = R_{12}^2/R_{11}, \quad (59)$$

a well knwon result of Onsager thermodynamics which has been obtained here without making reference to the dissipation function.

As a final check on the procedure, we can also write the L equations (60)

$$\begin{bmatrix} J_v \\ I \end{bmatrix} = 1/[(1/L_p) - (\beta^2/L_p^2)] \begin{bmatrix} 1/K & \beta/L_p \\ \beta/L_p & 1/L_p \end{bmatrix} \begin{bmatrix} \Delta P \\ E \end{bmatrix}$$

In the limit in which $I = 0$, $E = -\beta \Delta P$

so that

$$J_v = [(1/K) - (\beta^2/L_p)] \Delta P / [(1/K L_p) - (\beta^2/L_p^2)] = \Delta P/L_p \quad (61)$$

in agreement with the most basic linear phenomenology for volume flow in the presence of a hydrostatic pressure.

5.8 GOING BEYOND THE ONSAGER FORMULATION: NON-RECIPROCAL ENERGY CONVERSION

The above analysis, which follows mostly from my thesis work, <u>is not restricted to Onsager systems</u>, otherwise the Odum - Pinkerton-Kedem - Caplan (OPKC) approaches would contain the same information. In fact, they do not.

It has been shown above that one of the possible ways --and, probably, the most convenient way-- to describe energy conversion is by means of the hybrid coefficients **P** or **H** and that, in Onsager systems these have the following antireciprocal characteristics:

$P_{12} = -P_{21}$

or, equivalently,

$H_{12} = -H_{21}$.

These antireciprocities indicate that the Onsager systems for which the OPKC analysis holds have the characteristic that i) the ratio of output to input force when the output flow is zero is the same in magnitude to the ratio of input to output flow when the input force is zero (with a similar expression holding when forces and flows are exchanged, in the H machine).

Suppose, however, that because of linearization, because the system is too far from equilibrium or because some additional input is unknown to the experimenter it is found that the reciprocity does not hold. In such cases, the OPKC analysis cannot be used (how is q to be defined ?). By contrast, this approach remains valid: the two individual q´s are defined and the maximum efficiency can be found without reference to whether the equations are reciprocal or not.

This is not only a "theoretical" construct, but a very practical consideration. Engineers are extremely familiar with such non reciprocal machines (the amplifiers). It is unfortunate, then, that Onsagerists who are familiar with the non reciprocal approach through my thesis work persist in forcing reciprocal equations on systems which do not obey them.

5.9 MEASUREMENTS AND PATHS

From a topological point of view, the manifold where processes take place is generated by taking the topological product of a chosen port variable with other independent port variables, e.g.

denotes the sets of points which are in both X_1 and X_2.
Measurements such as those described schematically in FIGURE 5 may
be considered paths in the resultant manifold which restrict the
process to a limited portion of the manifold. For example,
consider the dependent variable X_1 as a function of the
independent variables J_1 and J_2. When a measurement is performed
in which J_2 is zero and a flow source is applied to port 1, port 1
represents the path of the set of points lying on the line

J_1 = constant

while port 2 represents the linear path

J_2 = zero.

If we solve the relevant R equations one can find several ways of
expressing both X_1 and X_2 in terms of J_1--e.g., plot
X_1 vs. X_2 using J_1 as a parameter or use a three dimensional
cartesian frame of reference X_1, X_2 and J_1. In either case, the
locus of allowable points is a line. In general, the set of all
values found for a given measurement defined in one of the boxes
of FIGURE 5 represents a line in the plane determined by one
dependent and one independent variable--i.e., the plane for which
the remaining independent port variable is zero.

Although this "planar" measurements are the ones generally
discussed, it is obvious that there are, even in the linear domain,
other possible planes which are different from the ones determined
when one of the variables is zero. For example, J_2 could be
related to J_1 in a "parasitic" fashion so that

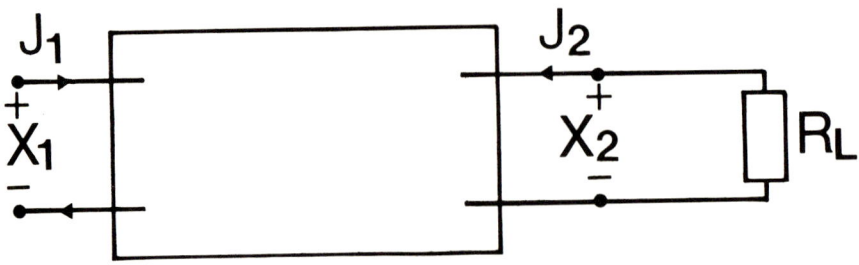

Fig. 5.5   Two port with linear load used in the text to
calculate maximum efficiency characteristics.

$$J_2 = r_{21} J_1$$

or, given that X_2 is a linear function of both J_2 and J_1 by

$$X_2 = -R_L J_2$$

in which we have purposely chosen the negative sign in front of the constant R_L (the load resistance). Thus by attaching the load resistance R_L to port 2, we have established a different linear path for the process.

A question of practical interest becomes: which is the path for which the ratio

$$- X_2 J_2 / X_1 J_1$$

is a maximum ?-i.e., for which the system is most efficient. The answer is unique only if the path lies on a plane and not on an arbitrary curve. This corresponds to a constant load resistance.

5.10   TWO PORT WITH LINEAR LOAD

In order to know how energy conversion will take place it is necessary to know the forces, flows, characteristics of a system, and the allowable ranges of operation. In the case of a linear, time invariant two port system in which a linear load of the form

$$R_L = X_L / J_L \qquad (62)$$

specifies the "operating path" it is possible to obtain an expression which is the analog of Carnot's maximum efficiency, but is valid for the steady state case. This maximum efficiency turns out to be independent of the forces and flows and dependent only on the transfer characteristics between the ports. The maximum efficiency is obtained, as expected, for a particular path or linear load wich is also a function of the system. In network terms, the problem is one of matching the load to the system for maximum operation. In an approach which uses "scattering" matrices to separate "power" components into incident, reflected and absorbed components, the problem can be reduced to finding the load which reflects the least possible power back. We shall not take this road, however and the interested reader is referred to any of the network texts suggested (ref. 126).

The energy converting box may be described in terms of any two port parameters. For convenience, however, we describe only the output port in terms of both H and P parameters. To this end, the second equation in set (4) can be rewritten to yield

$$H_{21} J_1 = - X_2 H_{22} + J_2. \qquad (63)$$

Moreover, since $X_2 = X_L$ it follows that $X_2 = - R_L J_2$ and that

$$H_{21} J_1 = J_2 (H_{22} R_L + 1). \qquad (64)$$

After rearrangement we obtain

$$J_2/J_1 = H_{21} / (R_L H_{22} + 1). \qquad (65)$$

Similarly, we can use the output port of the P network to find the output force across the load R_L. This yields

$$X_2 = (P_{21} X_1)(R_L / R_L + P_{22}) \qquad (66)$$

and, after rearrangement, we obtain the ratio of output to input force when the linear load is present

$$X_2/X_1 = P_{21}/(1 + P_{22}/R_L). \qquad (67)$$

Multiplying Eqs (66) and (67) we finally obtain an expression for the forward efficiency of energy conversion

$$e_{forward} = - (X_2 J_2 / X_2 J_2) =$$
$$= - P_{21} H_{21} /(1 + H_{22} R_L)[1 + (P_{22}/R_L)]. \qquad (68)$$

The maximum efficiency is found from the condition

$$\partial e/\partial R_L = 0 \qquad (69)$$

which yields a value for the optimum load for forward energy conversion

$$R_L^f (\text{max. efficiency}) = \sqrt{P_{22}/H_{22}}. \qquad (70)$$

Introducing Eq. (70) into (68) we obtain an expression for the maximum efficiency:

$$e_{max\ forward} = - P_{21}/H_{21}/(1 + \sqrt{H_{22} P_{22}})^2 . \qquad (71)$$

Using the conversions given in TABLE 1, the maximum efficiency can also be expressed in terms of **L** or **R** coefficients to obtain

$$e_{max\ forward} = \frac{(R_{21} R_{21} / R_{11} R_{22})}{[1 + \sqrt{1 - R_{12} R_{21} / R_{11} R_{22}}]^2} \qquad (72)$$

and

$$e_{max\ forward} = \frac{(L_{21} L_{21} / L_{11} L_{22})}{[1 + \sqrt{1 - L_{12} L_{21} / L_{11} L_{22}}]^2} \qquad (73)$$

respectively. Let the coupling coefficients q_{12} and q_{21} be defined by

$$q_{12} = - R_{12} / \sqrt{R_{22} R_{11}} \qquad (74)$$

and

$$q_{21} = - R_{21} / \sqrt{R_{22} R_{11}} \qquad (75)$$

or by

$$q_{12} = L_{12} / \sqrt{L_{22} L_{11}} \qquad (76)$$

and

$$q_{21} = L_{21} / \sqrt{L_{22} L_{11}} \qquad (77)$$

in terms of L coefficients. Introducing the coupling q's into Eqs.(72) and (73) it follows that a single expression gives the maximum efficiency in terms of either R or L parameters:

$$e_{max(forward)} = q_{21}^2 / [1 + \sqrt{1 - q_{12} q_{21}}]^2 \qquad (78)$$

In the special case considered by Kedem and Caplan in which the system shows Onsager type ssymmetry, $q_{12}=q_{21}=q$, in which q is the degree of coupling and the expression reduces to

$$e_{max} = q^2 / [1 + \sqrt{1 - qq}]^2, \qquad (79)$$

in agreement with the result of Kedem and Caplan.

In addition to the inherent generality of including non-reciprocal systems, the advantage of expressing e_{max} in the form given by Eq(78) is the directionality shown by the numerator q_{21}^2, which clearly distinguishes forward from reverse energy conversion. We can utilize the same line of reasoning to find the reverse maximum efficiency. In that case, we define

$$e_{reverse} = - X_1 J_1 / X_2 J_2 \qquad (80)$$

and the maximum efficiency becomes

$$e_{max(reverse)} = q_{12}^2 / [1 + \sqrt{1 - q_{12} q_{21}}]^2 \qquad (81)$$

and the optimum load for reverse energy conversion is

$$R^{reverse}_L(optimum) = \sqrt{H_{11}/P_{11}} .$$

This equation makes sense by noting that flipping of the P network leads to H parameters and vice versa. An interesting result is that a reciprocal system has the same maximum efficiency whether it is operated in forward or reverse modes, but this is not true of a non reciprocal system. Note that even in the reversible case the optimum loads for maximum forward and reverse efficiencies are not the same.

5.11 EXPERIMENTAL DETERMINATION OF THE OPTIMUM LOAD

An interesting practical result follows from the network analysis. Given that the load resistance for optimum efficiency is

$$R^{forward}_L = \sqrt{P_{22}/H_{22}} \qquad (82)$$

it follows that only two measurements are required to find the value of the optimum load using P and H coefficients:

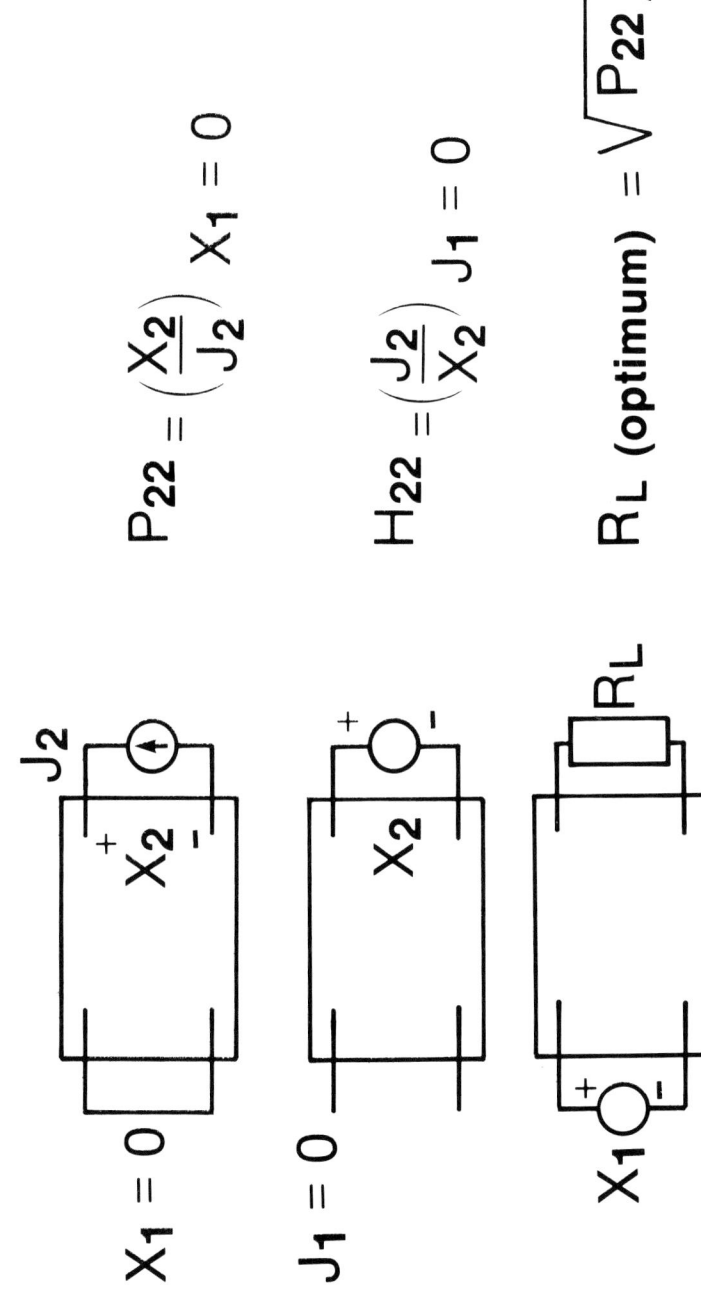

FIGURE 5.6 The two practical measurements shown yield the optimum linear load for a two port with constant coefficients.

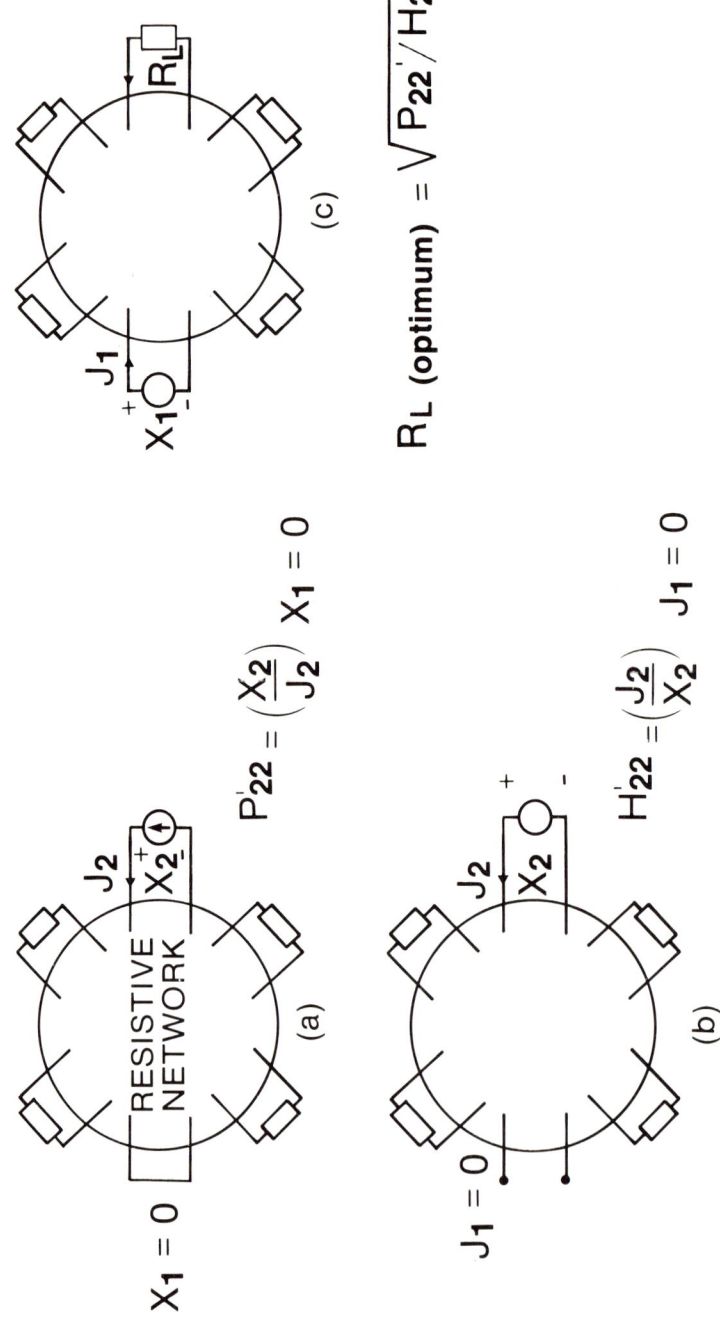

FIGURE 5.7 A multiport treated as a two port when n-1 outputs are resistive loads.

(1) The ratio of output force to output flow when the input is short circuited (level flow),

$$P_{22} = (X_2/J_2) \quad \text{with } (X_1 = 0) \tag{83}$$

(2) The ratio of output flow to output force when the input is open circuited (static head)

$$H_{22} = (J_2/X_2) \quad \text{with } (J_1 = 0). \tag{84}$$

5.12 EXPERIMENTAL DETERMINATION OF OPTIMUM EFFICIENCY

The optimum efficiency requires the two measurements given above plus two more:

(3) The ratio of output force to input force when the output flow is zero,

$$P_{21} = (X_2/X_1) \quad \text{with } (J_2 = 0) \tag{85}$$

and

(4) The ratio of output flow to input flow when the output force is zero,

$$H_{21} = (J_2/J_1) \quad \text{with } (X_2 = 0). \tag{86}$$

The above four measurements define the optimum operating values of a linear, time invariant system completely, in the forward mode, whether reciprocal or not.

In the case of reverse operating mode the optimum linear load requires the following measurements:

(5) The ratio of input force to input flow when the output force is zero,

$$H_{11} = (X_1/J_1) \quad \text{with } (X_2 = 0) \tag{88}$$

and

(6) The ratio of input flow to input force when the output is open

$P_{11} = (J_1 / X_1)$ with ($J_2 = 0$).

Note that, unlike the situation considered by Onsager´s phenomenology, which involves either resistive or conductive coefficients only hybrid coefficients are introduced here. This can be shown to be the simplest form of expressing the energy conversion characteristics of such a system, as both R and L representations are much more involved and contain less information.

5.13 THE CENTRAL ROLE OF THE HYBRID PARAMETERS IN ENERGY CONVERSION

Curiously, the seminal work by Odum and Pinkerton onenergy conversion in Onsager systems and its biological implications was given, indirectly, in terms of the hybrid coefficients described here, even though these authors failed to make a distinction between the Onsager and the hybrid coefficients.

Thus, Odum and Pinkerton rewrite the Onsager equations

$$J_1 = L_{11} X_1 + L_{12} X_2$$

$$J_2 = L_{21} X_1 + L_{22} X_2$$

in the form

$$J_1 = (1 + c f^2) X_1 - c f X_2 \quad (89)$$

$$J_2 = -c f X_1 + c X_2 \quad (90)$$

because, as they show, the l, c and f parameters appear in the expressions for maximum output power and maximum efficiency, which becomes in their terms

$$E_{max} = 2 [1 - (1/cf^2) (\sqrt{1 + cf^2/1} - 1) - 1 . \quad (91)$$

If one identifies l, c and f with P_{11}, $(1/P_{22})$ and P_{21}, their equations readily reduce to the hybrid coefficient expressions. TABLE 1.1 can be used to verify that the coefficients $(1 + c f^2)$,

- c f and c used in Odum and Pinkerton's phenomenological equations are, indeed, L_{11}, L_{12} and L_{22}. Moreover, it follows from the above discussion that

$$q = f / \sqrt{1/c} = P_{21} / \sqrt{P_{11} P_{22}}, \qquad (92)$$

thus showing the compatibility of the two formulations.

5.14 NON CYCLIC ENERGY CONVERSION IN THERMOSTATICS

The efficiency of energy conversion is calculated in thermostatics only for cyclic processes. We can now ask, using the above model, what is the most efficient path to follow, say, in the P-V plane when the heat bath supplies a differential amount of heat TdS reversibly. Assume that over a short distance the derivative (dV/dP) is relatively constant so that it can therefore be expressed as a load conductance g_L.

The conductance is attached to the output port, while the temperature source (a battery) is attached to the input port to represent the constant temperature bath. Defining the local efficiency for the infinitesimal distance travelled by

$$e_{local} = (- PdV / T\, dS)$$

and using the value of q

$$q^2 = g_{12}\, g_{21} / g_{11}\, g_{22},$$

in which the g_{ij}'s are defined from the free energy network of Chapter 2, it follows that the optimum path when the independent variables are S and T is given by

$$g_L = (dV/dP)_{optimum}$$
$$= [\, (\partial V/\partial P)_S\, (\partial V/\partial P)_T\,]^{1/2}.$$

5.15 THE COUPLING PARAMETER Q

The maximum efficiency found,

$$e_{max(forward)} = q_{21}^2 / [\, 1 + \sqrt{1 - q_{12}\, q_{21}}\,]^2, \qquad (93)$$

may be rewritten, after expanding and dividing numerator and denominator by $(2 - q_{12} q_{21})$ and multiplying and dividing the numerator by R_{12} and rearranging terms inside the square root, we obtain

$$e_{max(forward)} =$$

$$= (R_{21}/R_{12}) [q_{21}q_{12}/(2 - q_{12} q_{21})]$$

$$/ [1 + \sqrt{1 - (q_{12} q_{21})^2 / (2 - q_{12}q_{21})^2}]. \qquad (94)$$

Defining "Super Q", by

$$Q = 2R_{21}R_{12} / 4R_{11}R_{22} - 2R_{12}R_{21} \qquad (95)$$

$$= q_{21}q_{12} / (2 - q_{12}q_{21})$$

we obtain an expression for maximum efficiency of the form

$$e^f_{max} = (R_{21}/R_{12}) Q / (1 + \sqrt{1 - Q^2})$$

which is more concise than the previous result and summarizes all the possible parameter representations. Thus, from conversion TABLE 1 we obtain the equalities

$$Q = 2 |R_{21}R_{12}| / (4 R_{11}R_{22} - 2 R_{12}R_{21}) \qquad (96)$$

$$Q = 2 |H_{21}H_{12}| / (4 H_{11}H_{22} - 2 H_{12}H_{21}) \qquad (97)$$

$$Q = 2 |L_{21}L_{12}| / (4 L_{11}L_{22} - 2 L_{12}L_{21}) \qquad (98)$$

and

$$Q = 2 |P_{21}P_{12}| / (4 P_{11}P_{22} - 2 P_{12}P_{21}). \qquad (99)$$

Thus, Q has the same form, regardless of the coefficients chosen to define it. Moreover,

$$(R_{21}/R_{12}) = (L_{21}/L_{12}) = -(H_{21}/H_{12}) = -(P_{21}/P_{12})$$

so that the efficiency may be expressed in terms of any of the phenomenological coefficients, including the hybrid coeficients, as

$$ef^f_{max} = (n_{21}/n_{12}) Q / (1 + \sqrt{1 - Q^2}) \qquad (100)$$

with Q given by

$$Q = 2n_{21}n_{12} / 4n_{11}n_{22} - 2n_{12}n_{21} \qquad (101)$$

in which n_{ij} represents any of the phenomenological coefficients.

In addition to the value of unification, this expression has an additional property which will be considered in Chapter 6 : the maximum efficiency for the oscillatory case has the same form as the steady state when given in terms of an impedance defined Q.

As an example of the use of Q, we refer back to the Kedem Katchalsky equations used before in Q form, to obtain

$$Q = (1-\sigma)^2 \bar{c}_s^2 / [2w\bar{c}_s/L_p - (1-\sigma)^2\bar{c}_s^2] \qquad (102)$$

which can then be introduced into equation (100) to obtain the maximum efficiency in the reciprocal case. The optimum load will depend on whether the osmotic flow or the volume flows are used as driving processes; they will be two distinct quantities, even in the reciprocal case ($R^{forward}_L$ and $R^{reverse}_L$).

It should be stressed again that the present analysis also applies to the general non reciprocal case.

5.16 ENERGY CONVERSION IN MULTIPLE FLOW, LINEAR SYSTEMS

Caplan (ref. 42,43) extended the treatment of energy conversion in Onsager systems to multiple port situations. He defined a new degree of coupling given by

$$r_{ij} = -R_{ij}/\sqrt{R_{ii} R_{jj}} \qquad (103)$$

or, equivalently, by
$$l_{ij} = L_{ij}/\sqrt{L_{ii} L_{jj}}.$$

Caplan then considered the efficiency of energy conversion in the following situations:

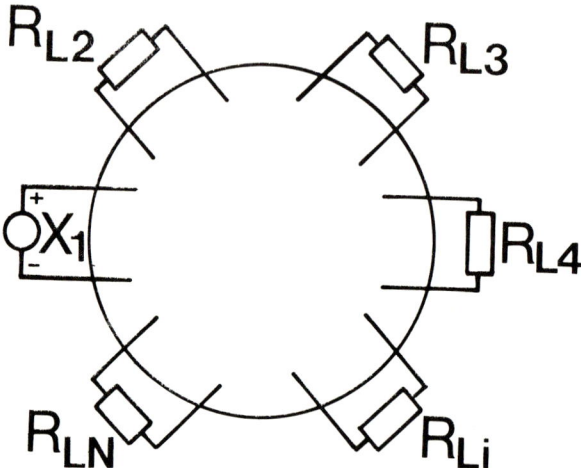

Fig. 5.8 A multiport network in which each port, except for one input, is terminated in a linear load R_{Li}.

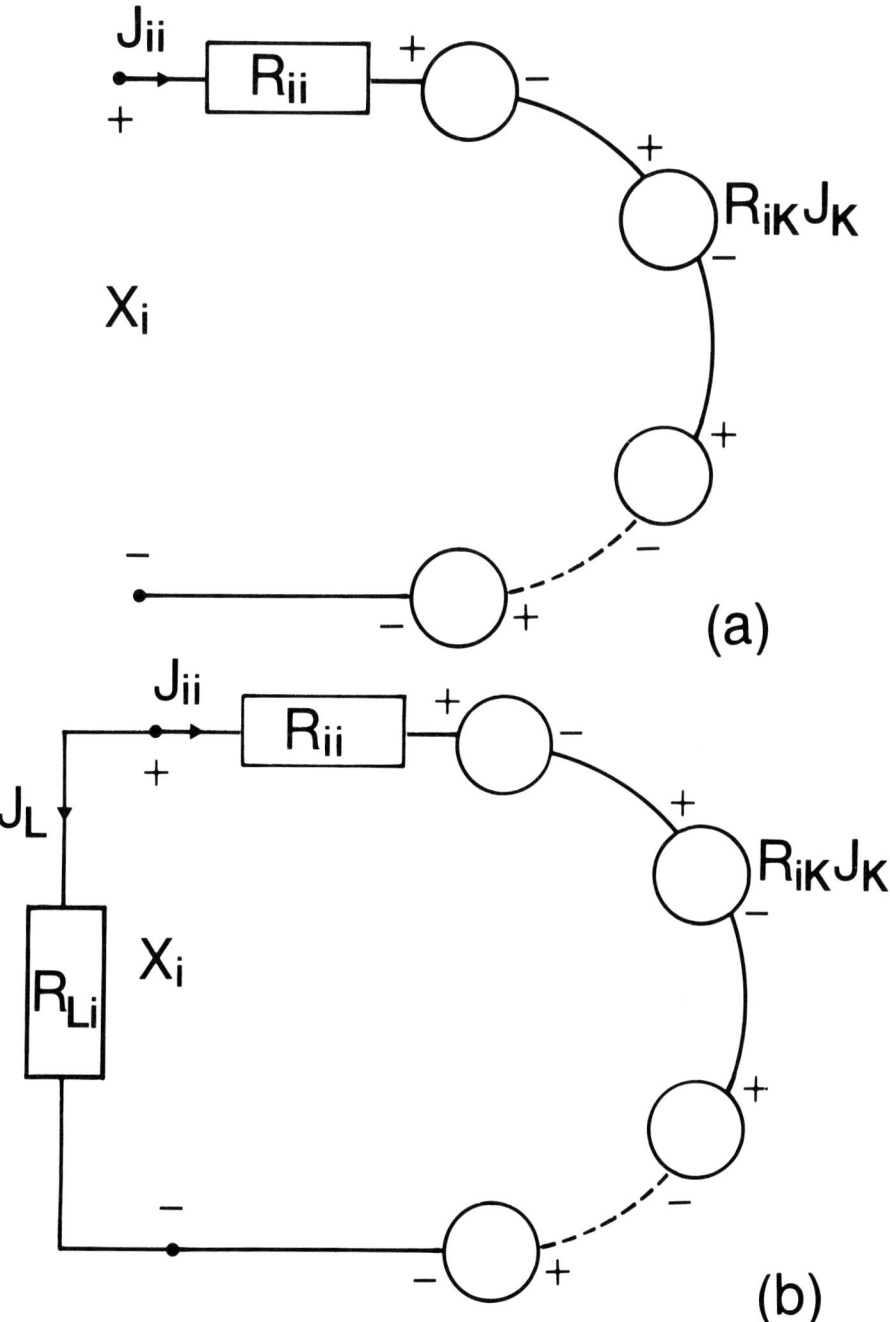

Fig. 5.9 Representation of each port in the multiple port situation.

(i) all processes (except for the input and output) are at level flow ($X_i = 0$)

(ii) all processes (except for the input and output) are in static head ($J_i = 0$)

(iii) all processes (except for the input and output) are in a mixed situation in which some are in static head while some work at level flow.

As Caplan pointed out, these various situations correspond to either short circuiting or open circuiting the terminals of an equivalent network. I considered, in addition to those, some which Caplan refers to as parasitic in which the ignored outputs are linear functions of the input quantities--these are, in network terms, resistive loads (Fig. 5.8) .

Consider the multiple flow system as a linear balck box with n terminal ports ; of these. we arbitrarily choose an input port (labeled 1) and an output port (2). We are now interested in the efficiency of energy conversion in the general linear case in which all terminals, except for the input port, have linear dissipative loads--i.e., a boundary condition of the type

$$X_i / -J_i = R_{Li} .\qquad(104)$$

The negative sign appears, as before, because J_{Li}, the flow across the resistive load, is defined as coming out of the network, whereas J_i, the flow conjugate to X_i is going into the network. The maximum efficiency for the multiport so defined is analogous to the case of the two port network, provided that proper equivalent coefficients are given. The optimum load resistance becomes

$$R_L = \sqrt{P'_{22} / H'_{22}}\qquad(105)$$

whose value can be determined by means of the experiments shown in FIGURE 7, while the maximum efficiency is

$$e_{max} = (n_{12}/n_{21}) \, Q' / [1 + \sqrt{1-Q'^2}]\qquad(106)$$

in which Q' is now calculated using equivalent coefficients.

In order to give an explicit determination of multiple port parameters, we start with a disjoint nxn network in which each force X_i is given by the sum of a conjugate term (a resistance) and (n-1) non-conjugate terms (controlled sources), as shown in FIGURE 9. The basic branch equation represented by the network is

$$X_i = R_{ii} J_i + \sum_{k \neq i} R_{ik} J_k \quad . \qquad (107)$$

When a load resistance R_{Li} is placed across port i, the situation is that shown in FIGURE 9b; the total resistance seen by the port is now $(R_{Li}+R_{ii})$ while the force across port i is given by

$$X_i = - R_{Li} J_i \; .$$

The general K.V.L. equation for the resistively loaded ports now becomes

$$0 = (R_{Li} + R_{ii}) J_i + \sum R_{ik} J_k \quad (i=1,2,\ldots,n) \qquad (108)$$

while the matrix equation representing the phenomenological equations for the network has the form

$$\begin{bmatrix} X_1 \\ X_2 \\ 0 \\ | \\ 0 \\ | \\ 0 \end{bmatrix} = \begin{bmatrix} R_{11} & R_{12} & R_{13} & \cdots & R_{1i} & \cdots & R_{1N} \\ R_{21} & R_{22} & R_{23} & \cdots & R_{2i} & \cdots & R_{2N} \\ R_{31} & R_{32} & R_{33} & \cdots & R_{3i} & \cdots & R_{3N} \\ | & | & | & & | & & | \\ R_{j1} & R_{j2} & R_{j3} & \cdots & R_{ji} & \cdots & R_{jN} \\ | & | & | & & | & & | \\ R_{N1} & R_{N2} & R_{N3} & \cdots & R_{Ni} & \cdots & R_{NN} \end{bmatrix} \begin{bmatrix} J_1 \\ J_2 \\ J_3 \\ | \\ J_j \\ | \\ J_j \end{bmatrix}$$

in which ports 1 and 2 are the input and output ports whose boundary conditions have not been specified. The vectors and matrix operators can be partitioned as follows:

$$\begin{bmatrix} X_{io} \\ --- \\ 0 \end{bmatrix} = \begin{bmatrix} R_{io} & | & R_1 \\ ----- & | & ---- \\ X_{io} & | & X_{io} \end{bmatrix} \begin{bmatrix} J_{io} \\ --- \\ X_{io} \end{bmatrix} \qquad (109)$$

in which

X_{io} = input output force vector, $\begin{bmatrix} x_1 \\ x_2 \end{bmatrix}$

0 = the zero vector (all $n-2$ entries are zero)

J_{io} = the vector of input and output flows = $\begin{bmatrix} J_1 \\ J_2 \end{bmatrix}$

J_L = the vector of flows at all ports except for 1 and 2.

R_L = the $(n-2) \times (n-2)$ matrix whose entries are: $(R_{Li} + R_{ii})$ for a diagonal element and R_{ik} for an off diagonal element.

R_1 = the $(n-2) \times 2$ matrix which relates input and output forces to all other ports --i.e., ports other than 1 or 2-- when the input and output flows are zero:

$$\begin{bmatrix} x_1 \\ x_2 \end{bmatrix} = \begin{bmatrix} R_{13} & R_{14} & \cdots & R_{1i} & \cdots & R_{1N} \\ R_{23} & R_{24} & \cdots & R_{2i} & \cdots & R_{2N} \end{bmatrix} \begin{bmatrix} J_3 \\ J_4 \\ J_i \\ J_N \end{bmatrix} \quad (110)$$

R_2 = the $2 \times (n-2)$ matrix which gives the vector 0 in terms of input and output flows when all other flows are zero:

$$\begin{bmatrix} 0 \\ 0 \\ | \\ 0 \end{bmatrix} = \begin{bmatrix} R_{31} & R_{32} \\ R_{41} & R_{42} \\ | & | \\ R_{N1} & R_{N2} \end{bmatrix} \begin{bmatrix} J_1 \\ J_2 \end{bmatrix} \quad (111)$$

R_{io} = the input-output matrix
$\begin{bmatrix} R_{11} & R_{12} \\ R_{21} & R_{22} \end{bmatrix}$

which gives the input force vector interms of the input-output flow vector when all other flows are zero.

With these definitions we may write two equations which describe the loaded phenomenological equations:

$$X_{io} = R_{io} J_{io} + R_1 J_L$$

and

$$0 = R_2 J_{io} + R_L J_L .$$

J_L can be eliminated to yield

$$X_{io} = (R_{io} - R_1 R_2 R_L^{-1}) J_{io}$$

and defining R' by

$$R' = R_{io} - R_1 R_2 R_L^{-1}$$

the last equation reduces to

$$X_{io} = R' J_{io}.$$

Thus, the resistance matrix R' defines the overall parameters searched for the multiport energy conversion problem.

5.17 ENERGY CONVERSION FOR ATTACHED IDENTICAL TWO PORTS

It is clear from the above discussion that the degree of coupling is an invariant property --i.e., a characteristic which does not vary with the "frame of reference" or variables chosen to describe the process. This makes geometrical sense, as q is a measure of an angle between two vectors. Given this property of q, it follows that whenever n identical converters are attached in such a way that each entry in the total phenomenological matrix $A_{ik(total)}$ equals n times the corresponding entry in a single converter, the coupling coefficient remains invariant.

The situations in which each converter form can be used are summarized in FIGURE 10.

These constant efficiency attachements were previously considered in Chapter 2 to determine the equilibrium conditions of a thermodynamic phase.

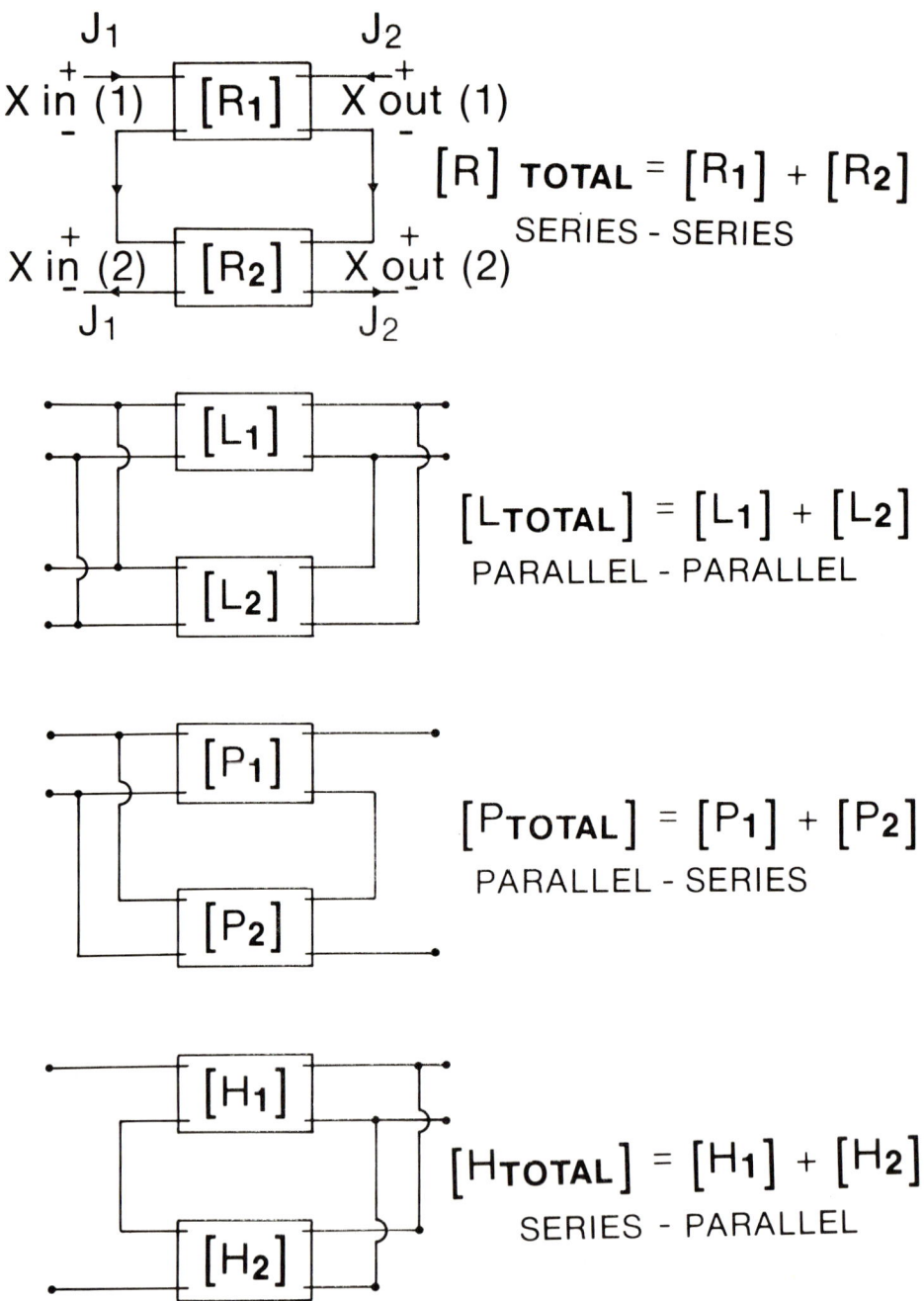

FIGURE 5.10   Attachement of **R, L, P** and **H** ports to obtain global parameters.

Chapter 6

TIME BEHAVIOR AND EVOLUTION OF NON-EQUILIBRIUM THERMODYNAMIC AND KINETIC SYSTEMS

6.1   INTRODUCTION: DISSIPATIVE VS STORING CONSTITUTIVE LAWS

Up to this point, we have only considered network descriptions of thermodynamic and kinetic systems in which linear resistances and sources are present. In this chapter the idea is to extend the treatment of both kinetic and thermodynamic networks to the non-equilibrium situation.

It has been shown above that resistive networks give rise to bilinear invariants of the form

$$\Phi = R_{ik} J_i J_k \qquad (1)$$

These expressions have the form of Raleigh's dissipation function and act as potentials in the sense that the forces can be found by taking the derivative

$$X_i = (1/2)(\partial \Phi / \partial J_i) \qquad (2)$$

in the direction of a given force J_i. In the special case in which the dissipation expression is reciprocal, the resultant metric can be equated with a distance in a euclidean space of n dimensions, provided that the R_{ik}'s are constant. (Otherwise, if the resistances are not constant it can always be locally imbedded in a euclidean space). All such metrics are isomorphic to resistive networks--i.e., they correspond to situations in which the internal forces are functions of local velocities (or the voltage is proportional to the through current, in the case of the electrical resistance). These cases are seen typically in frictional systems with a frictional coefficient proportional to the velocity and in which all inertial accelerations are fast decaying transients. The laws of diffusion provide one such example in which collision times are short and the ratios of the masses to

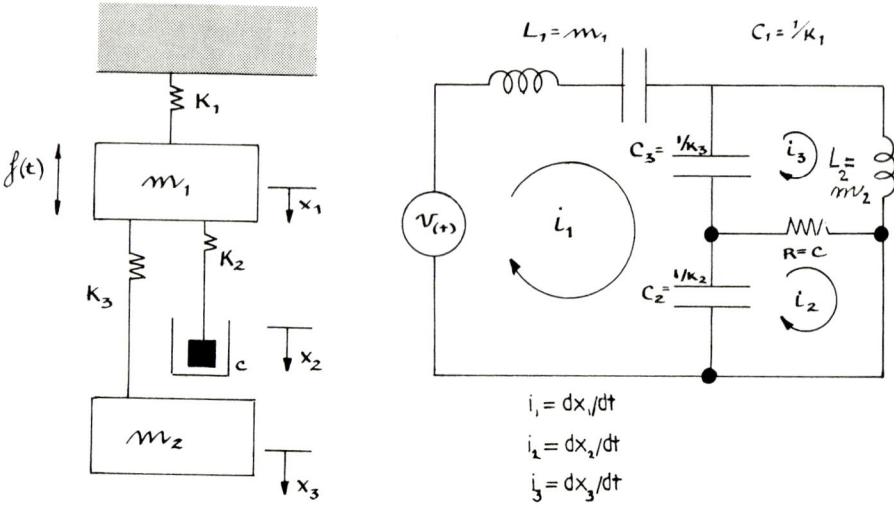

FIGURE 6.1   A classical example of the use of networks to simulate dynamical systems.

the frictional coefficients is small.

In a more general treatment, other types of coordinates must be included those are: 1) the ones leading to " potential energy" storage, which show the delays in the build up of a force caused by the size of the compartment where the increase in an intensive variable, such as mass density, is taking place and 2) the inertial coordinates, which lead to "kinetic" energy storage.

At first sight, the way to extend the thermodynamic and kinetic steady state treatment to nonequilibrium situations is rather obvious. As it is well known, R, L, C networks have been used to simulate many dynamical systems by analog computers -- as shown , for example , in FIGURE 1 . It would therefore make perfect sense to reverse the analogy and use the mechanical system to point in the direction of an appropriate dynamical dissipation. The immediate suggestion that one derives from network theory is: in order to extend steady state thermodynamics to non-equilibrium processes simply add inductors and capacitors.

Moreover, Lagrange's "equations of motion" have been available for networks for a long time (ref. 88). Guillemin showed their relationship to Kirchhoff's laws (ref. 88). The primary contribution in this field is, of course, the neglected work of Kron (ref.125) who extended the approach to non-linear and non-holonomic systems .

It would then appear that one can simply add the appropriate capacitors and inductors-- to represent potential and kinetic "energy" storage and proceed to apply Lagrange's equations to obtain a proper dynamic dissipation.

In the case of thermodynamics this approach works, but <u>in the case of kinetics it will only work if a network is first found such that both K.V.L. and K.C.L. are obeyed</u> . That is the system is first transformed into a system in which the forces can be found from potentials which have reciprocal gradients. This is a requirement that follows from the very derivation of Lagrange's equations : in order for Lagrange's equations to hold Newton's third law must also hold in the form

$$F_{ik} = -F_{ki},$$

or, alternatively, in the form of D'alambert's principle-- leading to matching of gradients in the case of a velocity dependent potential,

$$\nabla \phi_{ik} = \nabla \phi_{ki}.$$

Either of these statements is consistent with the network connectivity requirements. In the case of disconnected--i.e., non reciprocal networks-- holonomic Lagrangians would not be expected to be valid, as there is no energy like invariant.

We shall first derive Lagrange's equations from Tellegen's theorem, to show the connection between the two, and then obtain a dynamic dissipation for thermodynamic systems. The resultant representations will then be used to consider dynamic energy conversion in thermodynamic systems and dynamic coupling in reaction diffusion processes.

6.2 LAGRANGE'S EQUATIONS AND INVARIANCE

As is well known, Lagrange's equations arise in an effort to generalize Newton's equations of motion,

$$d/dt \; \partial T/\partial(dx_i/dt) \quad + \quad \partial V/\partial x_i = 0 \qquad (3)$$

equivalently,

$$m\,(d^2/dx_i^2) = F_i \qquad (4)$$

to non cartesian frames of reference. The solution is rather simple in its final result, at least for conservative systems, as it is found that by replacing both the kinetic energy T and the potential energy V by the Lagrangian L, the following equation is obtained

$$d/dt \; \partial L/\partial(dq/dt) \; - \; \partial L/\partial q_i \; = 0, \qquad (5)$$

which has the same form for all frames of reference.

In a general situations involving both conservation and dissipation the equations become

$$d/dt \; (\partial T/\partial q_i) \; - \; \partial T/\partial q_i \; + \; \partial U/\partial q_i \; + \; \partial D/\partial q_i = F_i \qquad (6)$$

in which D is the dissipated energy and F is the force.

6.3 MECHANICS AS AN ANALOG COMPUTER FOR NETWORKS

The kinetic energy term gives a functional relation between the velocities and momenta -- or, locally, between the momentum and the velocity of the mass whose momentum is being measured--,

$$p_j = p(v_j) \qquad (7)$$

and the second gives displacements which depend on a force (or vice versa),

$$x_i = x(F_i). \qquad (8)$$

(The reader may be more familiar with the common terminology of Mechanics $q_i = x_i$ and $\dot{q}_i = v_i$). In the special case of linear constutive laws we can write for each coordinate the familiar equations $p = mv$ and $x = kF$. In order to express these in terms of forces and velocities (or voltages and currents, etc), we differentiate to obtain the familiar expressions

$$dp/dt = m\, dv/dt + v\, dm/dt \qquad (9)$$

and

$$dx/dt = k\, dF/dt + F\, dk/dt. \qquad (10)$$

Of course, when m and k are constant this reduces to

$$F = dp/dt = m\, dv/dt \qquad (11)$$
and
$$v = dx/dt = k\, dF/dt \qquad (12)$$

respectively. In an ideal linear inductor, the magnetic flux is
$$\phi = L\, i \qquad (13)$$
leading to Lenz's law
$$v = d\phi/dt = L\, di/dt \qquad (14)$$
while in an ideal capacitor the constitutive law is
$$q = C\, V, \qquad (15)$$
which yields
$$i = dq/dt = C\, dV/dt. \qquad (16)$$
This is all well known and rather trivial. There are several points to highlight, however:

1. The above differential laws are valid <u>only</u> for linear inductors and capacitors (or masses and conservative forces) in which the coefficients of prportionality are time invariant and, in addition, magnetic fluxes (momenta) are proportional to the currents (velocities) and capacitive voltages (forces) are proportional to the charges (displacements). If this was not the case, incremental inductances and capacitances could be defined by

$$L = d\phi / di \quad (m = dp/dv, \text{ in the mechanical case}) \qquad (17)$$
and
$$C = dv / dq \quad (1/k = dF/dx, \text{ in the mechanical case}). \qquad (18)$$

2. The inductive and capacitive laws reflect "delays" in the coordinate "lengths". This can be seen by writing the equations in

integral form

$$i = (1/L)_0 \int^t V \, dt \quad \quad (\text{or } v = 1/m \int F \, dt) \quad \quad (19)$$

expressing the fact that it is impossible to build up an instantaneous current, unless the voltage is a delta function (or, in the mechanical case, that it is impossible to build up an instantaneous velocity unless the force is infinite) and

$$v = (1/C) \int i \, dt \quad \quad (\text{or } F = k \int v \, dt) \quad \quad (20)$$

expressing the fact that a capacitor cannot build up voltage instantaneously. Thus, in one case the flow coordinate is delayed relative to the force and in the other case the force coordinate is delayed relative to the flow.

3. The products v x i (or JX, or F x) lead to the stored forms

$$C \, Q^2/2 \quad \quad \text{and} \quad \quad L \, i^2/2 , \quad \quad (21)$$

which are not dissipated in time and can be retrieved.

6.4 INSTANTANEOUS DISSIPATION AND STORAGE

We can now consider the more general case away from the steady state, in which forces and flows are functions of time. In such non equilibrium situations, extensive quantities may not appear to be instantaneously conserved and, morevoer, forces may not appear to be instantaneously balanced. Clearly, Onsager's treatment does not provide any indication of how to find an instantaneous dissipation. We can, however, turn to Tellegen's theorem, which will work when the network obeys K.V.L. and K.C.L. For such cases, Tellegen's theorem states that

$$\sum_{\text{terminals}} X_i(t) \, J_i(t) = \sum_{\text{network}} X_k(t) \, J_k(t) \quad \quad (22)$$

in which $X_i(t)$ and $J_i(t)$ must be specified by the constitutive relations inside the system or network. In the case of non steady state systems there will be in general sources of divergence of the form

$$- \text{div } J_i = C \, dX_i/dt \quad \quad (23)$$

which can be viewed as a capacitive flow, J_c, such that the constitutive relation for the capacitor holds. In that case, K.C.L. is obeyed. Thus a capacitor serves to account for a temporary violation in the conservation laws for extensive quatities (mass, charge, etc).

Similarly, we can account for temporary violations of force balance (Newton´s third law) by means of inductors whose instantaneous current is delayed relative to the force. In fact, Newton´s second law is just one such example of how to account for a force which does not seem to be acting instantaneously, or appears to be temporarily imbalanced. FIGURE 2 shows the two

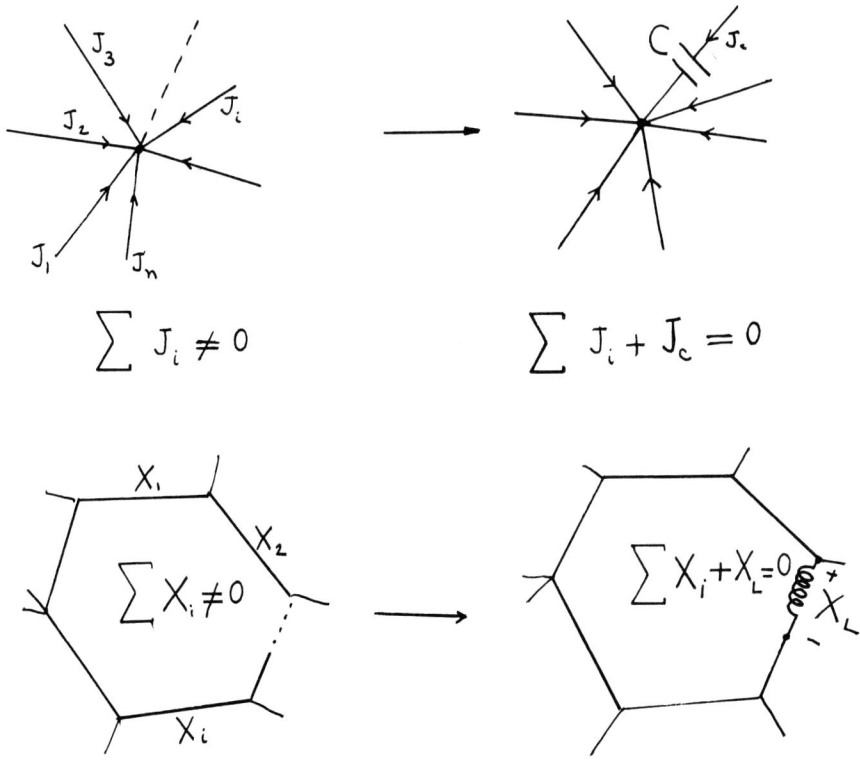

FIGURE 6.2   Temporary violations of K.V.L. and K.C.L. in the steady state can be accounted for in terms of capacitive and inductive effects.

situations in which K.V.L. and K.C.L. violations are accounted for by inductors and capacitors, respectively.

There are, of course, limitations to how far one can go with this approach. The primary assumptions are that forces which are not of frictional character have a build proportional to the amount of a source (imbalanced) extensive quantity entering the region, while imbalanced forces are proportional to the rate of change of the local, conserved extensive quantity.

These situations correspond, as described above, to the capacitive equations

$$J = C \, dX/dt \qquad (24a)$$

or the inductive (inertial) equations

$$X = L \, dJ/dt. \qquad (24b)$$

Given that both K.V.L. and K.C.L. are obeyed when capacitors and inductors are introduced, we can write, using Tellegen's theorem

$$\Phi = \sum_{\text{all } R} J_R X_R + \sum_{\text{all } C} J_C X_C + \sum_{\text{all } L} J_L X_L$$

in which the summations are taken for all capacitors, resistors and inductors. Introducing the constitutive equations for each of the elements, we obtain

$$\Phi = \sum_{\text{all } R} R_R J_R^2 + \sum_{\text{all } C} [1/2] \, C \, d X_C^2 /dt + \sum_{\text{all } L} [1/2] \, L \, d J_L^2 /dt \qquad (25)$$

The first term accounts for the steady state dissipation and corresponds, as in the thermostatic case, to the manifold imbedding in a higher dimensional ambient manifold in which orthogonal axes can be defined; this is an amount of dissipation, entropy, etc., which cannot be returned to the ports. The second and third terms are storage terms which are returned to both the frictional (resistive) elements or the ports, depending on the network.

These storage terms are the analogs of a time rate of change of kinetic "energy" and "potential energy".

Denoting these terms, for obvious reasons, by T and V the total dissipation becomes

$$\bar{\Phi}(t) = \bar{\Phi}(s) + d[T + V]/dt, \qquad (26)$$

in which $\bar{\Phi}(s)$, the amount dissipated by the resistances is the dissipation which would obtain if the system was in a steady state, while the second term is the rate of change of the stored dissipation.

An interesting question is: <u>will the stationary state be reached through a unique path?</u> The answer is in the affirmative if all the elements are positive and can be obtained through the direct use of Tellegen's theorem (although it is also a result of the Fundamental theorem on differential equations showing that there is a unique solution in that case).

Consider an initial state in which there is no storage of any kind and some later time at which both dissipative and storing forces are at work. Tellegen's theorem becomes, in incremental form:

$$\sum_{ports} \delta x_o \, \delta J_o = \sum_{network} \delta x_i \, \delta J_i, \qquad (27)$$

in which increments represent differences between the values of both sets of variables in both sets of experiments. Replacing the forces and flows by their explicit constitutive expressions (capacitive, resistive, or inductive) we obtain (ref. 26).

$$\sum \delta x_o \, \delta J_o = \sum R_R (\delta J_R)^2 + \qquad (28)$$
$$+ \sum d/dt[\, C_C (\delta x_C)^2(t))\,] + \sum d/dt[\, L_L (\delta J_L)^2\,]$$

in which the left hand side is zero because the boundary conditions are constant,

$$\sum \delta x_o \, \delta J_o = 0. \qquad (29)$$

If Eq. (28) is now integrated from some arbitrary time τ at which the conditions are known to t and if all elements (R, L and C) are positive it follows that all time variations must be zero.

This means that, with constant boundary conditions, there will be only one set of internal forces and flows for a given time.

6.5 LAGRANGE'S EQUATIONS ARE IMPLICIT IN TELLEGEN'S THEOREM

We now show that Tellegen's theorem includes Lagrange's equations as a special case.

[Given that Lagrange's equations are derived from the assumption that physical displacements are linear functions of some independent generalized coordinates and that Newton's third law acts either directly on the system as

$$F_{ik} = -F_{ki} \qquad (30)$$

or through the mediation of a potential gradient which leads to the balancing of internal coordinates (D'Alambert's principle), it is not surprising that network "equilibrium" equations can also be expressed in terms of Lagrange's equations. In fact James Jeans gave explicit forms for these and the classical textbook by Guillemin shows the relationship with K.V.L. and K.C.L.]

We presently show that these follow directly from Tellegen's theorem which, from a mechanical point of view, simply states the above continuities in forces and coordinate displacements.

We assume that the generalized coordinates q_a are inputs or outputs of a multiport network having conjugate forces X_a. The generalized coordinates are network "charges" -- distance, mass, charge, etc--, so that flows are defined by

$$\dot{q}_a = dq_a/dt = J_a. \qquad (31)$$

These charges can also be interpreted, in a microscopic model, as fluctuations in macroscopic extensive variables. Whether a system is actually a series of attached masses, pendulums, etc. or a single flow such as in the case of heat diffusion or a set of Onsagerian forces and flows, they will be representable, as discussed, using resistors capacitors (force storage elements) and inductors(flow storage elements). Usually, the network is given and each dependent coordinate can be expressed as functions of the independent generalized cordinates or "charges",

$$q_k = (q_1, q_2, q_3, \ldots q_{n'}).$$

We can also consider a second network having the same resistances, capacitors and inductors, in which the variables have been incremented by some arbitrary amounts. Using Tellegen's theorem, it follows that the forces in the first network are orthogonal to the incremented flows in the second network; this can be expressed in vectorial form as

$$X \cdot \Delta J^T = 0 \qquad (32)$$

whence it follows that the sum of products of forces times their conjugate increments at the inputs equals the same incremental form taken inside the network:

$$\sum_{\text{input variables}} X_o \Delta J_o = \sum_{\text{network variables}} X_N \Delta J_N \qquad (33)$$

in which we have followed the usual convention of reversing the direction of flows at the inputs and outputs, as done in Chapter 1. The sums of products inside the newtwork are of three kinds:

Resistive Elements

$$\sum_{\text{all resistances}} X_r \Delta J_r = \sum R_k J_r \Delta J_r = \sum (R/2) \Delta J_r^2 = \sum \Delta (R/2) J_r^2 \qquad (34)$$

Capacitive elements

$$\sum_{\text{all capacitors}} X_c \Delta J_c = \sum (q_c/C) \Delta \dot{q}_c \qquad (35.a)$$

Inductive elements

$$\sum X_L \Delta J_L = \sum L_L (\partial J_L/\partial t) \Delta J_L \qquad (35b)$$

Tellegen's theorem can now be expressed using these explicit forms. In addition we will keep all inputs constant excep for input J_a, which shows a variation ΔJ_a. Tellegen's theorem now reduces to

(35 c)
$$X_a \delta J_a = \Delta \sum (R/2) J_r^2 + \sum (q_c/C) \Delta \dot{q}_c + \sum L_L \Delta J_L \partial J_L/\partial t$$

Dividing through by the increment $\Delta J_a = \Delta \dot{q}_a$, we obtain

$$X_a = \Delta \sum [(R/2)J_r^2 / \Delta J_a] + \sum (q_c/C) \Delta \dot{q}_c / \Delta \dot{q}_a + \sum L_L (\Delta J_L / \Delta J_a)(\partial J_L / \partial t). \quad (36)$$

We now consider each term individually. The first term is simply the flow derivative of the dissipation function, or a similar analog, we label it D.E. (for dissipated "energy"). The second term can be first changed by dropping the dots--i.e., cancelling time derivatives which appear both in numerator and denominator> In the differential limit, we obtain

$$\sum (q_c/C) \Delta \dot{q}_c / \Delta \dot{q}_a = \sum (q_c/C) \partial q_c / \partial q_a = \partial \sum [(1/2) q_c^2/C] / \partial q_a. \quad (37)$$

The last term in the summation defines the "potential energy" or an analog, we label it V. Finally the third term in Equation (36) can also be expressed as

$$\sum L_L (\partial J_L / \partial t)(\partial J_L / \partial J_a) = \sum \partial^2 [(1/2) L_L J_L^2] / \partial J_A \partial t - \sum \partial [(1/2) L J_L^2] / \partial q_a. \quad (38)$$

Again, the last term is an analog of the kinetic energy and may be denoted by T, so that the above equation reduces to

$$\sum L_L (\partial J_L / \partial J_a)(\partial J_L / \partial t) = \partial^2 T / \partial J_a \partial t - \partial T / \partial q_a. \quad (39)$$

Finally, Equation (40) can be rewritten in the more familiar generalized version of Lagrange's equation including kinetic and potential energy (or analogs) and a dissipation potential:

$$X_a = \partial DE / \partial \dot{q}_a + \partial V / \partial q_a + \partial^2 T / \partial \dot{q}_a \partial t - \partial T / \partial q_a. \quad (40)$$

In the special case in which the potential energy is not a function of the generalized flows, we can write

$$X_a = \partial DE / \partial \dot{q}_a + \partial^2 L / \partial \dot{q}_a \partial t - \partial L / \partial q_a. \quad (41)$$

in which the Lagrangian L = T - V has been introduced. If the system is conservative, the equations reduce to

$$X_a = d(\partial L / \partial \dot{q}_a)/dt - \partial L / \partial q_a, \quad (42)$$

in agreement with Classical Mechanics.

6.6 EFFECTS OF COMPARTMENTAL STORAGE AND KINETIC TRANSIENTS

In Chapter 3 I treated the kinetic problem from a purely steady state point of view, assuming that the effective affinity of the chemical reaction could be sensed instantaneously by the diffusional coupling and that the baths are infinite, so that no variation in the boundary concentrations of A and B takes place. In my thesis work I pointed out that finite compartments behave like capacitors and used this result to introduce Fick's second law. Moreover, I introduced capacitors to represent chemical relaxation effects (see reference for further details). I should also point out that this type of modelling was very familiar to other workers, even though my results were independently obtained. This includes work by Krohn (ref. 128) and other, more recent efforts (refs. 55, 56). Moreover, a secondary reference from the book on Einstein's collected papers on Brownian motion (ref.70) indicates that Lorentz's wife (of the Lorentz transformation) had done extensive modelling on many types of such phenomena, especially as they relate to diffusion and Brownian motion. Unfortunately, that reference is inaccessible to me so that I have not investigated the matter further.

I review the transient corrections for completeness. In transient situations, there are sources of divergence. If these sources are conservative--i.e., if the dissipated energy or entropy can be accounted for and recovered at a latter time, these sources can be accounted for in terms of capacitive storage, as described above. Furthermore, these capacitive effects may represent a real physical capacitor in which there is a constitutive relation of the form $q = Cv$ representing the amount of stored extensive quantity --e.g., charge -- per unit of force measured. In general, the capacitive equation will have the form

$$(x/F) = C, \qquad (43)$$

in which x is a displacement, F is the force and C is the capacitance. The simplest such storage is shown by any region which serves to define a diffusional potential and in which the flow has not reached the expected steady state value. A "compartment", for example, is a special case of this situation. Given that the concentration of any species i is given by

$$c_i = n/V \qquad (44)$$

in which n is the number of moles, it is clear that a compartmental capacitor is simply the volume of the storage area for the Fick diffusional problem in which the concentrations are the proper potentials. Note that in the case in which the potentials are concentrations multiplied by kinetic constants (or, in general, polynomials) this effect must be taken into account, so that, in the example given above, the capacitor for the right hand compartment as seen by A is V/k_1, while the capacitor seen by B would be V/k_2. Each of these are assumed to attach between potential $k_1[A]$ and the "reference" potential or "ground" and between $k_2[B]$ and the ground. The total capacitance seen by the $k_2[B] -- k_1[A]$ port on the right must be obtained by adding the series capacitors. This part is rather trivial but the literature has an example in which the author failed completely to either note the fact that the volume is a function of the compartment, not the species considered, leading to serious errors in interpretation.

For large volumes, the capacitor is infinite and no compartmental effects are seen. Incidentally, the same approach to compartmental storage can be used if the forces are <u>thermodynamically</u> defined, in which case the capacitor must be

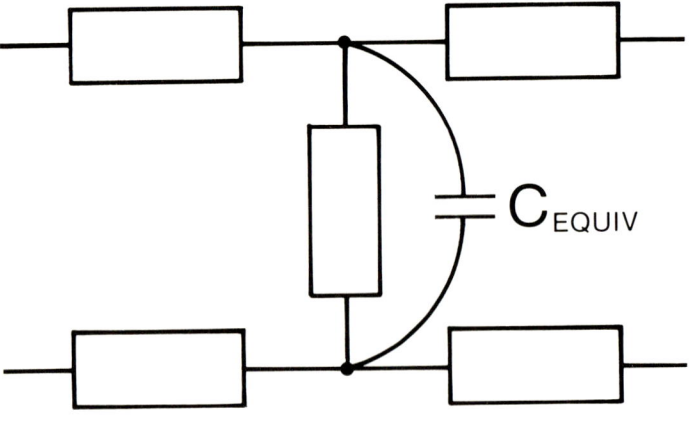

FIGURE 6.3 Electrical network representation of reaction diffusion coupled to compartmental effects.

defined accordingly

Given the force- displacement definition of a capacitor, the capacitor is also allowed to be a nonlinear function of the displacement. A second situation in which capacitive storage appears is when the observed rate of a process--e.g., a chemical reaction -- does not fit the steady force strength. In that case the defficiency in rate can be fictitiously assumed to flow into a storage process which both temporarily prevents the force from reaching an instantaneous value and which also can account for the instantaneous flow rates that enter the reactive region but have not yet reacted. This reflects the relaxation involved in the process. It is seen that the integral expression of the capacitive law now accounts for the observed situation and that a capacitor in parallel with the chemical resistance will account for the observed effects. The complete pseudoelectrical representation is given in FIGURE 3. The solution for the delayed chemical rate is simply given by an exponential of the form

$$J_c = J_c(\infty) [1 - \exp(-t/\tau)]. \qquad (45)$$

At infinite time, the rate reaches its steady value J_c, in which the time constant is given by

$$\tau = RC$$

in which the equivalent resistance which charges the capacitor has the value

$$G_{equivalent} = 1/R_{equiv} = 1 + 1/b, \qquad (46)$$

in which b is again the Thiele modulus. The overall time constant is then

$$\tau = C/(1 + 1/b).$$

6.7 TRANSMISSION LINE EQUATIONS DERIVED FROM THE CONNECTED TOPOLOGY

Analytically, the connected topology description presented for the chemical reaction diffusion coupling problem in Chapter 3 allows to express the network laws as a differential equation by considering the right hand and left hand side boundaries to be at x and $(x + \Delta x)$, repectively. Because of the cascaded nature of the problem --in which diffusive and reactive regions are attached in some kind of linear ordering--, it is convenient to express each region as a local vector at x which is a function of the vector at x+dx (again, I use the differential sign to represent a finite difference in general)

$$F(x) = A(x) F(x+dx), \qquad (47)$$

in which the vector **F** represents all the measurable potentials and flows at a point x. If a matrix **A** can be given as a function of position, it is clear that the matrix A_{total} which appear in the "integrated" form of the above equation, say, for a membrane of thickness L,

$$F(0) = A_{total} F(L), \qquad (48)$$

is simply found by multiplying the individual **A** matrices in the order in which they appear in the membrane:

$$A_{total} = A(x_0) A(x_1)\ldots\ldots A(x_L). \qquad (49)$$

If one atttempts to subdivide the diffusion-reaction process into unit operations, there is some indeterminacy in the definition of the problem. This is very clear, as we have seen, from the topological point of view, but it also surfaces in analytical descriptions. This observation prompted Bunow and Aris to introduce three separate matrices corresponding to the chemical reaction, diffusion (inert region) and an integrated reactive region in which both reaction and diffusion take place. In the topological, or network formulation, the unit transmission operation can be considered to be the whole elementary slab and it is not necessary to integrate over the region of the membrane first. The transmission matrix **A** is given by (50)

$$\begin{bmatrix} J_A(0) \\ J_B(0) \\ \hline \phi_A(0) \\ \phi_B(0) \end{bmatrix} = \left[\begin{array}{cc|cc} 1 & 0 & G & -G \\ 0 & 1 & -G & G \\ \hline 1/P_A & 0 & 1 & 0 \\ 0 & 1/P_B & 1 & 0 \end{array}\right] \begin{bmatrix} J_A(L) \\ J_B(L) \\ \hline \phi_A(L) \\ \phi_B(L) \end{bmatrix}$$

in which the $\phi_i(x)$ are the potentials of A or B at any point in the membrane. We consider here the possibility of many distributed reactions, so that, in general, G, P_A, P_B and the potentials will vary with position. The potentials have been left in an undeterminate form to stress that they could be either thermodynamic or kinetic potentials. In the particular case of the kinetic description with first order linear kinetics we can make the identifications

$$\phi_A(x) = k_1 [A](x), \qquad (51)$$

$$\phi_B(x) = k_2 [B](x) \qquad (52)$$

and G= 1. Transmission matrices are particularly useful if a numerical answer is desired for a multilayered membrane containing several different regions, but in the case in which the membrane consists of many homgeneous regions, we can use the methods of differential calculus to obtain a closed matrix solution as done by electrical engineers to treat transmission lines.

Denote by a(x) the generalized afinity at x, given by

$$a(x) = \phi_A(x) - \phi_B(x). \qquad (53)$$

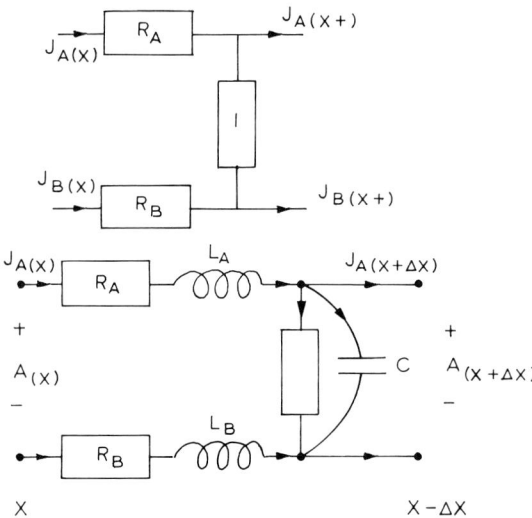

FIGURE 6.4 Basic unit box used to derive " transmission line" equations for the diffusion reaction problem.

Application of Kirchhoff's current law to the network shown in FIGURE 4 yields

$$J_A(x+) - J_A(x) = J_B(x+) - J_B(x) = -G\ a_x = -J_c \qquad (54)$$

while K.V.L. gives

$$a(x+) - a(x) = -R_A\ J_A(x) + R_B\ J_B(x). \qquad (55)$$

By taking the differential limits we obtain

$$\partial J_A/\partial x = -\partial J_B/\partial x = -G\ a = -J_c \qquad (56)$$

and

$$\partial a/\partial x = -\hat{R}_A\ J_A(x) + \hat{R}_B\ J_B(x), \qquad (57)$$

in which $\hat{R}_A$ and $\hat{R}_B$ are the generalized diffusive resistances per unit length. Equations (54) and (57) readily yield

$$\partial^2 a/\partial x^2 = G\ (\hat{R}_A + \hat{R}_B)\ a \qquad (58)$$

in which G is the total chemical conductance in a unit length slab x. Clearly, if G= 0,

$$\partial^2 a/\partial x^2 = 0 \qquad (59)$$

and the afinity decays linearly with x,

$$a = a_1\ x + a_2,$$

while, in general,

$$a(x) = a_+ \exp\{x\ \sqrt{G(R_A + R_B)}\} + a_- \exp\{-x\ \sqrt{G(R_A + R_B)}\} \qquad (60)$$

in which a_+ and a_- are constants determined by the boundary conditions. The + and - signs have been chosen to denote the fact that the solution has the nature of a superposition of a forward and a backward diffusing solutions. This result is in agreement with other approaches. Note that if $R_A + R_B = 0$, the problem is only defined for the integrated slab.

With the introduction of inertia and chemical relaxation, partial differential equations can be given explicitly using Kirchhoff's laws and the constitutive equations for the apropriately placed capacitors and inductors.

For example, in FIGURE 4 inductors representing the inertia of A and B have been placed in series with the diffusional resistances, while a capacitor representing the relaxation of the chemical reaction has been placed in parallel with the chemical resistance. The partial differential equations which follow directly from Kirchhoff's laws are

$$\partial J_A / \partial x = - G_{chem} A - c \, \partial A / \partial t , \qquad (61)$$

$$\partial J_B / \partial x = - G_{chem} B - c \, \partial B / \partial t , \qquad (62)$$

and

$$\partial a / \partial x = (R_A J_A - R_B J_B + L_A \partial J_A / \partial t - L_B \partial J_B / \partial t . \qquad (63)$$

Equations (61), (62), and (63) have a form very similar to the well studied transmission line equations. The solutions to these equations have been studied extensively and ref. 134 should be consulted for natural frequencies, approximations, reflecting properties, etc. Tellegen's theorem can again be used to determine the energy constraints imposed on the forces and flows at x and (x+ dx). The difference between input and output dissipations can then be expressed as a function of x in the form

$$- \partial \Phi / \partial x = \Phi_{diss} + d(T + U) / dt \qquad (64)$$

which is an analog of Poynting's theorem for the chemo-diffusional line. It should be clearly stressed that all of the above is only applicable once a connected network has been given--i.e, when the continuity imposed by Kirchhoff's laws has been established.

The same situation holds for the application of Lagrange's equations, as it has been shown above that they are a special case of Tellegen's theorem. There are several reports in the literature where both the integration and Lagrange's equations have been used in systems which are not connected and in which they are not valid in holonomic form.

6.8 DISSIPATION FUNCTION AND EFFICIENCY IN THE STEADY STATE

We define, as above, an average dissipation as found from the instantaneous dissipation

$$\Phi = (T\, d_i S/dt) = X(t)\, J(t) \qquad (65)$$

in which $X(t)$ and $J(t)$ are the conjugate quantities and a simple system of one force and one flow is first considered. The time varying force and flow may be replaced by their complex forms to yield

$$\Phi(t) = ||X||\ ||J||\ \text{Real}\ \exp j(\phi_x + wt)\ \text{Real}\ \exp j(\phi_j + wt) \qquad (66)$$

in which ϕ_x and ϕ_j are the phase angles for X and J respectively. This expression yields the average dissipation

$$\Phi(t)^{average} = ||X||\ ||J||\ \cos(\phi_x + \phi_j) = \qquad (67)$$

$$= ||X||\ ||J||\ \cos(\phi_j - \phi_x),$$

which can also be expresseed in terms of exponentials by

$$\Phi(t)^{average} = (1/2)\ \text{Real}\ [\ X\ J^*\ ] = \qquad (68)$$

$$= (1/2)\ \text{Real}\ [\ J\ X^*\ ].$$

The starred quantitites are complex conjugates of the original complex amplitudes. It can further be shown that

$$\text{Imaginary}\ \Phi = 2w(T_{average} - V_{average}). \qquad (69)$$

Since only average dissipation is calculated in the following derivation, the subscript "average" is omitted. Also we shall drop the 1/2 factor because only ratios of dissipations are found. The treatment follows closely references 53 and 216 .

6.9 GENERALIZED TWO PORT CONVERTER

Using the above expressions, we write

$$\Phi(t)^{input} = \text{Real} [X_1 \, J_1^*] = \tag{70}$$

$$= \text{Real} [J_1 \, X_1^*]$$

for the complex input power going into the converter and

$$\Phi(t)^{output} = \text{Real} [X_2 \, J_2^*] =$$
$$= \text{Real} [J_2 \, X_2^*]$$

in which we have denoted the output power or dissipation by a negative sign, as before, so that when O is positive, entropy, dissipation or power is delivered to the external world.

Given that we deal with time variations, we set up the equations in terms of the complex inputs and ouputs and the respective impedances,

$$X_1 = Z_{11} J_1 + Z_{12} J_2 \tag{71}$$

$$X_2 = Z_{12} J_1 + Z_{22} J_2 \tag{72}$$

We now assign, following the central reference, the following values for J_1 and J_2:

$$J_1 = 1 + j\,0 = 1 \tag{73}$$

and

$$J_2 = (L + j M) K; \tag{74}$$

in which the parameter K is defined by

$$K = - Z_{21} / 2 \, \text{Real} \, Z_{22} \tag{75}$$

This choice of variables is introduced without loss of generality as all possible load conditions can be simulated by changing the variables L and M. Furthermore, the restriction that one of the input quantities be unity posses no problem, as only ratios of

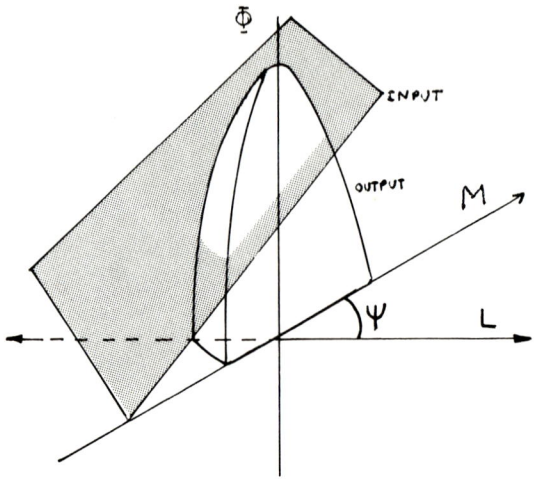

Fig. 6.5 Input and output parabola and the L, M plane. The quantities plotted are input and output dissipation as a function of L and M (after refs. 53 and 182).

quantities will be calculated--e.g., efficiency . With this choice of variables, the input and poutput dissipations become

$$\phi_{in} = r_{11} - L\, a/2\, r_{22} + Mb/2r_{22} \qquad (76)$$

and

$$\phi_{out} = L\, ||z_{21}||^2/\, 2\, r_{22} + [\, L^2 + M^2\,]\, ||z_{21}||^2/\, 4\, r_{22} \qquad (77)$$

in which r_{11} and r_{22} are the real parts of z_{11} and z_{22}, respectively, and a and b are the real and imaginary parts of the product of cross impedances,

$$z_{12}\, z_{21} = a + j\, b. \qquad (78)$$

Equations (76) and (77) represent a plane and a paraboloid of revolution , respectively, in a cartesian system with axes L, M and ϕ . These two surfaces are depicted in FIGURE 5 and replotted in two dimensions in FIGURE 6 . The two surfaces have the following general properties:

1. The axis of revolution of the paraboloid is the line L= 1, M= 0;these values of M and L give the maximum output dissipation o_{out}

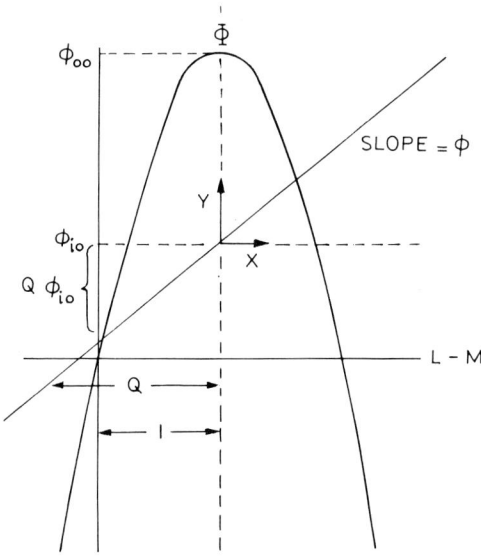

Fig. 6.6 Two dimensional projection of Fig. 6.5.

2. The point at which the input plane intersects the axis of revolution of the paraboloid has been labelled o_{io} and a cartesian coordinate system x,y has been defined with origin at o_{io}. Note that x is given in the direction of an angle   , which is, in turn, specified by the inclination of the input dissipation plane.

3. For a given $\dot{\Phi}_{io}$, the inclination of the dissipation input plane in the two dimensional drawing is given by the product $Q\ o_{io}$ on the Φ axis.

We are interested in the behavior of the system when the output dissipation is positive--i.e., when dissipation is delivered to the external world. This restricts the value of x to
$$-1 < x < +1. \tag{79}$$
Furthermore, we require that whenever the output dissipation is positive the input power cannot be negative. From the two dimensional drawing, it is seen that this condition is met when

$||Q|| < 1.$

The efficiency is now given by the ratio

$e = \phi_o / \phi_i$

and we wish to express this equation in terms of two dimensional parameters. In FIGURE 6 the parabola has the equation.

$$\phi_o = \phi_{oo} (1 - x^2) \qquad (80)$$

while the input line is given by

$$\phi_i = \phi_{io} (Qx + 1). \qquad (81)$$

The efficiency then becomes, upon introducing (80) and (81),

$$e = \phi_{oo} (1 - x^2) / \phi_{io} (Qx + 1). \qquad (82)$$

The optimum x for maximum efficiency is found by differentiating equation (82) with respect to x; the resultant value of x, x_{opt} is given by

$$x_{opt} = [-1 + \sqrt{1 - Q^2}] / Q \qquad (83)$$

in which the positive sign has been chosen for the square root to obey the stability condition. The maximum efficiency is

$$e_{mx} = 2 P_{oo} K_G / P_{io} \qquad (84)$$

in which

$$K_G = 2 [1 - \sqrt{1 - Q^2}] / Q^2 \qquad (85)$$

It is now necessary to express all quantities in terms of the three dimensional parameters L, M, P introduced originally and then in terms of the two port parameters. As a first step in finding (1+ jM) we plot the input and output surface in the complex plane L, jM. The distance x can be expressed as $x \exp(\Psi j)$; the magnitude of the vector is x, while the angle it makes with the L axis is Ψ. The complex number $x \exp(j\Psi)$ may also be given by

$$x \exp(j\Psi) = (L + jM) - 1 \qquad (86)$$

in which the value -1 that appears in the second term is a correction for the fact that x is measured from L=1. Thus,

$$(L + jM) = 1 + x \exp(j\Psi). \qquad (87)$$

Equations (83) and (84) provide a relationship between $x_{optimum}$ and K_G:

$$x_{optimum} = -Q K_G / 2 \qquad (88)$$

which relates x_{op} to $L + Mj$, provided that the angle Ψ is known. To this end, we look for the intersection of the input dissipation plane with the L, M plane--i.e., the plane O=0--, this is found by setting equation (76) equal to zero. The resultant line is given by the equation

$$M = (La/b) + [-2 r_{11} r_{22} / b]. \qquad (89)$$

A line perpendicular to the line described by (89) makes an angle Ψ with the L axis, as shown in FIGURE 5. From this plot, we obtain

$$\tan \Psi = -(b/a). \qquad (90)$$

Moreover, the negative product of the complex impedances Z_{21} and Z_{12} is

$$-(Z_{21} Z_{12}) = -a - jb, \qquad (91)$$

which has a complex conjugate expression given by

$$-(Z_{21} Z_{12})^* = -a + jb. \qquad (92)$$

Therefore, the angle Ψ can also expressed as

$$\Psi = \text{argument } -(Z_{21} Z_{12})^*. \qquad (93)$$

In the special case in which the product $(Z_{21} Z_{12})$ is real, =0 and x coincides with the L axis. The values of all quantities may now be specified in terms of the phenomenological parameters Z_{ik}, as done in reference 53

6.10 OPTIMUM LOAD IMPEDANCE

The load impedance Z_L is given by

$$Z_L = -X_2 / J_2 \tag{94}$$

$$Z_L = [Z_{21}/(L+jM)] + Z_{22}. \tag{95}$$

In particular, the optimum load impedance is given by

$$Z_L(\text{optimum}) = -Z_{22} + \{4 \text{ Real } Z_{22}\}/[2 - QK_G \exp(j\ )]. \tag{96}$$

The coefficient Q is found from similar triangles in FIGURE 6,

$$Q P_{io} = P_{io}/Q. \tag{97}$$

Given that Q is the perpendicular distance from L=1, P=0, M=0 to the line resulting from the intercept of the input plane with the L,M plane, it can be found by knowing the distnace from L=1 to the L value for which the input plane intercepts the L axis. Simple trigonometry on the angle Ψ leads to the final result

$$Q = 2 (P_{oo} Z_{12} / P_{io} Z_{21}). \tag{98}$$

Finally, the ratio ϕ_{oo}/ϕ_{io} is found by setting L=1 and M=0 in equations (76) and (77), which yields

$$\phi_{oo}/\phi_{io} = ||Z_{21}||^2/[4 \text{ Real } Z_{11} \text{ Real } Z_{22} - 2 \text{ Real } Z_{12} Z_{21}]. \tag{99}$$

6.11 RELATION TO THE STEADY STATE RESULTS

When no energy storage takes place in the system, the impedances are real. In this restricted case, the efficiency parameter Q becomes

$$Q = ||Z_{12}|| \ ||Z_{21}||/[2 Z_{11} Z^{22} - Z^{12} Z_{21}]. \tag{100}$$

This expression may be simplified further by introducing the generalized parameters q_{12} and q_{21}' defined by

$$q_{12}' = Z_{12} / \sqrt{Z_{11} Z_{22}} \tag{101}$$

$$q_{21}' = Z_{21} / \sqrt{Z_{11} Z_{22}} \tag{102}$$

to obtain

$$Q = ||q_{21}' q_{21}'|| / (2 - q_{21}' q_{12}'). \tag{103}$$

As expected, when the system is symmetric --i.e., it is described by reciprocal phenomenological equations the maximum efficiency predicted by Kedem and Caplan is obtained again.

The requirement that $||Q||<1$ imposes a restriction on q, in the steady state case:

$$q^2 < 1 \tag{104}$$

for the symmetric case and

$$||q_{12}' q_{21}'|| < 1 \tag{105}$$

for the asymmetric case with energy storage. In terms of impedance coefficients the last expression becomes

$$||Z_{12} Z_{21}|| < Z_{11} Z_{22}, \tag{106}$$

which is not equivalent to imposing positive definiteness to the phenomenological matrix, because Equations (105) and (106) do not exclude the possibility that the input dissipation be smaller than the output dissipation--e.g., there could be an amplification step or sources not included in the description, but the analysis would still be valid from the point of view of the included ports. The general condition here is derived not from positive definiteness, then, but from stability considerations in the energy storing elements (the requirement that they be positive).

Graphical techniques borrowed from the design of transmission line and transistor engineering are discussed elsewhere (ref. 182). Various charts analogous to the familiar Smith charts used by engineers to obtain optimum loads are obtained.

Appendix

COMMENTS ON THE STATIONARITY OF THE STEADY STATE AND THE "PRINCIPLE" OF MINIMUM ENTROPY PRODUCTION

The usual route taken in non-equilibrium thermodynamics to demonstrate the stationarity of the steady state is to use a variational (Eulerian or Hamiltonian) approach. The local entropy production is first integrated to find the total entropy production, Φ. The application of a first variation then shows that for the case in which $\delta\Phi = 0$ the Euler equation obtained leads to conservative flows --i.e., div $J_i = 0$. Moreover, differentiation of the total entropy with respect to time shows that the entropy production decays with time as it reaches the stationary state.

In the case of a single flow, for example heat conduction, the problem is clear cut and is, in fact, no more than a variation on the Lapace's equation theme. Thus, given the linear heat flow problem having the phenomenological equations

$$J_q (x,y,z,t) = L_{qq} \, \text{grad} \, (1/T), \qquad (1)$$

the first variation of the global entropy integral (given a solid which does not expand) yields

$$\Phi = \delta \int^V [L_{qq} \, \text{grad}^2 \, (1/T) \, dv] = 0 \qquad (2)$$

and leads to the Euler equation

$$\text{div grad} \, (1/T) = 0 \qquad (3)$$

or, equivalently, to div $J_q = 0$, showing that the state of minimum entropy production is indeed a stationary state in which there are no sources or sinks of heat. Moreover, differentiation of the entropy integral with respect to time in the neighborhood of the steady state but in the prescence of sources of the form

$$\text{div} \, J_q = - c \, \partial T / \partial t \qquad (4)$$

--i.e., capacitors--, yields

$$\partial \Phi/\partial t = -2 \int^V (c_V/T^2)(\partial T/\partial t)^2 \, dv. \quad (5)$$

Given that c_V is assumed to be positive, the stability of the steady state follows in the form

$\partial \Phi/\partial t < 0$
or
$\partial \Phi/\partial t = 0.$

In network terms the stability of the stationary state can either be considered to be a result of Tellegen's theorem or looked upon as a result of the use of operational calculus --or, more rigorously, the theory of distributions in functional analysis-- : in a network containing only positive resistances, capacitors and inductors, the poles and zeroes of the system function lie on the left hand side of the complex frequency plane. As a result, these systems are stable.

From a physical point of view, positive capacitors discharge if they are charged in the presence of dissipative structures (resistances) and positive inductors oppose changes in flow. As a result, any changes forced on the system will be met with forces and flows which tend to counteract the disturbance, in agreement with the dynamical idea of LeChatelier's principle. On a more interesting vein, it has been conjectured that the steady state distribution of flows yields a minimum in the evolutionary history of the system--i.e., that all systems tend to a minimum of entropy production, so that the steady state is, in a sense, a state of dynamic equilibrium.

The additional assumption is that the source of divergence conserves entropy. In network terms the claim that there is an evolutionary minimum of entropy production does not appear to be valid if both capacitive and inductive branches are introduced. Thus, for example, a network consisting of a resistance and a capacitance in series has minimum dissipation in the steady state, but a resistance in series with an inductance --as would represent, for example, the acceleration of a particle in the presence of both a driving force and a frictional force -- has <u>maximum</u> dissipation in the stationary state.

Chapter 7

PIECEWISE NON-LINEAR NETWORKS: STEADY AND OSCILLATORY

7.1. INTRODUCTION

Most interesting natural processes are not in the steady state, but exhibit, instead, transitions which either decay or result in sustained oscillations. While passive linear systems can give rise to sustained oscillations--such as the case of the familiar L-C oscillator--, these will eventually decay because of the presence of unavoidable parasitic energy dissipation in frictional elements. Most real systems capable of sustained oscillations, therefore, maintain these through active processes which neutralize the parasitic resistances by subtraction. These processes involve non linearities and feedback so that the usual difffferential equation approaches have limited applicability. Onsager´s theory is clearly inoperative in these ranges because it can not include large "signals" , non linearities or non-equilibrium. Unfortunately, there are no clear suggestions of what would constitute a proper dissipation function in such cases or , for that matter, proper forces and flows.

The network formalism leads to a natural extension of the Onsager theory in the non linear realm. First, we assume that the topological characteristics of conservation and continuity and conservation imposed by Kirchhoff´s laws still hold and that it is the branch --or overall-- box constitutive relations that change. Given Kirchhoff´s laws, Tellegen´s theorem holds and it then follows that the external dissipation can be equated with the internal dissipation in the system.

The general non linear problem is of course forbidingly complex. We therefore consider only two specific examples to illustrate the applicability of the technique to some general situations:

1. Thermodynamic models of muscle contraction (ref.182),which illustrates how a non linearity accounted for in terms of switching mechanisms leads to a unique dissipation function;

2. The Teorell oscillator., which is a classic example leading to oscillatory coupled phenomena.

Both of these problems may be represented by non-linear resistances. The approach to non linearity taken in this chapter consists in splitting this resistance into regions of constant slope (positive or negative) and then introducing force-flow characteristics which can effect the transition from one part of the curve to another. These are basically basic ideas which are in an embryonic stage and I will not treat them with outmost rigor or detail.

A more general approach to dealing with non-linear thermodynamic problems will be suggested in the closing chapter.

7.2 THERMODYNAMIC MODELS OF MUSCLE CONTRACTION

A phenomenological model of muscle contraction must predict not only the macroscopic coupling between hydrolysis of high energy compounds and the contractile process, but also the non-linearity known as the Hill equation

$$(P + a)(V + b) = (P_o + a)(V_m + b)a, \qquad (1)$$

in which P is the output tension and V the velocity of contraction during tetanic stimulation. P_o and V_m are the maximum tension obtained under isometric conditions (v→0) and the maximum velocity in the absence of a load, respectively.

The Hill non linearity presents a serious obstacle to the Classical irreversible thermodynamic analysis, because Onsager thermodynamics cannot consider such processes. Thus, thermodynamic models of muscle contraction have been given in which the phenomenological linearity is kept at the energy conversion step, while the non linearity is arbitrarily assigned to either a regulatory or to a switching mechanism.

Caplan (ref.43) considers the muscle as a mechanochemical energy converter having the dissipation function

$$\Phi = J_{chem} A + P V, \qquad (2)$$

in which J_{chem} is the reaction rate at the input of the converter and A is the chemical affinity of high energy compund hydrolysis.

The resultant linear Onsager phenomenology for this model is

$$A = R_{11} J_{chem} A + R_{12} V$$
$$P = R_{21} J_{chem} A + R_{22} V \qquad (3)$$

--in which P is the tension applied by an external load on the muscle, rather than the tension developed by the muscle. In Caplan's model, these equations are linear. An auxiliary nonlinear equation is given for a regulator step in the form

$$A_{reg} = A(J_{chem}). \qquad (4)$$

The basic clue in the derivation of a network representation for the system is obtained through an additional equation given by Caplan for the total input affinity:

$$A_{total} = A_{regulator} + A_{converter} ; \qquad (5)$$

which is a form of Kirchhoff's "voltage" law. Since the chemical rates at the converter input and at the regulator output reaction are the same, K.C.L. also follows. The resultant network is shown in FIGURE 1 (ref.182) in which it is assumed that the phenomenological equations are reciprocal, as done by Caplan, thus leading to a connected resistive model.

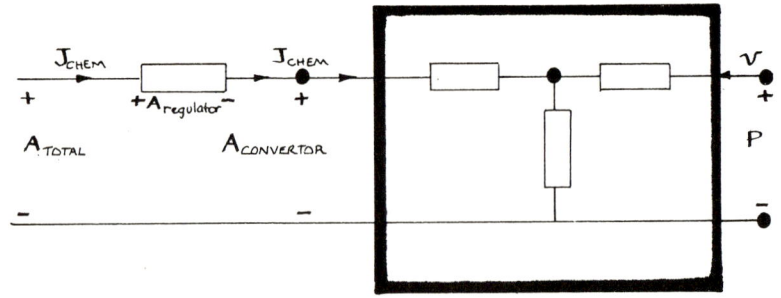

FIGURE 7.1 Onsager network and non linear resistance ("regulator") used to represent the Caplan model of muscle contraction in network terms.

Eqs. (1) and (5) translate into a (non linear) resistance in series with the **T** network representing the Onsager conversion step. A second model was proposed by Bornhorst and Minardi (45) in which they suggested that the Hill non linearity appears indirectly, through the involvement of an increasing number of energy converters -- assumed to be the individual sarcomeres. From the network viewpoint, the linear energy converting steps are simply two ports which connect together through some time switching process. These converters may be added using matrices, following the techniques discussed in Chapter 5. The addition of successive boxes changes the resistances in time so that the Hill curvature is mimicked. This process could be thought of as consisting of , say, many **H** converters as shown in FIGURE 2 . Each additional energy converter has the same input affinity and the same velocity as the previous ones while the total chemical flow and the total tension add up. Thus, the number of elements

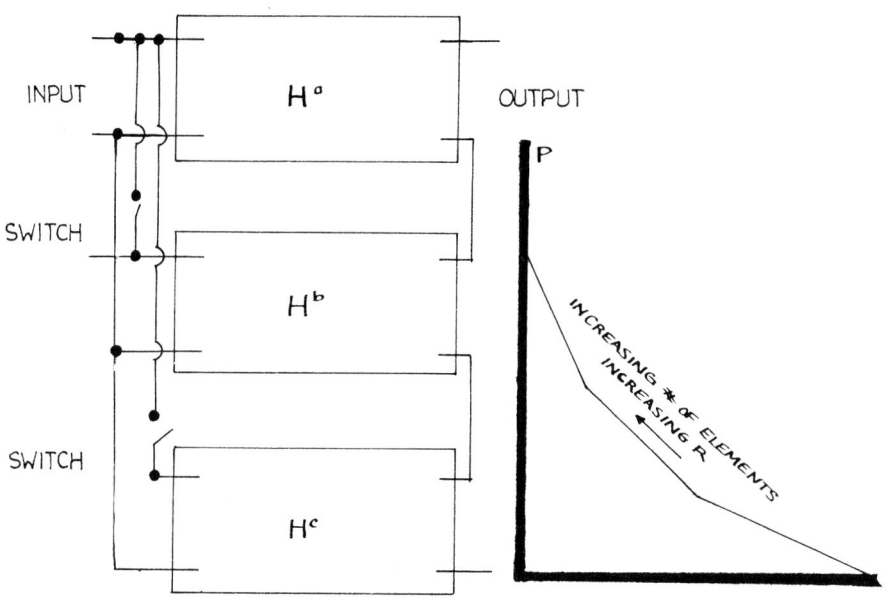

FIGURE 7.2 The non linear Hill curve can be synthesized by switching on identical **H** converters as the velocity of contraction increases.

that take part in the process increases with increasing tension --
i.e., decreasing velocity--, so that a non linear resistance is
synthesized. Moreover, the maximum total efficiency is the same as
that of a single converter. This follows readily by calculating q
for each converter and for the overall system , because

$$q_{unit} = H_{12} / \sqrt{H_{11} H_{22}}$$

while the overall degree of coupling is given by

$$q_{total} = n H_{12} / \sqrt{n^2 H_{11} H_{22}} = q_{unit}.$$

The interesting -- and practical --part of the problem arises
in the context of heat dissipation. Does one model dissipate more
than the other? According to Bornhorst and Minardi their model
dissipates less, because the switching process does not dissipate.
This is incorrect in the light of Tellegen's theorem because if
both models have the same inputs and ouputs and if they obey
Kirchhoff's laws they must have the same dissipation.

The problem is a dynamical analog of Maxwell's demon. At first
it looks as though a switch will not dissipate because it is
either closed and has no force across, in the ideal case, or it is
open in which case there is no flow through it. In network terms
the switching element is the ideal diode, shown in FIGURE 3. It

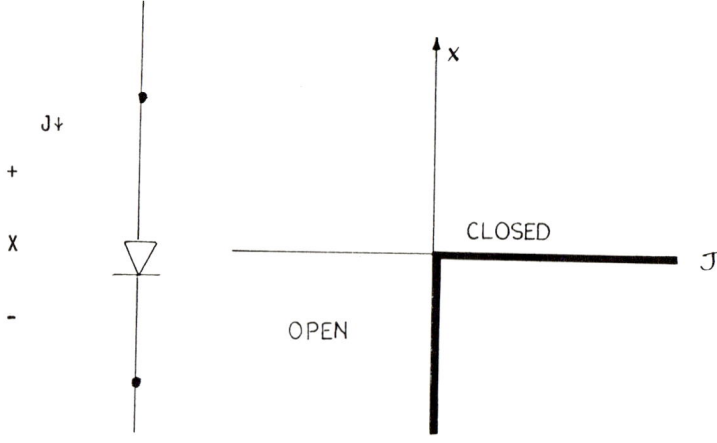

FIGURE 7.3. Network representation and ideal force-flow characteristic for the Ideal Diode. An ideal diode does not
dissipate as it is either open (X= 0) or closed (J=0).

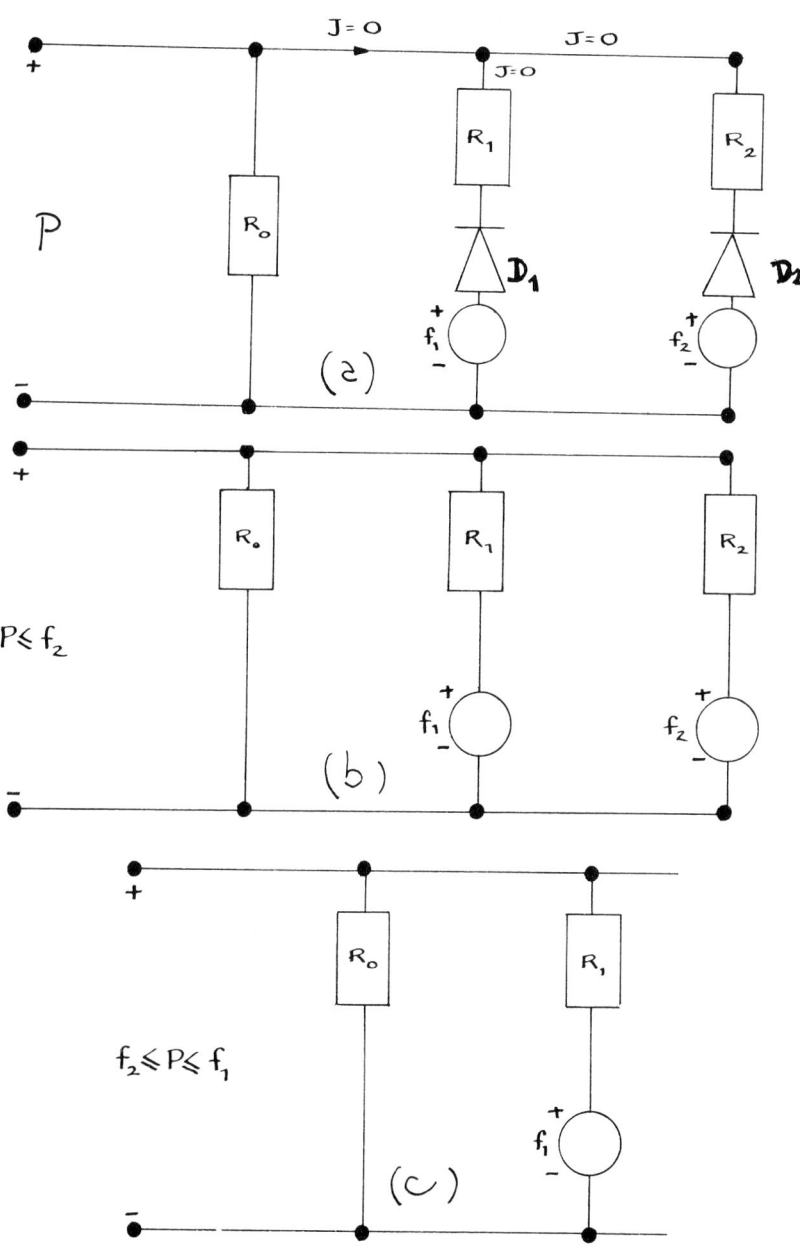

FIGURE 7.4 A resistance- ideal diode network in which the reference forces f_1 and f_2 keep the diodes closed until the input force P exceeds those values.

can be closely approximated analytically by the function

$$J = J_s [b \exp(aX) - 1]. \qquad (6)$$

The X-J characterisitc of the ideal diode is shown in FIGURE 3. Basically, it can be interpreted as stating that:

1. For any negative force, there is no through flow (OPEN) (a small amount equal to $-J_s$ in the above equation) and

2. For any positive flow the force is zero (CLOSED) -- about ln 1 for the equation given above.

Clearly, the diode *by itself* shows little or no dissipation. The problem consists in "programming" one or more diodes to switch at prespecified values of input forces or flows, in order to approximate any curve with the desired degree of precision.

This is achieved by providing internal reference levels in the form of force (voltage) or flow sources (currents). By combining sources, diodes and resistances one can achieve this approximation. The resultant states of the system can be calculated by allowing the diodes to be opened or closed, for various ranges of the input flows or forces (a standard network problem discussed in the textbooks).

The example of FIGURE 4 shows that an additional ("hidden") dissipation is required to provide the required memory function, because--in this example-- every time a diode closes there is an additional flow which goes both through a resistance and through the reference force. The dissipation at the source is the "hidden" dissipation.

Thus, if the reference forces in FIGURE 4 obey the inequality $f_1 > f_2$, the force across the diodes will keep them closed as long as the input force (pressure) P is smaller than f_2 (FIGURE 4 b). The resultant resistance seen at the input is the sum of R_o, R_1 and R_2 in parallel. As P increases and becomes larger than f_2 Diode 2 opens (FIGURE 4c) so that the resistance increases. Finally, when P exceeds f_1 Diode 1 opens and the resistnace increases even further. If the direction of incresing flow is considered, the situation is reversed. As the flow is increased more resistances in parallel *and sources* are added, thus leading to the additional dissipation. These cannot be eliminated.

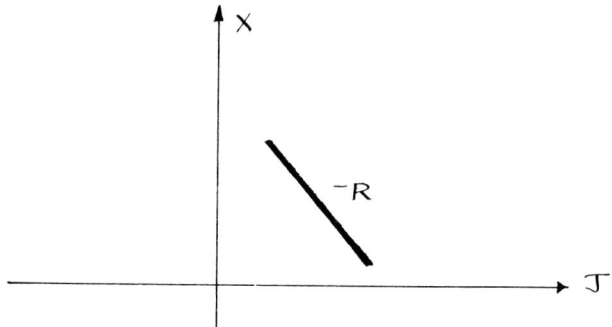

FIGURE 7. 5   The negative resistance curve.

7.3   CATASTROPHE THEORY VS. NONLINEAR NETWORK OSCILLATIONS

From an energy, or dissipation point of view , it is clear that the systems leading to self sustained oscillations must contain folding non linearities (ref. 275). A positive frictional process which is naturally occurring must be cancelled by a "negative" resistance having the form (refer to FIGURE 5)

X = - R J

Such a process is clearly energy suppplying and can therefore not go on into arbitrarily large or arbitrarily small values of J, because otherwise infinite sources of power would be required. Therefore, the most elementary type of X-J curve must fold at its end points .

This is a well understood process curve studied by Catastrophe Theory (C.T.). C.T. considers transitions which occur because of the presence of singularities in certain maps are surfaces which represent the possible energy states of the system. Taken in loose terms, C.T. fits the idea of cause and effect: the total energy of the system is parametrized by the "causes" --a control space, C-- while the displacements of the system-- or behavior space, X -- appears as heights in the energy surface. The types of possible transitions follow from the dimensionality of the control space, which reflects the degree of the polynomial in the energy plot. In particular, most analyses focus on the behavior in the neighborhood of the region, at critical points exhibiting maxima and minima. Transitions are obtained by considering the total energy of the system, $V = V(a,b,c...,x)$ in which x is a displacement variable and a, b, c,...are forces acting on the system. By looking at the derivatives

$$\partial v/\partial x = 0, \quad \partial^2 v/\partial x^2 = 0 \quad \partial^3 v/\partial x^3 = 0, \qquad (7)$$

the critical charactersitics of the energy surface can be analyzed.

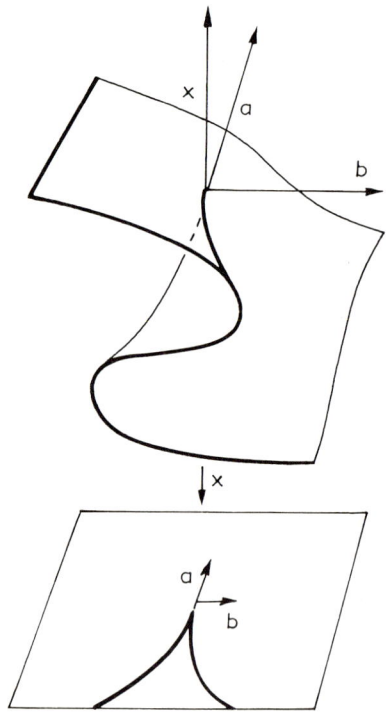

FIGURE 7.6 (a) The Cusp Catastrophe showing the folded maxima and minima . (b) Projection of the cusp in control space C=2 showing the bifurcation set. (after ref. 267)

FIGURE a shows the overall transitional behavior of a characteristic system, while FIGURE 6b shows the projection of the motion in a two dimensional space (a,b). This type of catastrophe, which can be approximated by a quartic potential of the form

$$V = (1/4) x^4 + a x + b x^2,$$

is called a **cusp** catastrophe and it appears when the dimension of the control space is C=2. A simpler type of catastrophe is the Fold Catastrophe which appears when the potential is cubic, of the form

$$V = (x^3 / 3) + a x. \qquad (8)$$

The dimension of the fold catastrophe, which is involved in all other higher catastrophes, is C= 1 and its graphical

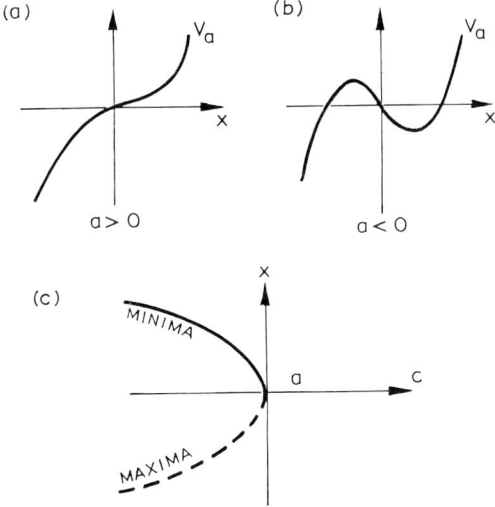

FIGURE 7.7 The Fold Catastrophe (ref. 267), in which the potentials are described by cubic functions, has the graph shown in (a) or (b), depending on the sign of the coefficient a in the expression $V_{(a,x)}$ = ax + $x^3/3$. The mainfold and map are shown in FIGURE (c).

representation is shown in FIGURE 7 . This is precisely the type of curve involved in a realizable non linear negative resistance. Other higher order catastrophes have been discussed by Woodcock and Poston (ref.267) . Essentially each higher order catastrophe serves to organize a lower one.

The analytical approaches involve either studying the bifurcation behavior in the control space, or plotting stereoscopic computer projections through surfaces of various dimensions. Although C.T. allows studying transitions without solving differential equations, it removes some valuable conceptual elements from the analysis of practical problems.

For example, we may wish to retain the direct visualization of cause-effect interactions of the force flow type , as treated in network thermodynamics. For that reason, it is desirable to merge

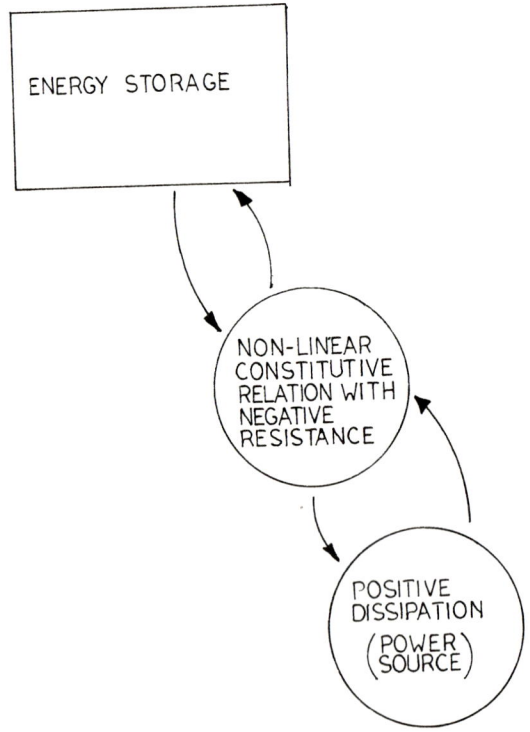

FIGURE 7. 8   Non linear oscillations can be described in mechanical or network terms in terms of three components: a source of power, an energy storage element and a negative resistance.

networks with catastrophes; this can be done by splitting contributions of energy supplying and storing elements (In fact, this form of analysis predated catastrophe theory).

In network terms, physical non linear oscillations include the following three elements:

First, a non linear curve in the X J plane which contains one or more regions of negative resistance. By analogy with C.T., this curve will be topologically isomorphic to a fold potential. We restrict our analysis to the case of one force and one conjugate flow, but it is clear that there must be at least one additional input not considered which provides the required source of energy. The typical curve is shown in FIGURE 9 . Incidentally, neither the + R nor the - R regions need be linear. We actually define the negative resistance as an incremental, monotonically decreasing function.

Second, there is an energy storing element, linear or non linear, whose constitutive relation either reflects the physical impossibility of changing a flow or a force instantaneously. In the first case, the element is able to store energy in a flow (inductor) ; in the second case, the element stores energy in a force (capacitor).

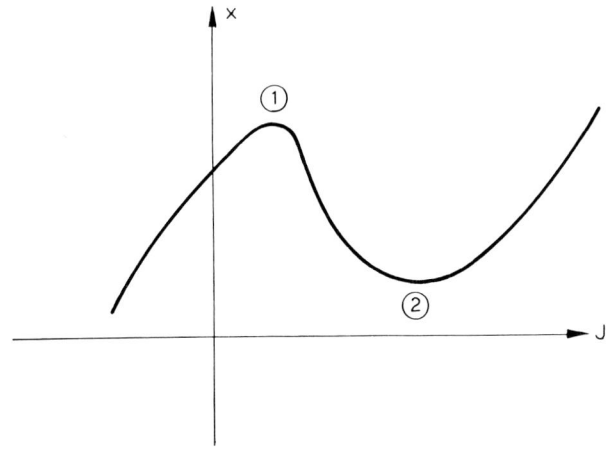

FIGURE 7.9 The typical nonlinear resistance containing a region of negative resistance is isomorphic to the fold catastrophe .

Third, there is a source of constant or pulsed power (dissipation) which is capable of shifting the non linear, negative resistance curve in the X-J plane, either permanently or for an arbitrary, specified time.

7.4 THE NEGATIVE RESISTANCE

The key element in all of this is the negative resistance which has the following properties (refer to FIGURE 10) a) it obeys the law $R = -X/J$, if linear; b) when connected in series with a positive resistance of the same absolute value it cancels out; c) when connected to a capacitor it <u>charges</u>, rather than discharges, the capacitor ; d) the negative resistance generates dissipation in the amount $-R J^2$.

Thus, a negative resistance is another device which is logically consistent with Onsager thermodynamics, except that it violates the condition $\phi > 0$.

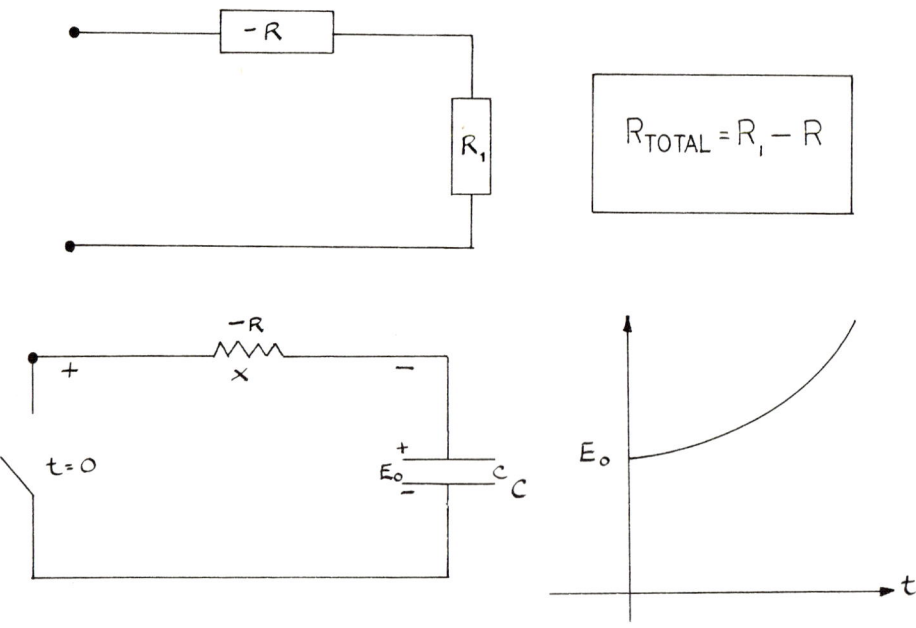

FIGURE 7.10 Some basic characteristics of the negative resistance: it subtracts from a positive resistance in series and it leads to building exponentials when connected to a capacitor (ref. 275).

7.5 THE TEORELL OSCILLATOR

T.Teorell (refs. 231, 232, 234, 235, 236) invented and analyzed in detail the behavior of a membrane system which exhibits sustained oscillations of the relaxation type. Since the initial description by Teorell, many other studies followed (refs. 45, 46, 68, 69, 78, 80, 238, 239). Caplan and Mikulecky (refs. 45, 46) have given a very comprehensive review of various attempts to obtain a complete theoretical description.

Basically, the system consists of a porous, negatively charged membrane separating two compartments kept at different salt concentrations--hence, of different electrical conductivity --, as shown in FIGURE 11. When a current of sufficient strength is passed across the membrane, bulk flow motion takes place because of electroendosmosis ; the resultant difference in water level between the two chambers provides a restoring force which resets the system back to its initial values--except for a slow drift effect in the concentration differences between the two compartments. Teorell observed oscillations in bulk flow, hydro-

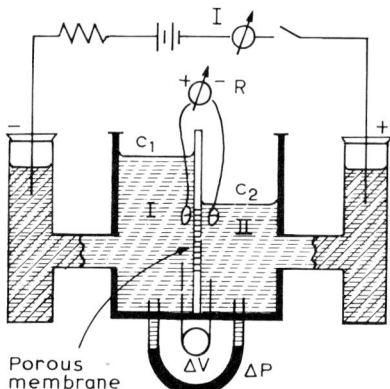

FIGURE 7.11 Basic set experimental set up for the Teorell oscillator (after ref. 231).

static pressure differences, voltage and membrane resistance which depend on the strength of the current, the geometry of the system, the physical characteristics of the membrane and the direction of the current; moreover, both stable and astable oscillations were seen, as shown in FIGURE 7.12 .

The main problem consists in predicting the oscillatory behavior. As Caplan and Mikulecky point out, all models attempt to

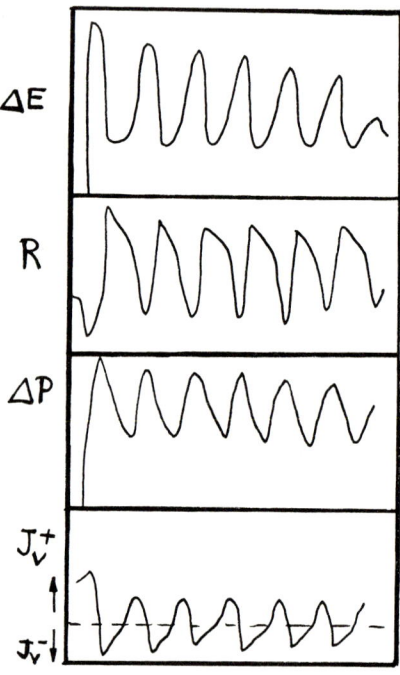

FIGURE 7.12 The oscillations in potential, resistance, hydrostatic pressure and water flow seen in the Teorell oscillator. The direction of positive flow defined by Teorell is reversed in the text. Also note that the resistance is plotted upside down. (The dashed line represents the resistance in the right hand compartment as a function of time . (After ref. 235).

explain the non linearities and the observed time delays. Franck has analyzed the required feedback in order to obtain some of the non linearities (refs. 235).

Teorell himself gave a simple model, which he simulated by analog computer. In the model proposed by Teorell, the concentration profile inside the membrane was assumed to be distributed according to the Manegold and Solf expression previously derived

$$(c_o - c_x) = (c_o - c_L)[1 - \exp(a J_v x)][1 - \exp(a J_v)]. \quad (9)$$

Teorell used this information to find the total resistance to the current, R, across the membrane by integration

$$R = \int_0^L [1/f\, c(x)]\, dx \quad (10)$$

in which f is a conductance function and L is the membrane thickness. The resultant overall resistance is, in this approximation, very non linear and can be depicted by the sigmoidal curve shown in FIGURE 13. Following Caplan and Mikulecky, I have reversed the direction of bulk flow because it contradicts the port definition of direction and it appears to contradict a Saxen relation.

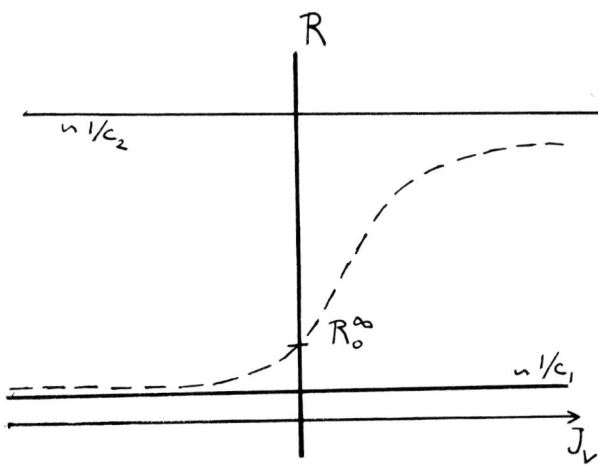

FIGURE 7. 13   Sigmoidal resistance function used by Teorell on the basis of the Manegold and Solf convection diffusion equation (after ref. 235).

The additional assumptions made by Teorell were:

- There is a delay in the value of the measured resistance relative to the value which would be obtained if the flow J_V was constant (defined as J_V^{∞}),

$$dR/dt = - k (R - R^{\infty})J_V. \qquad (11)$$

The presence of energy storage is introduced by allowing hydrostatic pressure differences to develop between the chambers. This storage is represented as

$$J_V = q\, d\Delta P/dt \qquad (12)$$

by Teorell and as

$$qJ_V = - d\Delta P/dt$$

by Caplan and Mikulecky. We retain the Teorell defintion, because q can be directly identified with a capacitor. For the Teorell "capacitor" with cross section of the membrane chambers equal to $1 cm^2$ the geometric factor q is $q = (1/981)\ cm^3/dyne$, the pressure is expressed in $dynes/cm^2$ and the resultant flow is measured in cm/sec (a velocity).

- The use of approximate phenomenological equations relating the dependent variables J_V and I to the voltage across the membrane and dP. By neglecting the cross coefficients between I and dP, Teorell found the electrical current using Ohm´s law directly.

Thus, the Teorell analysis focuses on the changes in the concentration profile <u>inside</u> the membrane and on the delays between the bulk flow and the steady state concentration profiles. While Teorell ascribed the non linearity to distortion of the concentration profile inside the membrane, Kobatake and Fujita (ref. 122) attributed the non linearity to concentration dependence of the phenomenological equations and then integrated these equations along streamlines in order to obtain global equations. Caplan and Mikulecky critized the absence of streaming currents in the analysis and proceeded to give complete phenomenological equations, à la Onsager, including the possible reciprocities. Mikulecky, Theford and DeSimone (ref. 121) then analyzed the system from the point of view of Catastrophe Theory, using the equations developed by Mikulecky and Caplan as a basis. In their analysis, the oscillator was seen as a cusp catastrophe in the concentration profile and as a bifurcation in the diffusion convection equation. They also provided a simplified analysis which gave the current voltage characteristics in terms of variations between two extreme values of concentrations.

The present model is a hybrid between C.T. and Network theory, but it has the advantage that the various processes--e.g., energy storage, negative feedback, etc. --can be identified with specific structures. Particularly interesting is the prediction of reaching certain states in the negative resistance region by experimental measurements.

A simple physical interpretation of the process is as follows. The (ionized) salt distributes along the capillary according to the equilibrium requirements of concentration and charge; the ions in the capillary also induce a charge in the glass thus leading to bulk flow for an appropriate value of the potential difference.

In the system considered--which has a positive hydrostatic pressure bias--flow will naturally occur in the positive direction until the potential reverses that situation leading to a negative

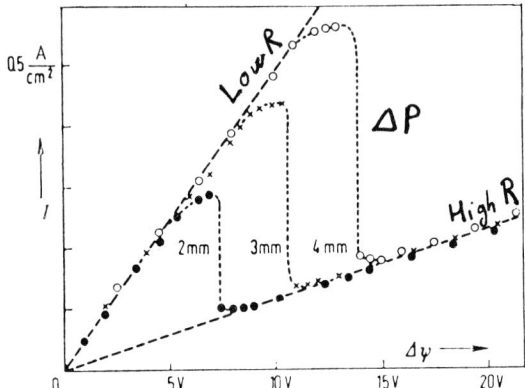

FIGURE 7.14   Experimental current voltage relations for the Teorell like system consisting of a glass filter placed across a NaCl concentration difference, as a function of the pressure difference (ref. 155).

bulk flow. It is clear that the presence of the bulk flow must affect the concentration gradients inside the capillaries, which can be ascertained by measuring the resistance inside the capillary.

There are basically two resistance values seen: a small resistance, for positive bulk flow, and a large resistance for negative bulk flow --i.e., during endosmosis. These cases correspond to high and low concentrations inside the capillary, respectively. To a first approximation one can then assume, that the salt concentration inside the capillary jumps between two extreme levels, as shown in FIGURE 14.

A further property of this transition is Kobatake's observation (ref. 155) that there is no folding for which the resistance is actually negative. This is illustrated in FIGURE 15.

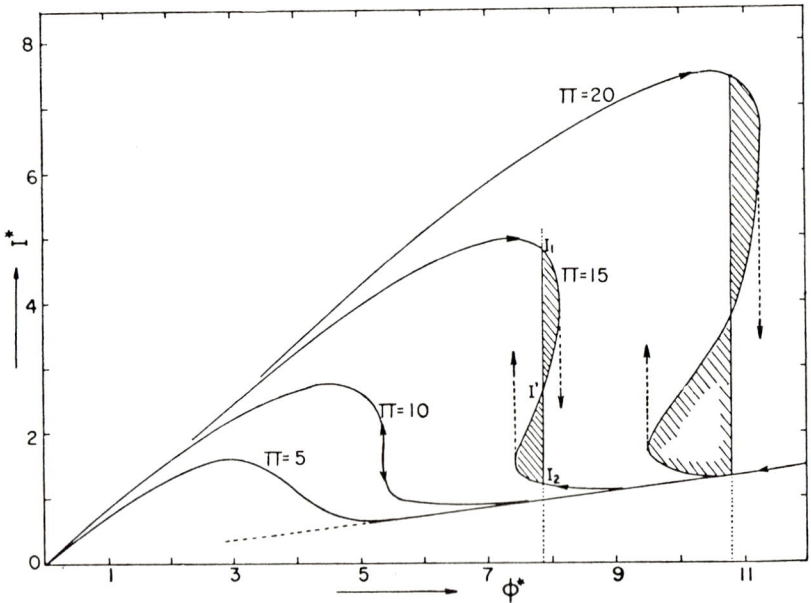

FIGURE 7. 15   The theoretical curves for the electrical membrane current vs. potential difference at various pressure differences predicts the negative resistance regions shown; however, only transitions between two steady states are observed (after ref. 155).

7.6 STATIC NETWORK REPRESENTATION AND DYNAMIC ANALYSIS

In the network analysis the time varying and the static processes can be separated, so that the coupling processes can be considered independently of the hydrostatic storage.

We assume that, to a first appproximation, positive charges are excluded from the membrane so that the membrane resistance is determined, as did Teorell, purely from the bulk flow through integration of the Manegold and Solf equations given in Chapter 3. The resultant resistance has the sigmoidal dependence on the volume flow depicted in FIGURE 16 and it can adequately be approximated by a three piece resistance:

$$R = R_{low} \qquad (J_v > 0)$$

$$R = R_o - aJ_v \qquad (J_v = 0),$$

and

$$R = R_{high} \qquad (J_v < 0),$$

If the flow is assumed, to a first approximation, to be linearly related to the hydrostatic pressure difference ΔP, the electrical potential and a constant biasing osmotic force

$$J_v = L_p (\Delta P - bE - \Delta \pi),$$

it follows from these equations and Ohm's law that, in the negative resistance region the voltage is given by

$$E = [I R_o / (1 - bL_p a)] - aL_v [\Delta P - \Delta \pi]/(1 - bL_p a).$$

If $bL_p > 1$ the i-E slope

$$R_o / (1 - bL_p a)$$

is negative. For R_{high} and R_{low} the i-E curves are simply lines going through the origin. The resultant i-E curve and the effect of changing ΔP is shown in FIGURE 16.

From the above equations we obtain

$$J_v = L_p \Delta P/(1 - L_p abI) - L_p (\Delta \pi + b R_o I)/ (1 - L_p a bI)$$

This line has a slope

$L^P/(1 - L^P ab I)$,

which is negative for $abI > 1$. The ΔP intercept is

$\Delta P = \Delta \pi + bI R^o$.

In the saturation ranges, $E = I R^{low}$ or $E = bI R^{high}$, so that

$J^V = L^P (\Delta P - \Delta \pi - bI R^{low})$

and

$J^V = L^P (\Delta P - \Delta \pi - bI R^{low})$

and are lines of slope L^P and intercepts $\Delta P = bIR^{low}$ and $\Delta P = bI R^{high}$, as shown in FIGURE 16.

Clearly, the hydraulic conductivity may have two different values for the two positive branches of the resistance curve.

The complete oscillator consists of the network given above--i.e., one of three possible networks or ranges -- and an external difference in water pressure that appears as gravitational energy storage in the height between two columns of water outside the membrane.

This hydrostatic "capacitor" has the differential equation given above,

$-J^V = q \, d\Delta P/dt,$ \hfill (13)

in which the flow entering the capacitor is the flow leaving the membrane, so that they have opposite signs (we use the membrane flow J^V) and the hydrostatic pressure difference is the same for the capacitor and the membrane. The capacitor equation requires that whenever the flow J^V is negative--i.e., into the capacitor and out of the membrane-- the pressure buildup be positive, $P>0$, while if the flow is positive the pressure difference is negative. As a result of these restrictions the instantaneous position in the system in the $J^V - \Delta P$ plane can only have the directions shown.

Note that the system cannot go from the positive into the

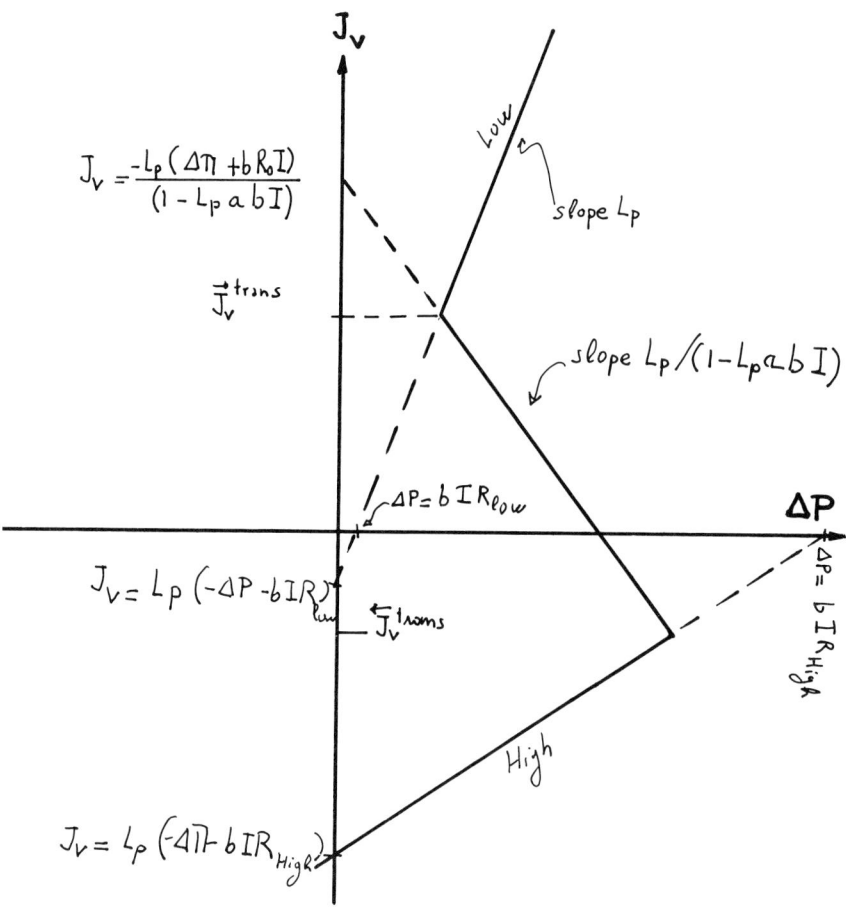

Fig 7.16 A possible simplified view of the negative resistance and three piece curve in terms of physical parameters, as described in the text.

negative resistance region, as it would require the capacitor to violate its constitutive law. Moreover, the capacitive equation can also be integrated to yield

$$P(t) = \int_0^t (1/q) J^V \, dt \qquad (14)$$

which indicates that the force cannot change instantaneously across the capacitor (if one excludes pulses). As a result, the only consistent solutions are build ups or decays of force along the positive resitance regions (where the capacitor sees sources in series with fictitious resistances) and jumps between the two branches of the resistance curve, as shown in FIGURE 17.

The resultant equations are simply exponential buildups or decays in which the system tends to go to asymptotic force values. For example, if the hydraulic capacitor C charges from an initial discharged state (force = 0) to a final asymptotic force value ΔP^{oo} in series with a resistance R, the force ΔP can be expressed as

$$\Delta P(t) = \Delta P^{oo} [1 - \exp(-tL^P/RC)], \qquad (15)$$

in which $\Delta P^{oo} = R^{high} b + \text{delta Pi}$. This equation follows readily by solving the differential equation for the RC net.

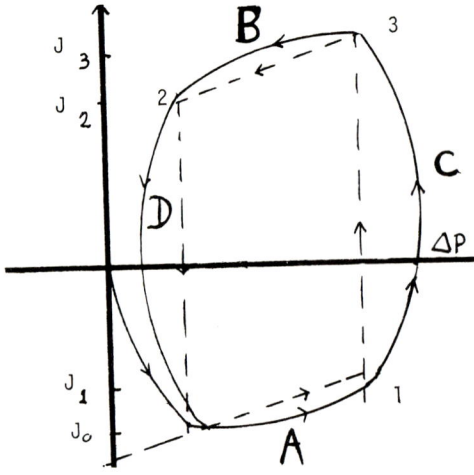

Fig. 7.17 Path described by the membrane system in the P-J^V plane.

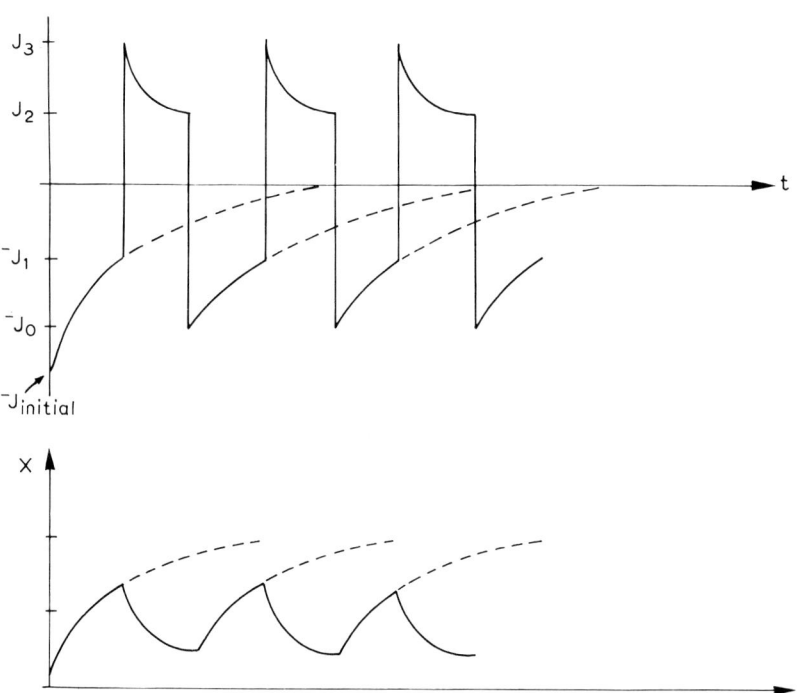

Fig. 7.18   Resultant relaxation oscillator curves for flow and pressure difference (force X).

(ref. 275). Moreover, when the intial time is t^o, instead of t=0, the equation simply changes to

$$\Delta P(t) = \Delta P^{oo} [1 - \exp(t^o - tL^P/C)],$$

which has the same slope but is shifted in the t axis.

Capacitive discharge is treated analogously. A capacitor which discharges from ΔP^o to $\Delta P = 0$ and which is in series with a linear resistance R will show the exponential decay

$$\Delta P(t) = \Delta P^o [\exp(-tL^P/RC)], \quad (16)$$

in which $P^o = R^{low} b_I + \Delta \pi$
while, if the decay starts at $t = t^o$, the force is specified by

$$\Delta P(t) = \Delta P^o [\exp-(t-t^o)L^P/C].$$

Moreover, if the resistance is non linear a piecewise linear solution may be given. The flow is obtained from the constitutive equation by differentiating the force,

$$J^V = (q \Delta P/RC) [\exp- tL^P/C]. \quad (17)$$

In the transition regions, the motion is instantaneous and the capacitor never sees the negative resistance. The resultant simplified waveforms for this model are plotted in FIGURE 17. These are typical relaxation oscillator waveforms, as expected.

Referring to the curves given, we can define several descriptive times in these waveforms. The time for the initial build up from $J^{initial}$ to J^1 is given by

$$T^{1'} = R^1 C \ln [J^V(initial) / J^3] = R^1 q \ln \Delta P^{oo}/\Delta P^{oo} - \Delta P^3$$

The transition time from 0 to 1 is given by

$$T^{1'} = R^1 q \ln [J^o / J^3] = R^1 q \ln (\Delta P^{oo} - \Delta P^2)/\Delta P^{oo} - \Delta P^3$$

and the time for the transition (3) to (2) is given by

$$T^2 = R^2 q \ln [J^1 / J^2] = R^2 \ln (\Delta P^3 / \Delta P^2).$$

7.7 INERTIAL DELAYS

The resultant shapes of the curves predicted do not correspond exactly with the experimental findings. In part, this is because non linearities have been neglected. However, the most relevant

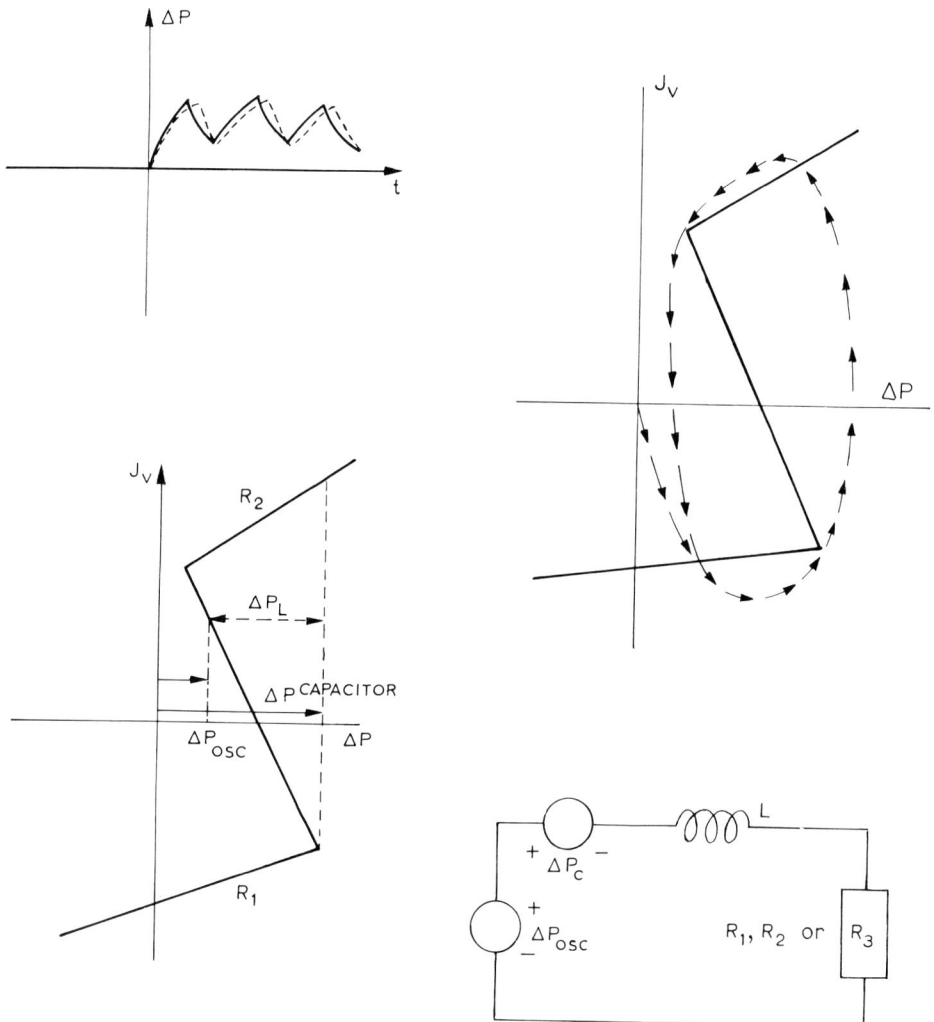

Fig. 7.19 (a) The flow/force curves have continuous derivatives when the inductor is included in the analyisis --i.e., the waveforms are "rounded". (b) The force across the inductor is the difference between the force seen by the non-linear resistance, for a given flow, and the constant force that would hold if only the capacitor where present. (c) The equivalent charging circuit when the inductive effects are included (after similar electronic equivalents in ref 275).

omission has to do with the presence of inertial delays.

Although to a first approximation the jumps can be assumed to be instantaneous, the more accurate model will show transition times between the two positive branches of the fold curve. These are introduced in terms of the network because of the presence of inductors--i.e., inertial delays. When inertial forces are included, the capacitor will "see" the negative resistance region, in series with an inductor during the transitions. The transition time is approximately given by, if the effect is small by (ref. 275)

$$T = T^p + 5 \, L/R, \qquad (18)$$

in which T^p is the total time the inductor sees the negative resistance region. In addition to the delay, a "rounding" in the waveforms results when the inertial corrections are small.

These effects are shown in FIGURE 18.

7.8   PHYSICAL REALIZATION OF THE NEGATIVE RESISTANCE

The negative resistance may be demystified by noting that the constitutive equation $R = -X/J$ may be viewed as a flow source controlled by X, as shown in FIGURE 20 a. If the same network is redrawn slightly, as shown in FIGURE 20 b, it looks as though there is a sensing flow and a feedback process.

The negative resistance becomes the two port (FIGURE 21.)

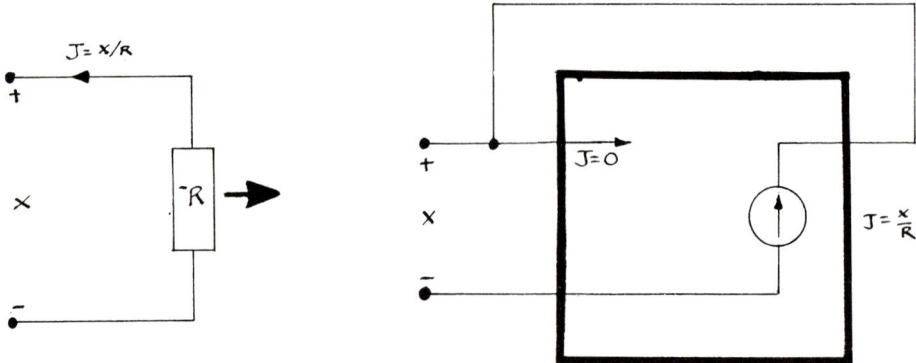

FIGURE 7.20 Ideal two port representation of a negative resistance

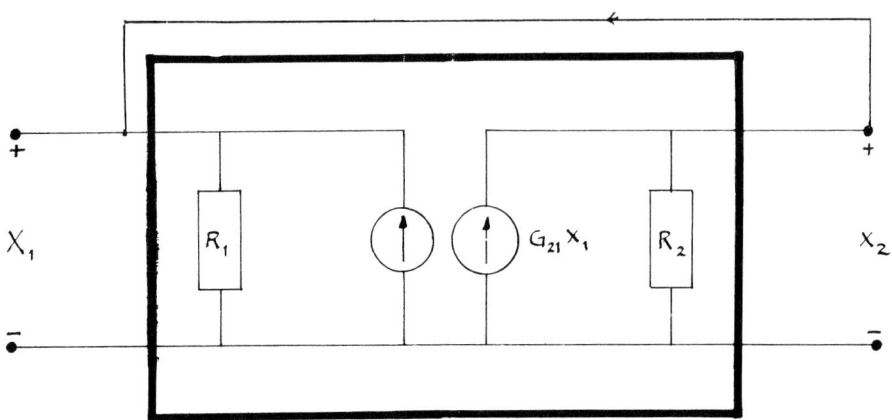

FIGURE 7.21 More realistic two port representation in which input and leakage flows are accounted for through the parallel conductances.

$$\begin{bmatrix} J^1 \\ J^2 \end{bmatrix} = \begin{bmatrix} 0 & 0 \\ -1/R & 0 \end{bmatrix} \begin{bmatrix} x^1 \\ x^2 \end{bmatrix}$$

In real life the input resistance would not be infinite, so that the sensing flow would not be zero and in general there will be other coefficients which enter in the picture. In such a model, the negative resistance is not a one port, but a two port

$$\begin{bmatrix} J^1 \\ J^2 \end{bmatrix} = \begin{bmatrix} G^{11} & G^{12} \\ G^{21} & G^{22} \end{bmatrix} \begin{bmatrix} x^1 \\ x^2 \end{bmatrix},$$

represented by the network in FIGURE 20, in which the coefficients G^{11}, G^{12}, G^{21} and G^{22} are assumed to have the limiting values 0, 0, $-1/R$ and 0.

Thus, a negative resistance is a feedback amplification process (i.e., non reciprocal in its energy converting characteristics)in which the input leakage is zero and which requires some source of dissipation to operate. Franck has given several examples of feedback processes which will act as negative resistances.

7.9 EXPERIMENTAL TEST OF UNSTABLE STATES

The actual presence of a negative resistance can be verified, in the J^V ΔP plane by attaching a variable source --e.g., a tube whose water level can be regulated -- and a positive resistance --i.e., a tube or membrane having positive hydraulic conductivity to scan the negative resistance states, as shown in FIGURE 22. If the overall resistance is positive, a steady state solution can then be obtained.

7.10 COMPLETE NETWORK MODEL

A complete network model which accounts for the three piecewise linear parts of the J^V ΔP curve is shown in FIGURE 23, in which the diodes and reference sources provide the required breaking points which switch the system from positive to negative resistance and viceversa. This network reflects the feed back process required to obtain the negative resistance region.

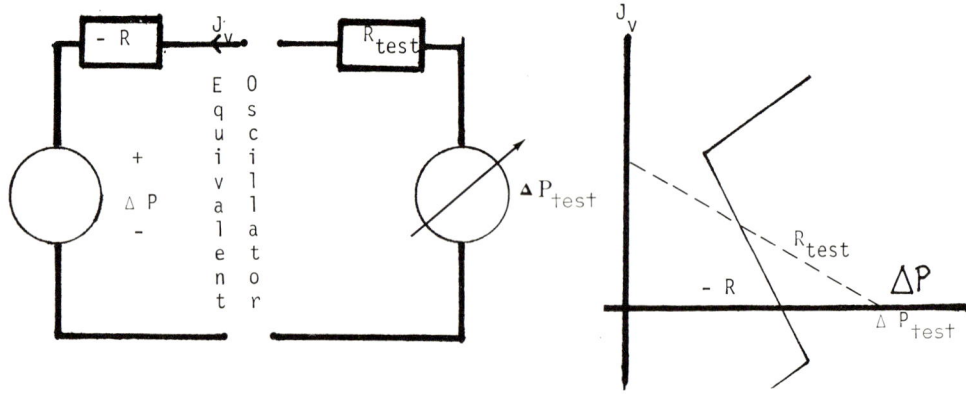

FIGURE 7.22 Schematic arrangement to view the negative resistance region as a steady state in the J^V-ΔP plane. A positive resistance and a hydrostatic source are placed in series.

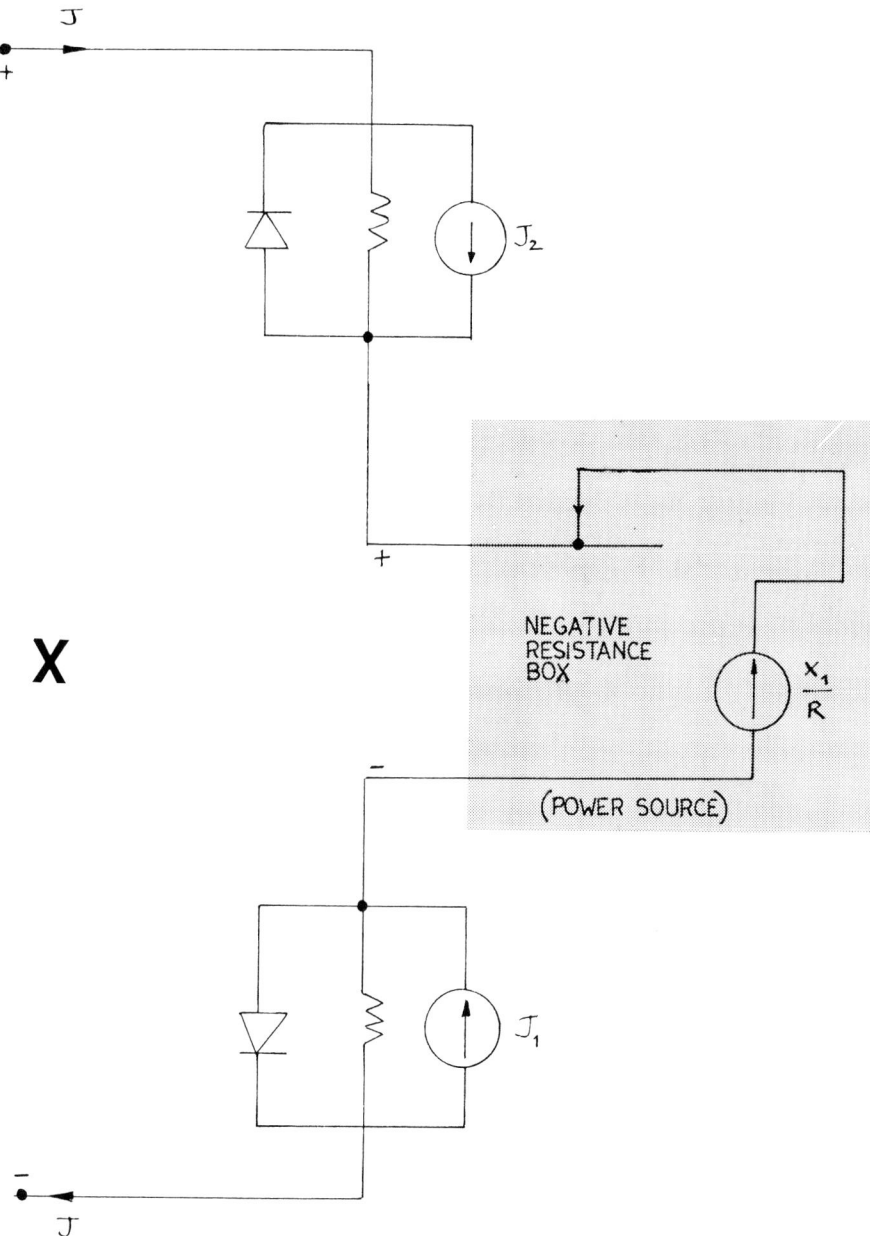

Fig. 7.23 The complete oscillator system may be approximated using three separate networks, as shown. The two ideal diodes give the breaking points in the non-linear curve.

7.11 DISSIPATION FUNCTION

The dissipation function, or some similar invariant cannot be calculated for the Teorell (or any other)oscillator on account of its non linear characteristics and nonequilibrium properties. Some authors have attempted to get out of these problems by assuming that the oscillator is always trying to reach some steady state which can be found by extrapolation of the direction of motion, say, in the J P or the I E planes. Such procedures have, of course, no physical justification.

The network technique assures that whatever dissipation is seen at the ports corresponds, instantaneously, to the sum of dissipations inside every element inside the system. In this particular case, the non linear "box" and the storing capacitor plus constant sources. For any flow in the network, the amount dissipated by the $J^V- \Delta P$ box is simply the product of the flow times the force read off directly from the non linear curve for that value of the flow--note that this does not work in reverse as there are three flows for every pressure. This will be the correct dissipation for the non linear box independently of whether we know its detailed internal mechanism or not.

The dissipation stored at any time in the capacitor and inductor is found using analogs of the stored energy.

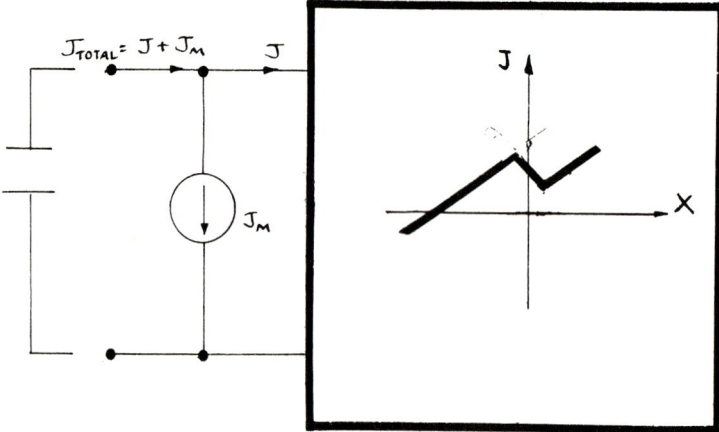

Fig. 7.24 Possible arrangement to trigger the oscillator with a pressure pulse.

7.12 ASTABLE OSCILLATIONS, TRIGGERING AND "ACTION" POTENTIALS

One of the potentially interesting application of this network approach is the anlysis of astable oscillations . This arises in cases in which the negative resistance region does not cross the X axis, in which case the system cannot oscillate. A fast pulse may trigger the system to go through a single oscillation, as shown in FIGURE 24. (Such triggers are seen, for example , during sudden increases in permeability in nerve membranes leading to depolarization.)

7.13 EXPERIMENTAL TESTS

In the original work by Teorell the following values were observed for the time transitions

$T^1 = 9.4$ min
$T^2 = 11.3$ minutes

leading to the two time constants $\dot{T}^1 = 13$ min and $T^2 = 16$ min. Using the known value of q one can then obtain the resistances R^1 and R^2 directly . Their inverses are the hydraulic conductivities for the two positive resistance branches :

$L^p(1) = 1.3 \times 10^{-6}$ cm^3/ dyn sec
and
$L^p(2) = 1.06 \times 10^{-6}$ cm^3/ dyn sec.

These values are in close agreement with the conductivities calculated by Mickulecky and Caplan.

Transition times are harder to read from the published experimental data. Using a value $(T-T^p)$ of about 5 min and $L^p(2)$ above in Eq. (18) gives an inertial inductor
$L = 6 \times 10^7$ dyn sec^2/ cm^3.
Another piece of information is the resistance of the membrane which varies as an inverse function of the concentration. The concentration is, in turn, a function of the volume flow as derived in Chapter 3. It is simpler, however, to assume that-- to a first approximation -- there are three relevant regions: i) when the flow is positive the concentration of the salt in the membrane is about that in the left chamber, ii) when the flow is negative

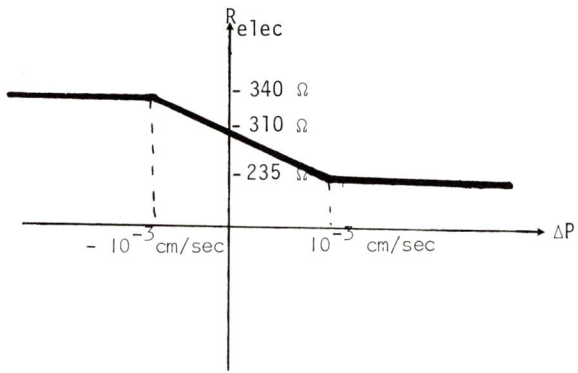

Fig. 7.25 Three segment piecewise resistance approximation.

the concentration is that of the right chamber and iii) when J=0 the concentration profile is that corresponding to pure diffusional flow. The resultant three part resistance curve is shown in FIGURE 25 when the right and left compartment resistances were 340 and 235 Ohms, respectively, and the resistance at zero flow was 310 Ohms. (The transition pressures were calcualted assuming that the observed curves contain an overshoot due to inertial delay) . In a given experiment these are only approximate values as the resistances change from cycle to cyle because of net transport of salt from the concentrated to the dilute compartment.

The total hydrostatic pressure may now be viewed as consisting of two components, to a first approximation : a hydrostatic component related to L^P given by the curve shown in FIGURE 26 (a) and the electrosmotic component shown in FIGURE 26(b). These lead to the total curve shown in FIGURE 26(c). Assuming that the electrosmotic factor has the form

$$\Delta_P \text{electric component} = L\ E$$

if cross coefficients are neglected, one obtains two values for L:

L = 300 dynes/ cm^2 volt   (large $J^V < 0$)
and
L = - 3 dynes/ cm^2 volt   (large $J^V > 0$) .

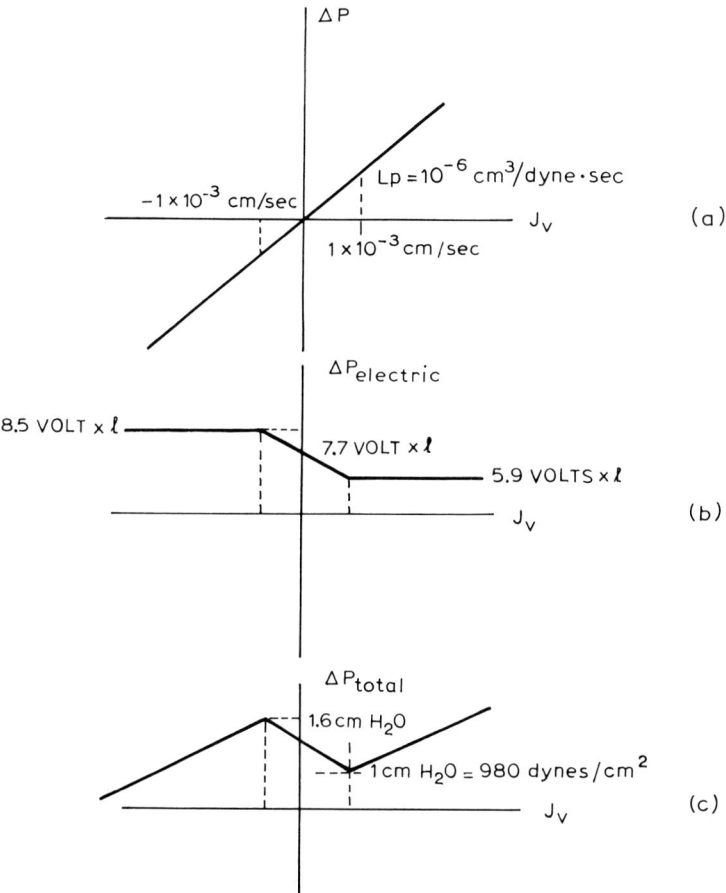

Fig. 7. 26 As a first approximation, the sigmoidal curve could be obtained by adding the hydrostatic and electroosmotic components found using Ohm's law. This is not an accurate representation as there is a three hundred fold change in the electroosmotic coefficient.

These two values reflect the concentration dependence of L; but a more detailed analysis is not justified here on the basis of this model.

The simplification in terms of two stable and one unstable state agrees with the findings of Theford et al. who have considered the range of validity of the assumption as well as the voltage current experiments performed at constant P in which the transition between the concentration regions are clearly seen. However, the approximation to bulk resistance is, of course, extreme as it neglects velocity gradients in the capillary porous plug--which are present when the capillary diameter is not much larger than the thickness of the double layer. In addition surface conductance effects can be substantial.

(Clearly, we have not deduced the electrosmotic component from first principles--such as calculation of potentials and surface conductivities--, but have proceeded in reverse to get some idea about the overall behavior of the system).

                              Milsok Truoa tulupin
                              Quuzt so rpuripiasco
                              [s mcru]
                              i trnuodsd asxu
                              st keipioauud
                              ycp Inn.

                              Amu Huianut, Prophets of
                              the Flower People, Year 4.

Chapter 8

PERSPECTIVES

   As with any other theory or model, the " pseudoelectrical" network paradigm in which lumped parameters are used has some distinct advantages over other techniques as well as some inherent limitations. I will quickly go over some of these side issues in this final chapter and will mention a few routes which I consider worth exploring to extend the range of application using this approach (some of these comments reflect work in progress).

   The general advantages and disadvantages are easy to summarize. On the positive side, network approaches provide a natural representation for the sequence: geometry, metrically induced topology, manifold and graph in the geometrical description of physical processes (ref.140-2) . As a result, they aid in the mechanistic understanding and decomposition of a process and the possible modelling of such processes. A major advantage over other geometric techniques is their capability of representing n dimensional systems by means of flat, or planar drawings while keeping the correct physical relations between forces and flows as the frames of reference change (covariance). The major disadvantage is that not all problems can be treated using networks. However, this is a criticism which can be made of any other particular technique used in physical chemistry. A secondary criticism may be the lack of rigor . This is more a reflection on my presentation than on the field itself. It is possible, for example, to give a much more formal, axiomatic view of this same ideas by defining abstract "machines" (ref. 24) . The little experience I have had in presenting some of this material seems to indicate , however, that a higher level of abstraction is neither desired by physical chemists nor by biologists, who are trying to figure out specific mechanisms. This may explain why the now classic work of Oster and Perelson on chemical reactions (ref. 177, 181) or the elegant thermodynamic work of Oster Perelson and Katchalsky have failed to find wide use (ref. 178).

In terms of the specific problems regarding the kinetic and thermodynamic models discussed here there are two immediate , practical advantages which surfaced in the present context : 1) the possibility of retaining the analyticity of transport imbedded processes and 2) providing a natural classification and extension scheme for thermodynamics.

The first part of this issue has been extensively discussed through examples in Chapters 3 and 4 . Although the networks used help to solve several integration problems in Membrane Science, their limitations are obvious, as it is not clear how a kinetic potential can be generalized to include more complex chemical networks and kinetics. For example, how would the generation of HBr from H_2 and Br_2 be anlayzed using this technique? The most natural route is, of course, to use differential flow rates and a Riemannian length. The resultant locally linear networks would then be represented again as connected resistive networks.

In the field of thermodynamic applications an immediate criticism is the absence of beta variables --which depend on odd functions of the velocity. I excluded those from the present analysis because of two reasons: 1) magnetic fields do not appear in the common transport problems of membrane science and 2) the purpose of analyzing the Onsager construct was to show how detailed balance -- and , therefore, K.V.L.-- was built into the theory of fluctuations of alpha variables given by Onsager. I see little point in generalizing the treatment using the same ideas as it turns out that a graph technique based on Lie algebras leads to a better model of the e.m. field in four dimensional form. This material will be published shortly.

A more serious problem, however, has to do with the actual physical coupling between different coordinates.

In this monograph the progression:" equilibrium state , reciprocal stationary state , non-reciprocal stationary state , non equilibrium state and irreversible process" was studied by using Kirchhoff's laws in resistive networks which were progressively "disconnected" through the use of non reciprocal two ports and with the subsequent addition of sources and energy storing elements. The central "glue" in this form of network thermodynamics --as in the case of Onsager thermodynamics or their variations (ref. 273)-- were Kirchhoff's voltage and current laws. The use of the former is justifiable because it corresponds to the principle of detailed balance. But the use of K.C.L. is not so evident . If a flow is conserved (which is required by Tellegen's theorem in a connected version of a physical description), what is it? In a chemical reaction it is certainly conservation of mass and there is no problem. But in the resistive models having a purely bilinear dissipation in the fluctuations we are invariably led to a resistive model which couples coordinates by "mixing" flows. Formally, this is correct, but physically there must be another way for one process to drive another, unless one falls back into the idea that entropy is transferred between ports (ref. 191) . The use of fluctuations does introduce a common coin between different coordinates but the requirement to be close to equilibrium leads to further inconsistencies, as discussed in the text. Therefore, the use of entropy as the "glueing fluid" --which is a standard technique in all irreversible theories -- , is not an entirely satisfactory devise . A more physically pleasing approach would use momentum exchange between processes or some momentum like field. This problem was indirectly considered in the corrections and extensions of Onsager's theory given by Casimir (ref. 47) and by Machlup and Onsager (ref. 128), in which a kinetic "energy" term of the form

$$[1/2] L_{ik} a_i a_k$$

was added to the entropy change calculated from microscopic quantities. This step corresponds to the introduction of an inductance network which includes mutual inductance coupling -- i.e. , the cross terms which allow coupling between coordinates .(By contrast, the self inductances used in Chapter 6 only lead to consideration of inertial terms) . In these networks, the inductors couple through the exchange of a field -- the magnetic flux that goes thorugh a coil integrated through an area. This is a conserved quantity, o , which is exchanged among various

coordinates . In the physical system momentum is exchanged among coordinates and can be assumed to be conserved under appropriate conditions. The resistances now provide a frictional path that accounts for the losses in each coordinate. The restriction of fluctuations being small can be completely removed and replaced by the requirement that the exchanges of momenta among various processes be statistical in nature. (Thus, we are led to the Langevin equation). Furthermore, if the fluctuations are seen as bursts of noise pulses,the resultant voltage (force) is given by:
$X = L \, di/dt = L \, d(u_o)/dt = L \, X$ and the constant coefficient equations are recovered in the form of average variables. Prof. Wyatt and his MIT associates have studied some of these problems in certain detail (ref 268-9). In the case of a lumped parameter network , the Lagrangian equations including dissipation, potential and kinetic energies follow, as shown above, from Kirchhoff´s laws (ref. 125). This can be shown either by using tellegen´s theorem, or by equating virtual works at the inputs and the system, as done by Biot (ref 22). I have also utilized non linear Lagrangians on R, L networks to analyze Brownian motion evolution (ref. 190). These routes suggest a direct generalization of thermodynamics to non linear problems in which there are sources of curvature, as suggested by Ruppenier (209) and by Hertz before him .

Specific relations between the network formalism and statistical mechaniccs have been pointed out throughout the monograph and are rather obvious. I am currently exploring the use of networks to extend some of the statistical mechanical ideas to the non linear domain. The work of Propp (re. 198) and that of Baur et al.(17) are also very relevant in this area.

On a different level, biological applications abound (refs. 115, 143-158, 241- 6). These are mainly due to contributions of Prof Mikulecky and associates at the University of Virginia ,as well as some of my own efforts (ref. 182). But beyond the superficial modelling there are more basic questions to be answered.

One of the very interesting possibilities is the updating of the ideas of structure and function. These two concepts present a dicotomy as they are ususally looked upon as the" container " (anatomy) which holds the dynamic process (function). In fact , it is more realistic to consider the interactions between the two by looking at structure as the sum total of various static-dynamic interactions which can be summarized by K.V.L. and K.C.L. (conservation of mass, etc) and which define a network of forces and flows (representing both structure and function).

A further problem related to biological structure is : given a global response of an association of cells, say, is it possible to decompose the observed behavior into constituent elements in order to infer what the basic properties of the individual cells of a specific tissue, say. This is a problem which has bneen analyzed by both Mikulecky and by Rapundalo , but it is not completely solved even for the simple case in which one is only trying to study the diffusional and storing properties of the cells.

In fact, the area of global decomposition of component processes may be one of the most promising aspects of the network paradigm. The reason is that these networks , unlike the superficial modelling which can be achieved with other mathematical techniques (e.g., linear or non linear programming) is a true representation of the allowable physical processes which can lead to that particular phenomenology. Given a certain superficial behavior, it is possible to decompose the global equations into subprocesses which are consistently "hooked up" so that they agree both with observed phenomenology and with the orthogonality requirements of forces and displacements.

There is a close parallel between this aproach and the idea of "observables" in quantum mechanics . The operators of quantum mechanics lead to observables when it is possible to decompose the description into a set of eigenvectors in a Hilbert space. These vectors have the property that they displace in orthogonal directions and , moreover, each displacement is proportional to an eigenvalue scalar. In the case of the networks considered here a process is a valid observable whenever it is possible to decompose it into orthogonal directions (eigenvectors) which remain parallel to themselves ; these are the internal branches of a network obeying K.V.L. and K.C.L. Conversely, given an observable, it is possible to find consistent decompositions in terms of network branches. Of course, in the cases in which the constitutive processes are known, the global behavior can be found by attaching subprocesses using Kirchhoff´s laws -- that was the route taken in the representation of reaction diffusion processes, for example.

The idea of the consistent physical representation has not been used to its full power either here or elsewhere, as much of the network use has been either pure modelling or the building of global equations from local laws.

But the demonstration of the full power of these techniques awaits the application to more complex situations in which the global phenomenology is known --e.g., the representation of a thermodynamic surfaces --to obtain global Equations of State.

REFERENCES

1. R.Abraham, and J.Marsden, **Foundations of Mechanics,** Benjamin, NY.,1967.

2. R.B.Adler, L.J.Chu, and R.M.Fano, **Electromagnetic Energy Transmission and Radiation.** John Wiley and Sons, New York, 1960.

3. M.A.Akivis, V.V.Goldberg, **An Introduction to Linear Algebra and Tensors,** Dover Publications, Inc. N.Y. 1972.

4. P.S.Alexandrov, **Combinatorial Topology, Vol. 1** Graylock Press, Baltimore, MD 1956.

5. A.A.Andronow, and C.E.Chaikin, **Theory of Oscillations,** Princeton,N.J.:Princeton University Press,1949

6. E.J.Angelo, **Electronic Circuits** McGraw-Hill Book Co.,Inc., 1958.

7. R. Aris, Indust. Engng. Chem. Fundamentals, 3 (1964), 28.

8. R. Aris, **Arch. Rational Mech. Anal., 19** (1963), 81

9. R. Aris, **Arch. Rational Mech. Anal., 27** (1968), 356.

10. R. Aris, Indust. Chem.Engng. J., 2 (1971), 140.

11. R.F.W.Bader and T.T. Ngvyen-Dang, **Adv. Quant. Chem., 14**(1981),63.

12. A.Baggeroer, **State Variables,and Communication Theory** M.I.T. Press, Cambridge, Mass.,1970.

13. A.T. Balaban(ed.), **Chemical Applications of Graph Theory** Academic Press, London, 1976.

14. N.Balabanian and T.A Bickart, **Electrical Network Theory,** John Wiley and Sons, N.Y., 1969.

15. K. Balasubramanian, Computer Generation of the Characteristic Polynomials of Chemical Graphs, **Journal of Computational Chemistry, Vol. 5** No 4, (1984) 387.

16. B. Baumsalg and B. Chandler, **Theory and Problems of Graph Theory** ,Mc-Graw Hill Book Co. NY, 1968.

17. M. Baur, J.R.Jordan, P.Jordan, and J.Mayer, Towards a Theory of Linear Nonequilibrium Statistical Mechanics, **Ann.of Physics, 35**(1965) 96.

18. V. Belevitch ,**Classical Network Theory**, Holden - Day, San Francisco, 1968.

19. M. Berkowitz et al., J. Chem. Phys., 79, 5563 (1983).

20. P.J.Bhattacparyya, **J. Biol** Chem. 253(1978),3848.

21. S. Bienstock and K.G. Gordon, Piecewise analytical Integration of chemical rate equations I.,**Jour.Chem. Phys, 77**(1982),2902

22 M.A. Biot, Thermodynamic Principle of Virtual Dissipation and the Dynamics of Physical-Chemical Fluid Mixtures Including Radiation Pressure. **Quart. Jour. of Applied Math.** (1982), 517.

23 K.B. Bird, W.E. Stewart and E.N. Lightfoot, **Transport Phenomena**, John Wiley & Sons, Inc. New York, N.Y. 1965.

24 L.S. Bobrow and M.A. Arbib, **Discrete Mathematics**, W.B. Saunders Co., Philadelphia, 1974.

25 L. Boltzmann, **Lectures on Gas Theory**, translation published by University of California Press, Berkeley. 1964.

26 A.G. Bose and K.N. Stevens, **Introductory Network Theory**, Harper and Row Publishers, New York, 1965.

27 R. Bott, DSc Thesis (Carnegie Institute of Technology, 1949)

28 R.K. Brayton, Nonlinear Reciprocal Networks, **SIAM-AMS Proceedings,** 3 (1971), 1.

29 R.K. Brayton and J.K. Moser, A Theory of Nonlinear Networks.I, **Quart. of Applied Math.**, April 1964, 1.; A Theory of Nonlinear Networks. II, **Quart. of Applied Math.**, July 1964, 81.

30 L. Brillouin, **Science and Information Theory**, Academic Press, New York, NY, 1962.

31 P.W. Bridgman, **The Nature of Thermodynamics**, Harvard University Press, Cambridge, Mass., 1941.

32 P.W. Bridgman, **Phys. Rev** 3 (1914), 273.

33 P.W. Bridgman, **A condensed Collection of Thermodynamic Formulas**, Harvard University, Cambridge, Mass., 1925.

34 S. Brush, **Kinetic Theory**, Pergamon Press, Oxford, 1965.

35 B. Bunow, Harvard University Thesis, 1971.

36 B.J. Bunow, Chemical Reactions and Membranes: A Macroscopic Basis for Facilitated transport, Chemiosmosis and Active Transport, **J. theor. Biol.** 75 (1978), 51.

37 B. Bunow and R. Aris, Diffusion and First Order Reaction in a General Multilayered Membrane, **Mathematical Biosciences** 26 (1975), 157.

38 B. Bunow and D.C. Mikulecky, in **Biophysics of Membrane Transport VI** (J. Kuczera et. al., eds.), Wroclaw, Poland, 1981.

39 A. Burks and J. Wright, Theory of Logical Nets, **Proc. of the I.R.E.**, Oct 1953, 1357.

40 R.G. Busacker and T.L. Saaty, **Finite graphs and Networks: An Introduction with Applications**, McGraw-Hill Book Co., New York, 1965.

41 S.R. Caplan, **J. Phys. Chem.**, 69 (1965), 3801.

42 S.R.Caplan, **J. theor. Biol. 10** (1966), 209.

43 S.R.Caplan, Autonomic Energy Conversion,**Biophys. J. 8** (1968),1167.

44 S.R.Caplan, **Proc. Natl.Acad.Sci. USA 81** (1981), 4314.

45 S.R.Caplan and A. Essig, **Bioenergetics and Linear Nonequilibrium Thermodynamics**, Harvard Unviersity Press, Cambridge, Mass.,1984.

46 S.R.Caplan and D.C.Mikulecky (J.A. Marinsky, ed.), **Transport Processes in Membranes**, Dekker, New York, 1966.

47 H.B.G.Casimir, **Rev. Mod. Phys.** , **17** (1945), 343.

48 R.Christensen, Entropy Minimax, A Non-Bayesian Approach to Probability estimation from empirical Data. **Proceedings of the IEEE Systems**, Man and Cibernetics Conference, Boston, Mass., Nov 5,1973.

49 R.Christensen, **Entropy Minimax Sourcebook** (5 volumes) Entropy Ltd.,Lincoln Mass.,1982-1985.

50 B.L. Clarke, in **Advances in Chem. Physics**, edited by I. Prigogine and S. A. Rice, **43**, p 1, Wiley, New York, 1980.

51 R. Clay, **Non-Linear Networks and Systems**, John Wiley and Sons, Inc., New York, 1971.

52 E.J. Corey, **Pure Appl. Chem.**, **14** (1967), 19.

53 A.Cote and J.B.Oakes, **Linear Vacuum Tube and Transistor Circuits**, McGraw-Hill Book Co., Inc., New York, N.Y., 1961.

54 H.S.M.Coxeter, **Regular Poytopes**, Dover Publications, Inc. New York, N.Y, 1963.

55 J.Crank , **The Mathematics of Diffusion**, Oxford University Press, London, 1956.

56 J. Crank and G.S. Park, **Diffusion in Polymers**, Academic Press, London, 1968.

57 P.Curie, **Ouevres** , Gauthier-Villars, Paris, 1908.

58 S.R.De Groot and N.G. van Kampden, **Physica 21** (1954), 39.

59 S.R.De Groot and P.Mazur, **Phys. Rev. 94** (1956), 218.

60 S.R.De Groot and P. Mazur, **Non Equilibrium Thermodynamics**, North Holland, Amsterdam, 1962.

61 J.A. deSimone, **J. theor. Biol. 68** (1977), 225.

62 J.A. deSimone and S.R.Caplan, **J. theor. Biol. 39** (1973), 523.

63 C.A. de Soer and G.F.Oster,Globally Reciprocal Stationary Systems, **Int J. Eng. Sci. 2** (1973), 141.

64 L.F.Del Castillo and E.A.Mason, Energy Barrier Models for Mem-

brane Transport, **Biophysical Chemistry, 9** (1979),111.

65 L.F.Del Castillo and E.A. Mason, Statistical Mechanical Theory of Passive Transport Through Partially Sieving or Leaky Membranes, **Biophysical Chemistry, 12** (1980), 223.

66 A.Dold and B. Eckmann, **Lecture Notes on Mathematics: Structural Stability, the Theory of Catastrophes, and Applications to the Sciences**, Springer-Verlag, Berlin 1975

67 L.Donati, **Rend.Acad.Sci. Bologna, 4** (1899), 29.

68 R.J. Donnelly, R. Herman and I. Prigogine,eds., **Non-Equilibrium Thermodynamics, Variational Techniques and Stability**, University of Chicago Pres, Chicago, Illinois, 1966.

69 H. Drouin, Verlag Chemie, GmbH (1969); also H.Drouin and E. Zimmermann, **GIT Fachzeitschr. Lab 12** (1968), 178.

70 A. Einstein, **Investigations on the Theory of Brownian Movement**, Dover Publications, New York, 1956.

71 N.R. Eyres, R. Hartree, D.R. Inghan, J. Jackson, R. Sarjant, R.J. Wagstaff, **S.M. Phil. Trans. Roy. Soc. A 240, 1**.

72 R. Fano, L.J. Chu and R. Adler, **Electromagnetic Fields, Energy and Forces**, John Wiley and Sons, Inc., New York, NY, 1963.

73 V.F. Fatushensko, Y.A. Chimadzev, V.S. Markin, **Biophysika 25** (1980).

74 E. Fermi, **Thermodynamics**, Dover Publications, New York, NY, 666 1936.

75 R. Field, R. Noyes, Oscillations in Chemical Systems IV. Limit Cycle behavior in Model of a real chemical reaction, **Jour. Chem. Phys, 60** (5) (1979)

76 R. Fieschi, S.R. deGroot and P. Mazur, **Physica**, 20 ,67.

77 D.D. Fitts, **Non-Equilibrium Thermodynamics**, McGraw Hill, 1962.

78 U.F.Franck, **Ber. Bunsengew. physik. Chem. 67** (1963), 657.

79 U.F.Franck, **Z. physik. Chem.N.F. 3** (1955), 183.

80 U.F.Franck, **Electrochemica Acta, 23** (1978), 1081.

81 P. Franklin,**Methods of Advanced Calculus**, McGrawHill Book Co., Inc., New York, NY, 1944.

82 R.M. Gallagher, **Finite Element Analysis Fundamentals**, Prentice- Hall, Inc., New Jersey, 1975.

83 C.W. Gear, **Commun. ACM, 14** (1971), 176, 185.

84 S.Giambo, P. Pantano and P. Tucci, An Electrical Model for the

KortewegdeVries equation, **American Journal of Physics**, 52 (1984), 283.

85 J.W.Gibbs, **The Scientific Papers of J.Williard Gibbs** ,Longmans and Green, London, 1906 (Reprinted by Dover Publications, N.Y., 1960.

86 P.Glansdorf and I.Prigogine, Sur les Proprietes Differentieles de la Production d´Entrophie,**Physica** 20(1954), 773.

87 P.Glansdorf and I.Prigogine, On a General Evolution Criterion in Microscopic Physics,**Physica** ,30 (1964) 351.

88 E.A.Guillemin, **Introductory Circuit Theory** , Wiley, New York, 1953

89 P.R. Halmos,**Finite Dimensional Vector Spaces**, D Van Nostrand Co., Inc. Princeton, NJ, 1958.

90 F. Harary, Graph Theory and Electric Networks **I.R.E. Trans. Circuit Theory** (1959),95.

91 F. Harary, **Graph Theory**, Addison Wesley, 1972 .

92 H.Hertz, **Ann. Phys., 36** (1888), 1.

93 B. Hess, E.M. Chance, H.Busse and B. Wurster in **Analysis and Simulation of Biochemical Systems** (H.C. Hemker and B.Hess, eds), 1972 .

94 Higgins, **J. Ind. and Eng. Chem.,** 59 (1967), 18.

95 F.B.Hildebrand, **Methods of Applied Mathematics,** Prentice-Hall, Inc., Englewood Cliffs, NJ, 1961.

96 T.Hill, **Free Energy Transduction in Biology**, Academic Press, New York, 1977.

97 T.Hill, The Linear Onsager coefficients for biochemical kinetic diagrams as equilibrium one-way cycle fluxes, **Nature,** 5878(1982), 84.

98 T.Hill, Studies in Irreversible Thermodynamics.IV. Diagramatic Representation of steady state Fluxes for unimolecular Systems., **J.theor. Biol., 10**(1966), 442.

99 H.R. Hirsh, Further Comments on Negative Solvent Pressure in Osmosis, **Journal of Biological Physics, 11**(1983).

100 H.R. Hirsh, On the Mechanism of Flow Through Porous Membranes, **Currents in Modern Biology,** 1(1967), 133.

101 H.R. Hirsh and C.C. Peck, The Onsager Reciprocal Relations in Transformed Coordinastes: Application to transmembrane Flow, **Currents in Modern Biology** , 1(1967), 235.

102 H.R. Hirsh and C.C. Peck, The Onsager Reciprocal relations in Transformed Coordinates: Restriction of the Rate of Entropy Production to Positive Definite Forms, **Currents in Modern Biology** , 2(1968), 133.

103 H.R. Hirsh and C.C.Peck, The Onsager Reciprocal Relations in

Transformed Coordinates: Arbitrary Flux and Displacement transformations, **Currents in Modern Biology**, 2 (1968),241.

104 J.M. Houston, Theoretical Efficiency of the Thermoionic energy Converter, **Journal of Applied Physics, 30** (1959),4.

105 G.J. Hooyman and S.R. deGroot, Phenomenological Equations and Onsager Relations, **Physica , 21** (1955), 73.

106 G.J. Hooyman, S.R. deGroot and P. Mazur, Transformation Properties of the Onsager Relations, **Physica , 21**, 360.

107 E. Huf and D.C.Mikulecky, Regulatory aspects of Na^+ Transport in complex Epithelia, **American Journal of Physiology,**1983.

108 H. Jahnke, **Dissertation, TH Darmstadt,** 1962

109 F.E.Jaumot,Jr., Thermoelectric effects, **Proceedings of the IRE, 46** (1958), No 3.

110 A. Katchalsky and P.F. Curran, **Non Equilibrium Thermodynamics in Biophysics**, Harvard University Press, Cambridge, Mass., 1965.

111 A. Katchalsky and O. Kedem, Thermodynamics of Flow Processes in Biological Systems, **Biophys. Journal, 2** (1962), 53.

112 O. Kedem and A. Katchalsky, Permeability of Composite Membranes, **Trans. Faraday Soc., 59** (1963), 1918,1931 and 1963.

113 O. Kedem and A. Katchalsky, **Transactions of the Faraday Soc., 59**(1968), 1918.

114 O. Kedem and S.R. Caplan, **Transactions of the Faraday Soc., 61**(1965), 1897.

115 D.B. Kell, **Biochem. biophys. Acta, 549**(1979), 55.

116 J.Kestin, **A Course in Thermodynamics**, Blaisdell Publishing Co., Waltham, Mass. 1966.

117 R.B. King, **Theoret. Chimica Acta, 60** (1982), 409.

118 E.L. King and C. Altman, **J. Phys. Chem., 60** (1956), 1375.

119 J.G.Kirkwood, R. Baldwin, P. Dunlop, L. Gusting and J. Kegeles Flow equations and frames of Reference for Isothermal Diffusion in Liquids, **Journal of Chem. Phys, 33** (1960), 1505.

120 I. Klotz, **Chemical Thermodynamics**, W.A.Benjamin, Inc., New York,1964.

121 Y. Kobatake, **Physica, 48**(1970), 301.

122 Y. Kobatake and H. Fujita,**J.Chem Physics 40** (1964), 2212

123 Y. Kobatake, N Takeguichi, Y. Toyoshima and H. Fujita, **J.Phys. Chem. , 69** (1965), 3981.

124 A.S. Kompaneyets, **Theoretical Physics**, Dover Publications, Inc, New York ,1962.

125 G. Kron, **Tensors for Circuits**, Dover Publications, Inc., New York, 1959.

126 F. Franklin Kuo, **Network Analysis and Synthesis**, John Wiley and Sons, New York, 1966.

127 C. Kuo-Chen and L.W. Min, **J. theor. Biol**, 91 (1981), 637.

128 S. Machlup and L. Onsager, Fluctuations and Irreversible Processes II. Systems with Kinetic Energy, **Phys. Rev.**, 91(1954), 1572.

129 E. Manegold, **Kapillarsysteme, Bd. 1** (1955), Strabenbau, Chemic and Technik, Verlagsgesellschaft Heidelberg.

130 E. Manegold and K. Solf, **Colloid Z**, 59 (1932), 179.

131 M. Mansfield, **Introduction to Topology**, D van Nostrand and Co., Princeton, NJ, 1963.

132 H. Margenau and G. Murphy, **The Mathematics of Physics and Chemistry**, D. van Nostrand and Company, Princeton, NJ, 1962

133 E.A. Mason, R. Wendt and E.H. Bresler, Similarity Relations for Membrane Transport, **Journal of Membrane Science**, 6, (1980), 283.

134 S.J. Mason and H.J. Zimmermann, **Electronic Circuits, Signals and Systems**, John Wiley and Sons, New York, 1960.

135 P. Mazur and S.R. deGroot, On Onsager´s Relations in a Magnetic Field, **Physica**, 19 (1954), 961.

136 P. Mazur and S.R. deGroot, Extension of Onsager´s Theory of Reciprocal relations II, **Phys. Rev.**, 94(1954), 218.

137 G.D. Mehta, T.F. Morse and M.H. Vaneshpajooh, Generalized Nernst-Planck and Stefan Maxwell equations for Membrane Transport, **Journal of Chem. Phys.**, 64 (1976), 3917.

138 J. Meixner, Thermodynamics of electrical Netowrks and the Onsager-Casimir Reciprocal Relations, **Journal of Math. Phys**, 4(1963), 154.

139 D. Menzel, **Mathematical Physics**, Dover Publications, Inc. NY, 1961.

140 P.G. Mezey, **Theoret. Chem. Acta,** 54, (1980), 95

141 P.G. Mezey, **Theoret. Chem. Acta,** 60, (1982), 409.

142 P.G. Mezey, **Theoret. Chem. Acta,** 62, (1982), 133.

143 D.C. Mikulecky, A Network Thermodynamic Approach to the Hill-King Altman approach to kinetics: Computer Simulation, in **Physical Methods in the Study of Cellular Biophysics** (M. Dinno, ed), Alan Liss Inc., 1984.

144 D.C. Mikulecky, A simple Network thermodynamic method for series-parallel coupled flows, **J. Theoretical Biol.**, **69**(19770, 511.

145 D.C. Mikulecky, System Identifiability Based on the Power Series Expansion of the Solution, **Mathematical Biosciences**, **41** (1978), 21.

146 D.C. Mikulecky, A Network Thermodynamic two-port element to represent the coupled flow of salt and current, **Biophys. J.**, **25** (1979), 323.

147 D.C. Mikulecky, **Biofluid Mechanics**, 2 (1980), 372.

148 D.C. Mikulecky, **Fed. Proc.**, **40** (1981), 372.

149 D.C. Mikulecky, in **Biophysics of Membrane Transport VI**, Wroclaw, Poland, 1981.

150 D.C. Mikulecky and T.U.L. Biber, **Fed. Proc.**, 39 (1980), 737.

151 D.C. Mikulecky and S.R. Caplan, The choice of reference frame in the treatment of membrane transport by non-equilibrium thermodynamics, **J. Phys. Chem.** 70 (1966), 3049.

152 D.C. Mikulecky and J.J. Feher, The relation between global and Local transport parameters in an epithelial membrane having an asymmetric transport mechanism, **Jour. theor. Biol.**, **88**(1981), 575.

153 D.C. Mikulecky, E.G. Huf and S.R. Thomas, **Biophys. J.**, **25** (1979), 87.

154 D.C. Mikulecky and L. Peusner, Network Thermodyanmics and Bioenergetics: Some useful new results and their implications, Biophys. Soc. Abstracts, 1985.

155 D.C. Mikulecky, W.A. Theford and J.A. deSimone, Local vs. global Reciprocity and Analysis of the Teorell Membrane Oscillator Using Catastrophe Theory, in **Biophysics of Membrane Transport IV** (S. Przestalski, J. Kuczera and J. Idzior, eds), Wroclaw, Poland, 1977, p 75.

156 D.C. Mikulecky and S.R. Thomas, Some network Thermodynamic models of coupled, dynamic, Physiological Systems, **J. Franklin Institute**, 308 (1979), 309.

157 D.C. Mikulecky and S.R. Thomas, Network thermodynamic Simulation of Nonlinear Dynamic Physiological Systems Using Circuit Simulation Programs, **Proceedings of the 1979 Summer Computer Simulation Conference**, Toronto, Canada, 1979.

158 D.C. Mikulecky, W.A. Wiegand and J.S. Shiner, **j. theor. Biol.**, **69** (1977), 471.

159 D. Miller, On the experimental verification of the Onsager reciprocal relations, in **Transport Phenomena in Fluids** (H. Hanley, ed.), M. Dekker, New York, 1969.

160 C.W. Misner, K.S. Thorne and J.A. Wheeler, **Gravitation**, W.H. Freeman and Co., San Francisco, 1973.

161 M.R. Moore, **Thesis**, Harvard University, 1971.

162 W. Moore, **Physical Chemistry**, Prentice Hall, Inc. Englewood Cliffs, NJ, 1964.

163 O. Nathason and O. Sinanoglu, The Metric Geometry of Near Equilibrium Thermodynamics, **J. Chem. Phys., 72** (1980), 3127.

164 J. Nutton et al, Quasistatic Processes as Step Equilibrations, **J. Chem. Phys., 83** (1985), 334.

165 H.T. Odum and R.C. Pinkerton, Time's Speed Regulator : The Optimum Efficiency for Maximum Power output in Physical and Biological Systems, **American Scientist, 53** (1955).

166 D.E. Oken, S.R. Thomas and D.C. Mikulecky, **Kidney International, 19** (1981), 359.

167 W.E. Olmstead and X.X. Handeisman, Diffusion in a Semi Infinite Region with Nonlinear Surface Dissipation, **SIAM Review, 18**(1976).

168 L. Onsager, Reciprocal Relations in Irreversible Proces-ses.I **Physical Review, 37** (1931), 405.

169 L. Onsager, Reciprocal Relations in Irreversible Processes, II. **Physical Rev., 38** (1931), 2665.

170 L. Onsager and S. Machlup, Fluctuations and Irreversible Processes, **Physical Rev., 91** (1953), 1505.

171 H.G. Othmer and L.E. Scriven, **Indust. Engng. Chem. Fundamentals 8** (1969), 302.

172 H.G. Othmer and L.E. Scriven, **J. theoret. Biol., 32** (1971), 502.

173 H.G. Othmer and L.E. Scriven, **J. theoret. Biol., 43** (1974), 83.

174 G.F. Oster and D. Auslander, Topological Representations of Thermodynamic Systems I. Basic Concepts, **Journal of the Franklin Institute,** (July 1971), 1.

175 G.F. Oster and A. Perelson, Systems, Circuits and Thermodynamics, **Israel Journal of Chemistry, 11,** (1973), 445.

176 G.F. Oster and A. Perelson, **Israel J. Chem, 11** (1973), 445.
G.F. Oster and A. Perelson, **Arch. Rational Mech. Analysis 55** (1974), 230.

177 G.F. Oster and A. Perelson, **Arch. Rational Mech. Analysis 57** (1974), 31.

178 G.F. Oster, A. Perelson and A. Katchalsky, **Nature, 234** (1971), 393.

179 G.F. Oster, A. Perelson and A. Katchalsky, **Quart. Rev. Biophys., 6** (1973), 1.

180 J.A. Pennline, J.S. Rosenbaum, J.A. DeSimone and D.C. Mikulecky, **Math. Math. Biosciences, 37** (1977), 1.

181 A. Perelson and G.F. Oster, **Arch. Ration. Mech. Analysis, 57**(1975), 31.

182 L. Peusner, PhD Thesis, **The Principles of Network Thermodyna-**

mics and **Biophysical Applications**, Harvard University, Cambridge, Mass, 1970.

183 L.Peusner, **Concepts in Bioenergetics**, Prentice Hall, Englewood Cliffs, NJ, 1974.

184 L.Peusner, Global reaction: Diffusion coupling and reciprocity in linear asymmetric kinetic networks, J.Chem. **Phys.** ,77 (1982), 5500.

185 L.Peusner, Topological derivation of nonlinear convection-diffusion equations using network theory, **Phys. Rev. A. 28** (1983), 6.

186 L.Peusner, Hierarchies of Irreversible Energy Conversion systems: A Network Thermodynamic Approach. I Linear Steady State Without Storage, **J. Theor. Biol.** 102 (1983), 7.

187 L.Peusner, Changes in reference frame with respect to barycentric velocity in isothermal diffusion: Duality between reference flows and potentials in mechanical equilibrium, **J.Chem. Phys., 80** (1984), 2727

188 L.Peusner, Electrical Network Representation of n-dimensional chemical manifolds, in **Graphs and Topology in Chemistry** , ed. R.B.King , Elsevier, Holland, 1984.

189 L.Peusner, Network Thermostatics, **J. Chem. Phys., 83,** 3 (1985), 1276.

190 Peusner, Network Representation Yielding the Evolution of Brownian motion with multiple particle interactions, **Phys. Rev. A., 32** (1985), 1237.

191 L.Peusner, Premetric Thermodynamics, **J. Chem. Soc. , Faraday Trans. 2, 81** (1985), in press.

192 L.Peusner, Hierarchies of Irreversible Energy Conversion systems: A Network Thermodynamic Approach.II . Network Derivation of Linear Transport Equations, **J. Theor. Biol.** (1985), in print.

193 L.Peusner, D.C. Mikulecky , B. Bunow and S.R Caplan, A Network Thermodynamic Approach to Hill and King Altman reaction diffusion kinetics, **J. Chem. Phys.** (1985), in press.

194 L.Peusner, Hierarchies of Irreversible Energy Conversion systems: A Network Thermodynamic Approach. III. Why are Onsager Equations Reciprocal? , **J. Theor. Biology** , in press.

195 L. Peusner, Topologically Consistent Thermodynamic and Kinetic Definitions of Reaction Diffusion Convection Potentials in the Presence of Reactive Layers, **J. Membrane Science** , accepted for publication.

196 F.C. Phillips, **An Introduction to Crystallography**, John Wiley and Sons, New York, 1963.

197 I. Prigogine, **Introduction to the Thermodynamics of Irreversible Processes**, Charles C. Thoms, Springfield, Illinois, 1955.

198 M. Propp, personal communication.

199 G. Raisbeck, **Information Theory: An Introduction for Scientists and Engineers**, MIT Press, Cambridge, Mass, 1964.

200 Lord Raleigh, **The Theory of Sound** (reprinted by Dover Publications, New York, 1945).

201 M.J. Randic, **Am. Chem. Soc.**, **97** (1975), 6609.

202 N. Rashevsky, **Mathematical Biophysics Physico-Mathematical Foundations of Biology**, Dover Publications, Inc, New York,

203 I.W. Richardson, On the Principle of Minimum Entropy Production, **Biophysical Journal, 9** (1969), 265.

204 I.W. Richardson, The Metrical Structure of Aging (Dissipative) Systems, **Journal of Theoretical Biology, 85** (1980), 745.

205 I.W. Richardson and R. Rosen, Aging and the Metrics of Time, **J. theor. Biol., 79** (1979), 415.

206 O.E. Rossler, **Z. Naturforsch, Teil A, 31** (1979), 259 ; (1976), 1664.

207 K.J. Rothschild, S.A. Ellias, A. Essig, and H.E. Stanley, Non-Equilibrium Linear Behavior of Biological Systems, **Biophys.J., 30** (1980), 209.

208 H. Rottenberg, S.R. Caplan and A. Essig, **Nature, 216** (1967), 611.

209 G. Ruppeiner, Thermodynamics: A Riemannian Geometric Model, **Phys. Rev. A, 20** (1979), 1608.

210 P. Salmon, S. Berry, Thermodynamic Length and Dissipated Availability, **Phys. Rev. Letters, 51** (1983), 1127.

211 J. Sandblom, A. King and G. Eisenman, Linear Network Representations of Multistate Models of Transport, **Biophysical Journal, 38** (1982), 93.

212 L.A. Santalo, **Vectores y Tensores**, Editorial Universitaria de Buenos Aires, Buenos Aires, 1961.

213 J. Schnakenberg, **Thermodynamic Analysis of Biological Systems**, Springer Verlag, New York, 1977.

214 Y. Schiffmann, Chemical triggering and Histeresis: the Privileged Position of Two dimensional Chemical Systems, **Physica, 114A** (1982), 74.

215 Y. Schiffmann, Non Equilibrium as a source of Unmixing, **Mathematical Biosciences, 64** (1983), 261.

216 L.G. Schimpf and J.G. Linvill, The design of Tetrode Transistor Amplifiers, **Bell Sys. Tech, J., 35** (1956), 813.

217 J. Serrin, Conceptual Analysis of the Classical Second Laws of Thermodynamics, **Archive of Rational Mechanics and Analysis**,

70 (1979), 355.

218 J. Serrin, An Outline of Thermodynamical Structure, in **New Perspectives in Thermodynamics**, Springer-Verlag, 1984.

219 D. Shear, **J. Chem. Phys.**, **48** (1968), 4144.

220 O. Sinanoğlu, L.S. Lee, Theor. Chem. Acta, 48 (1978), 287.

221 S. Smale, **J. Diff. Geom, 7** (1972), 1972.

222 A. Sommerfeld, **Thermodynamics and Statistical Mechanics**, Academic Press, New York and London, 1966.

223 D.M.Y. Sommerville, **An Introduction to the Geometry of n-Dimensions**, Dover Publications, Inc., NY, 1958.

224 I. Stakgold, **Boundary Value Problems in Mathematical Physics**, Macmillan, London, 1967.

225 A.J. Staverman, Non-Equilibrium Thermodynamics of Membrane Processes, **Trans. Farad. Soc., 48** (1952), 176.

226 A. Stern, Stability and Power Gain of Tuned Transistor Amplifiers, **Proc. of the IRE, 45** (1957), 335.

227 J.W. Stucki, **Prog. Biophys. Molec. Biol., 33**(1978), 99.

228 C.J. Suckling, K.E. Suckling and C.W. Suckling, **Chemistry Through models**, Cambridge University Press, Cambridge, 1978.

229 O. Sinanoglu, **J. Am. Chem. Soc., 97** (1975), 2309.

230 B.D. Tellegen, **Phillips Res. Reports, 7** (1952), 259.

231 T. Teorell, **Exp Cell Res. Suppl., 3** (1955), 339.

232 T. Teorell, **Faraday Soc. Discuss., 21** (1956), 9.

233 T. Teorell, Acta Soc. med. Upsaliensis, 62 (1957), 60.

234 T. Teorell, **Z.physik Chem. N.F.** 15 (1958), 385.

235 T. Teorell, **J.Gen. Physiology, 42** (1959), 831.

236 T. Teorell, **J.Gen. Physiology, 42** (1959), 847.

237 T. Teorell, Acta Soc. med. Upsaliensis 65 (1960), 231.

238 T. Teorell, **Arkiv fur Kemi 18** (1961), 401.

239 T. Teorell, **Biophys. J., 2** (1962), 27.

240 T. Teorell, **Ann. New York Acad. Sci., 137** (1966), 950.

241 K. Thakker, Pharmacokinetic-Pharmacodynamic Modelling and Simulation Using the Electrical Circuit Simulation Program SPICE 2, Conference on Clinical Pharmacology and therapeutics Aug 5, 1983, Washington, DC.

242 R. Thom, **Structural Stability and Morphogenesis**, Benjamin, New York, 1975.

243 R.Thomas, Logical Description Analysis and Synthesis of Biological and Other Networks Comprising Feedback Loops, in **Aspects of Chemical Evolution**, John Wiley and Sons, in press.

244 R.Thomas, Fully Asynchronous Logical Description of Networks Comprising Feedback Loops, **Lecture Notes in Biomathematics, 49** (1983), 784.

245 S.R.Thomas and D.C.Mikulecky, **Am.J.of Physiology, 235** (1978)

246 S.R.Thomas and D.C.Mikulecky, Transcapillary Solute Exchange, **Microvascular Research, 15** (1978), 207.

247 W. Thomson, **Proc. Royal Soc. Edin., 8** (1874), 325.

248 J.M.Thorman and Sun-Tak Hwang, Engineering Aspects of the Continuous Membrane Column, **Scientific Membranes:vol II, Hyper and Ultrafiltration uses**, Albin F.Turbak, ed.

249 K. Tomita and I. Tsude, **Phys. Lett. A, 71** (1979), 489.

250 C. Trusdell, **Rational Thermodynamics**, McGraw-Hill, New York, 1969.

251 J.J. Tyson, **J. Math. Biol., 5** (1978), 351.

252 J.J. Tyson and J.C.Light, **J. Chem. Phys., 59** (1973), 4164.

253 I.Ugi and P. Gillespie, **Angew. Chem. Int. Ed. Engl., 10** (1971), 915.

254 H.H. Ussing and K. Zerhan, **Acta Physiol. Scand., 23** (1950), 110.

255 van N.G.Kampen, Thermal Fluctuations in Nonlinear Systems, **Jour. of Math. Phys., 4** (1963), 190

256 W. Wakeeham and E.A Mason, Diffusion Through Multiperforate lining, **Ind. Eng. Chem Fundam., 18**(1979), 301.

257 J.Wei, **Ind. J. Chem. Phys., 36** (1962), 1578.

258 F. Weinhold, Geometric Representation of Equilibrium Thermodyanmics, **Acc. of Chemical Res, 9** (1976), 236.

259 F. Weinhold, The Metric Geometry of Equilibrium Thermodynamics IV Vector Algebraic Evaluation of Thermodynamic derivatives, **Journal of Chem. Phys., 63** (1975), 2496.

260 F. Weinhold, Thermodynamics and geometry, **Physics Today, March** (1976), 23.

261 J.N. Weinstein, B. Bunow and S.R. Caplan, **Desalintaion, 11** (1972), 341.

262 R.P.Wendt, E.A. Mason and E.H. Bresler, effect of Heteroporosity on flux equations for membranes, **Biophysical Chemistry, 4** (1976), 237.

263 H.V. Westerhoff in <u>Discussion Forum</u>, **TIBS**, August 1982, 275.

264 H.Weyl, **Space, Time, Matter**, Dover Publications, New York, 1922.

265 H.Weyl, **The Theory of Groups and Quantum Mechanics**, Dover Publications, New York ,reprint of 1922 book.

266 H. Weyl, **Rev. Math. Hispanoamericana, 153** (1923), 153.

267 A.E.R. Woodcock and T.Poston, **Lecture Notes in Mathematics: A Geometrical Study of the Elementary Catastrophes**, Springer-Verlag, Berlin, 1979.

268 J.L. Wyatt and L.O. Chua, A Theory of Nonenergenic N-Ports, **Circuit theory and Applications, 5** (1977), 181

269 J.L. Wyatt, L. Chua and G. Oster, Non Linear n-Port Decomposition via the Laplace Operator, **IEEE Transactions on Circuits and Systems, 9** (1978), 74.

270 J.L. Wyatt, Network Representation of Reaction Diffusion Systems far from Equilibrium, **Computer Programs in Biomedicine, 8** (1978),180.

271 J.L.Wyatt, D.C.Mikulecky and J.A. DeSimone, **Chem. Eng. Sci. 35** (1980), 2115.

272 W. Yourgrau and S.Mandelstam, **Variational Principles in Dynamics and Quantum Theory**, Dover Publications, Inc. New York, 1968.

273 W. Yourgrau, A. vander Merwe and G.Raw, **Treatise on Irreversible and Statistical Thermophysics** Dover Publications, New York, 1982.

274 E.C.Zeeman, **Catastrophe Theory in Biology** (A.Dold and B. Eckmann, eds.), Springer Verlag, New York, 1975.

275 H.J. Zimmerman and S.J. Mason, **Electronic Circuit Theory, John Wiley and Sons, Inc., New York, 1960.**

LIST OF SYMBOLS

A,	affinity of a chemical reaction,
$a_{(x)}$	affinity of a chemical reaction at x, in a membrane system
c_i,	molar concentration of component i, mol/cm^3
c_s,	molar concentration of solute, mol/cm^3
c_w,	molar concentration of water, mol/cm^3
$\bar{c}_i$,	average concentration of component i, mol/cm^3
Δc_i,	concentration difference of component i in a membrane system, mol/cm^3
D,	diffusion coefficient in a binary solution,
E,	internal energy, cal.
E,	electrical potential difference, Volts
e,	quantity of electric charge, Coulombs
F,	Helmholtz free energy
F,	Faraday 96,500 coulombs/mole
f,	external force
G,	Gibbs free energy
H,	enthalpy
I,	electric current, Amps = Coulombs/sec
J_i,	flow of component i, mol/cm^{2}sec

$\dot{J}_s$, solute flow, mol-cm^2

J_v, volume, flow, cm/sec

J_D, velocity of solute relative to solvent in a membrane system

J_{ch}, rate of chemical reaction, mol/cm^{3}sec

J_c, net chemical flow in a membrane system

J_i^D, net diffusional flow in a membrane system

k_i, k_{-i}, rate coefficients for the ith step in a chemical reaction, 1/sec.

K, equilibrium constant for a chemical reaction

L_{ik}, phenomenological coefficient relating the ith flow to the kth force

L_p, filtration coefficient of a membrane (hydraulic conductivity), cm/sec-mm Hg

n_i, number of moles of component i, dimensionless

P, permeability, cm/sec

P_{ik}, hybrid phenomenological coefficient

Q, heat, cal

R_{ik}, resistance coefficient relating the ith force to the kth flow

S, entropy, cal/°k

d_iS, entropy produced within a system, cal/°k

T, absolute temperature, °Kelvin

t, time, sec.

V, volume, cm^3

v, velocity of center of mass

v_i, velocity of component i

v_s, velocity of the solute in a binary system

v_w, velocity of water (solvent)

W, work, cal

X_i, thermodynamic force acting on component i

z_i, valence of component i, dimensionless

α_i, fluctuation in the ith parameter of a system from its equilibrium value

β, electroosmotic coefficient, Volts/atm.

γ_i, activity coefficient of component i

κ, electrical conductance, amp s/volt

μ_i, chemical potential of component i, cal/mole

μ_i^0, standard chemical potential of component i

$\tilde{\mu}_i$, electrochemical potential of component i

μ_s, chemical potential of the solute in a binary system

μ_w, chemical potential of water (solvent)

ν_i, stoichiometric coefficient of component i in a chemical reaction

$d\xi$, degree of advancement of a chemical reaction, dimensionless

$\Delta\pi$, osmotic pressure difference across a membrane, atm.

ρ, density, gm/cm^3

σ, local entropy production, cal/cm^3 °K. sec

σ_i, reflection coefficient for component i, dimensionless

τ, time interval, sec.

Φ, dissipation function in a discontinuous system, cal/cm^2.sec

ω_i, mobility of component i,

INDEX

Branch 8

Chain 8
___, zero 20
___, one 20
Cocontent 59
Content 59
Convection/reaction 178
Coupling coefficients
___ and hyperangle 120
Coupling parameters 269
Curl 15
Cycle, topological 22

Derivatives
___ partial, as port measurements 93
___ partial, thermodynamic 102
Diffusion
___ network representation 147
___ /reaction in symmetric membranes 167
Diode, ideal 312
Dissipation 59
___ in diffusion-reaction 150
___ instantaneous 294
Divergence 15

Efficiency 265
Entropy two ports 114
Equilibrium
___ mechanical 236
___ uniqueness of 132

Fluctuations, microscopic 195
___, as an incompressible fluid 205

Geometry 1
Gradient 14
___, discrete equivalent 14
___ in thermodynamics 76
Graph, definition 8
Graph, directed, 8
___ and orientability 17
Graph theory 3

Heat capacities
___ ratios of 110
Hybrid networks
___ in Onsager thermodynamics 250

Ideal gas two port 108
Incremental networks 231
Inner (dot) product
___, geometry of 119
___ and dissipation 63
___ of chains 21
Isomorphism 2

Jacobians 101

Kedem-Katchalsky equations 245

____non additive
    term 184
__network derivation 250
Kinetic networks
____and diffusion 153
____and thermody-
    namics 151
____first order 154
Kirchhoff's laws 10
___ as vector spaces 25

Laplacian 16
Legendre Transform-
    ation 58
Loop 8
___,balanced 225

Manegold and Solf
    equation 176
Metric 2
____network thermo-
    dynamic 104
Manifold 2
Manifold coordinates 2
Matrices
____ Cut set 56, 80
____ Mesh 74
____ Nodal 68
____ Tie set 56, 77
Maxwell's recipro-
    cities 98
Microscopic variab-
    les 199

Network
____definition 9
____equations, RLC 82
____,phase 91
Non-energenic 30

Onsager equations 191
Operator, boundary 21
____ coboundary 21
____ linear 45
____ non linear 45
Orthogonality
____topological 22

Passive processes
____and connectivity 170
Passivity 30
Pi network 48
____fluctuations 209
____thermostatic 106
Planar Network 1
Planar graphs 1
Potentials
____and functions of
    state 77
____and chemical
    rates 149
____chemical 140
Probabilities
____non equilibrium
    states 209
Pythagoras theorem 3

Reciprocity 30
____and connectivity 115
____general 46
____general thermos-
    tatic 125
____in Onsager thermo-
    dynamics 191
Resistance 27
____linear 28
____as load 34
____non-linear 323
Reversibility 212

___ microscopic 212
___ and detailed balance 212

Sources 29
State
___ in Onsager thermodynamics 196
___ transitions and metric 203
Superposition 30

T network 48
___ fluctuations 210
___ geometry of 118
___ incremental 232
___ thermostatic 106
Tellegen's theorem 17
___ complex form of, 84
___ continuous analog 23
___ incremental form 20
Tensors 14
Teorell oscillator 321
___ fundamental 116
Thermodynamics
___ Equation of state 129
___ metric 88,122
___ port variables 90
___ vectorial character 87
Thevenin's theorem 37
Thiele modulus 162
___ physical interpretation of 164
Topology 1
Transformation 60
Transport, facilitated 169
___ kinetic analysis 221
Tree 8

Two ports 37
___ attachment 111
___ energy 96
___ thermodynamic transformations 100
___ conversions among 40

Vector Spaces
___ affine 60
___ Banach 62
___ dimension of 52
___ euclidean 62, 201
___ Hilbert, in thermostatics 140
___ inner product 62
___ integrable square 142
___ linear 60
___ metric 61
___ properties of 25

RAYMOND H. FOGLER LIBRARY
DATE DUE

BOOKS ARE SUBJECT TO
AFTER TWO WEEKS